NERVOUS CONTROL OF THE EYE

The Autonomic Nervous System
A series of books discussing all aspects of the autonomic nervous system. Edited by **Geoffrey Burnstock**, *Autonomic Neuroscience Institute, Royal Free Hospital School of Medicine, London, UK.*

This book is part of a series. The publisher will accept continuation orders which may be cancelled at any time and which provide for automatic billing and shipping of each title in the series upon publication. Please write for details.

This book is due for return not later than the
last date stamped below, unless recalled sooner.

NERVOUS CONTROL
OF THE EYE

Edited by

Geoffrey Burnstock

Autonomic Neuroscience Institute
Royal Free Hospital School of Medicine
London, UK

and

Adam M. Sillito

Institute of Opthalmology
University of London
UK

harwood academic publishers
Australia • Canada • France • Germany • India • Japan
Luxembourg • Malaysia • The Netherlands • Russia
Singapore • Switzerland

Amsteldijk 166
1st Floor
1079 LH Amsterdam
The Netherlands

British Library Cataloguing in Publication Data

Nervous control of the eye. – (The autonomic nervous system
  ; v. 13)
  1. Eye   2. Autonomic nervous system
  I. Burnstock, Geoffrey   II. Sillito, A. M.
  612.8′4

ISBN   90-5823-018-X
ISSN: 1047-5125

Contents

Preface to the Series — Historical and Conceptual Perspective of The Autonomic Nervous System Book Series

The pioneering studies of Gaskell (1886), Bayliss and Starling (1899), and Langley and Anderson (*see* Langley, 1921) formed the basis of the earlier and, to a large extent, current concepts of the structure and function of the autonomic nervous system; the major division of the autonomic nervous system into sympathetic, parasympathetic and enteric subdivisions still holds. The pharmacology of autonomic neuroeffector transmission was dominated by the brilliant studies of Elliott (1905), Loewi (1921), von Euler and Gaddum (1931), and Dale (1935), and for over 50 years the idea of antagonistic parasympathetic cholinergic and sympathetic adrenergic control of most organs in visceral and cardiovascular systems formed the working basis of all studies. However, major advances have been made since the early 1960s that make it necessary to revise our thinking about the mechanisms of autonomic transmission, and that have significant implications for our understanding of diseases involving the autonomic nervous system and their treatment. These advances include:

(1) Recognition that the autonomic neuromuscular junction is not a 'synapse' in the usual sense of the term where there is a fixed junction with both pre- and postjunctional specialization, but rather that transmitter is released from mobile varicosities in extensive terminal branching fibres at variable distances from effector cells or bundles of smooth muscle cells which are in electrical contact with each other and which have a diffuse distribution of receptors (*see* Hillarp, 1959; Burnstock, 1986a).

(2) The discovery of non-adrenergic, non-cholinergic nerves and the later recognition of a multiplicity of neurotransmitter substances in autonomic nerves, including monoamines, purines, amino acids, a variety of different peptides and nitric oxide (Burnstock *et al.*, 1964; Burnstock, 1986b; Rand, 1992; Milner and Burnstock, 1995; Lincoln *et al.*, 1995; Zhang and Snyder, 1995).

(3) The concept of neuromodulation, where locally released agents can alter neurotransmission either by prejunctional modulation of the amount of transmitter released or by postjunctional modulation of the time-course or intensity of action of the transmitter (Marrazzi, 1939; Brown and Gillespie, 1957; Vizi, 1979; Fuder and Muscholl, 1995).

G. Burnstock — Editor of The Autonomic Nervous System Book Series

(4)   The concept of cotransmission that proposes that most, if not all, nerves release more than one transmitter (Burnstock, 1976; Hökfelt, Fuxe and Pernow, 1986; Burnstock, 1990a; Burnstock and Ralevic, 1996) and the important follow-up of this concept, termed 'chemical coding', in which the combinations of neurotransmitters contained in individual neurones are established, and whose projections and central connections are identified (Furness and Costa, 1987).

(5)   Recognition of the importance of 'sensory-motor' nerve regulation of activity in many organs, including gut, lungs, heart and ganglia, as well as in many blood vessels (Maggi, 1991; Burnstock, 1993), although the concept of antidromic impulses in sensory nerve collaterals forming part of 'axon reflex' vasodilatation of skin vessels was described many years ago (Lewis, 1927).

(6)   Recognition that many intrinsic ganglia (e.g., those in the heart, airways and bladder) contain integrative circuits that are capable of sustaining and modulating sophisticated local activities (Saffrey et al., 1992; Ardell, 1994). Although the ability of the enteric nervous system to sustain local reflex activity independent of the central nervous system has been recognized for many years (Kosterlitz, 1968), it has been generally assumed that the intrinsic ganglia in peripheral organs consist of parasympathetic neurones that provided simple nicotinic relay stations.

(7)   The major subclasses of receptors to acetylcholine and noradrenaline have been recognized for many years (Dale, 1914; Ahlquist, 1948), but in recent years it has become evident that there is an astonishing variety of receptor subtypes for autonomic transmitters (see *Pharmacol. Rev.*, **46**, 1994). Their molecular properties and transduction mechanisms are being characterized. These advances offer the possibility of more selective drug therapy.

(8)   Recognition of the plasticity of the autonomic nervous system, not only in the changes that occur during development and aging, but also in the changes in expression of transmitter and receptors that occur in fully mature adults under the influence of hormones and growth factors following trauma and surgery, and in a variety of disease situations (Burnstock, 1990b; Saffrey and Burnstock, 1994).

(9)   Advances in the understanding of 'vasomotor' centres in the central nervous system. For example, the traditional concept of control being exerted by discrete centres such as the vasomotor centre (Bayliss, 1923) has been supplanted by the belief that control involves the action of longitudinally arranged parallel pathways involving the forebrain, brain stem and spinal cord (Loewy and Spyer, 1990; Jänig and Häbler, 1995).

In addition to these major new concepts concerning autonomic function, the discovery by Furchgott that substances released from endothelial cells play an important role in addition to autonomic nerves, in local control of blood flow, has made a significant impact on our analysis and understanding of cardiovascular function (Furchgott and Zawadski, 1980; Burnstock and Ralevic, 1994). The later identification of nitric oxide as the major endothelium-derived relaxing factor (Palmer et al., 1988; *see* Moncada et al., 1991) (confirming the independent suggestion by Ignarro and by Furchgott) and endothelin as an endothelium-derived constricting factor (Yanagisawa et al., 1988; *see* Rubanyi and Polokoff, 1994) have also had a major impact in this area.

In broad terms, these new concepts shift the earlier emphasis on central control mechanisms towards greater consideration of the sophisticated local peripheral control mechanisms.

Although these new concepts should have a profound influence on our considerations of the autonomic control of cardiovascular, urogenital, gastrointestinal and reproductive systems and other organs like the skin and eye in both normal and disease situations, few of the current textbooks take them into account. This is largely because revision of our understanding of all these different specialised areas in one volume by one author is a near impossibility. Thus, this Book Series of 14 volumes is designed to try to overcome this dilemma by dealing in depth with each major area in separate volumes and by calling upon the knowledge and expertise of leading figures in the field. Volume 1, deals with the basic mechanisms of *Autonomic Neuroeffector Mechanisms* which sets the stage for later volumes devoted to autonomic nervous control of particular organ systems, including *Heart, Blood Vessels, Respiratory System, Urogenital Organs, Gastrointestinal Tract, Eye* and *Skin*. Another group of volumes will deal with *Central Nervous Control of Autonomic Function, Autonomic Ganglia, Autonomic–Endocrine Interactions, Development, Regeneration and Plasticity* and *Comparative Physiology and Evolution of the Autonomic Nervous System*.

Abnormal as well as normal mechanisms will be covered to a variable extent in all these volumes depending on the topic and the particular wishes of the Volume Editor, but one volume edited by Robertson and Biaggioni, 1995, has been specifically devoted to *Disorders of the Autonomic Nervous System* (*see* also Bannister and Mathias, 1992).

A general philosophy followed in the design of this book series has been to encourage individual expression by Volume Editors and Chapter Contributors in the presentation of the separate topics within the general framework of the series. This was demanded by the different ways that the various fields have developed historically and the differing styles of the individuals who have made the most impact in each area. Hopefully, this deliberate lack of uniformity will add to, rather than detract from, the appeal of these books.

G. Burnstock
Series Editor

REFERENCES

Ahlquist, R.P. (1948). A study of the adrenotropic receptors. *Am. J. Physiol.*, **153**, 586–600.
Ardell, J.L. (1994). Structure and function of mammalian intrinsic cardiac neurons. In: *Neurocardiology.* (Eds.) Armour, J.A. and Ardell, J.L. pp. 95–114. Oxford University Press: Oxford.
Bannister, R. and Mathias, C.J. (Eds.) (1992). *Autonomic Failure.* Third Edition. Oxford University Press: Oxford.
Bayliss, W.B. (1923). *The Vasomotor System.* Longman: London.
Bayliss, W.M. and Starling, E.H. (1899). The movements and innervation of the small intestine. *J. Physiol. (Lond.)*, **24**, 99–143.
Brown, G.L. and Gillespie, J.S. (1957). The output of sympathetic transmitter from the spleen of a cat. *J. Physiol. (Lond.)*, **138**, 81–102.
Burnstock, G. (1976). Do some nerve cells release more than one transmitter? *Neuroscience*, **1**, 239–248.
Burnstock, G. (1986a). Autonomic neuromuscular junctions: Current developments and future directions. *J. Anat.*, **146**, 1–30.

Burnstock, G. (1986b). The non-adrenergic non-cholinergic nervous system. *Arch. Int. Pharmacodyn. Ther.*, **280** (suppl.), 1–15.

Burnstock, G. (1990a). Co-Transmission. The Fifth Heymans Lecture - Ghent, February 17, 1990. *Arch. Int. Pharmacodyn. Ther.*, **304**, 7–33.

Burnstock, G. (1990b). Changes in expression of autonomic nerves in aging and disease. *J. Auton. Nerv. Syst.*, **30**, 525–534.

Burnstock, G. (1993). Introduction: Changing face of autonomic and sensory nerves in the circulation. In: *Vascular Innervation and Receptor Mechanisms: New Perspectives.* Edited by L. Edvinsson and R. Uddman, pp. 1–22. Academic Press Inc: San Diego, USA.

Burnstock, G., Campbell, G., Bennett, M. and Holman, M.E. (1964). Innervation of the guinea-pig taenia coli: Are there intrinsic inhibitory nerves which are distinct from sympathetic nerves? *Int. J. Neuropharmacol.*, **3**, 163–166.

Burnstock, G. and Ralevic, V. (1994). New insights into the local regulation of blood flow by perivascular nerves and endothelium. *Br. J. Plast. Surg.*, **47**, 527–543.

Burnstock, G. and Ralevic, V. (1996). Cotransmission. In: *The Pharmacology of Smooth Muscle.* (Eds.) Garland, C.J. and Angus, J. pp. 210–232. Oxford University Press: Oxford.

Dale, H. (1914). The action of certain esters and ethers of choline and their reaction to muscarine. *J. Pharmacol. Exp. Ther.*, **6**, 147–190.

Dale, H. (1935). Pharmacology and nerve endings. *Proc. Roy. Soc. Med.*, **28**, 319–332.

Elliott, T.R. (1905). The action of adrenalin. *J. Physiol. (Lond.)*, **32**, 401–467.

Fuder, H. and Muscholl, E. (1995). Heteroceptor-mediated modulation of noradrenaline and acetylcholine release from peripheral nerves. *Rev. Physiol. Biochem. Physiol.*, **126**, 265–412.

Furchgott, R.F. and Zawadski, J.V. (1980). The obligatory role of endothelial cells in the relaxation of arterial smooth muscle by acetylcholine. *Nature*, **288**, 373–376.

Furness, J.B. and Costa, M. (1987). *The Enteric Nervous System.* Churchill Livingstone: Edinburgh.

Gaskell, W.H. (1886). On the structure, distribution and function of the nerves which innervate the visceral and vascular systems. *J. Physiol. (Lond.)*, **7**, 1–80.

Hillarp, N.-Å. (1959). The construction and functional organisation of the autonomic innervation apparatus. *Acta Physiol. Scand.*, **46** (suppl. 157), 1–38.

Hökfelt, T., Fuxe, K. and Pernow, B. (Eds.) (1986). Coexistence of neuronal messengers: A new principle in chemical transmission. In *Progress in Brain Research*, vol. 68, Elsevier: Amsterdam.

Jänig, W. and Häbler, H.-J. (1995). Visceral-Autonomic Integration. In: *Visceral Pain, Progress in Pain Research and Management*, (Ed.) Gebhart, G.F. Vol. 5, 311–348. IASP Press: Seattle.

Kosterlitz, H.W. (1968). The alimentary canal. In *Handbook of Physiology,* vol. IV, (Ed.) C.F. Code, pp. 2147–2172. American Physiological Society: Washington, DC.

Langley, J.N. (1921). *The Autonomic Nervous System*, part 1. W. Heffer: Cambridge.

Lewis, T. (1927). *The Blood Vessels of the Human Skin and Their Responses.* London: Shaw & Son.

Lincoln, J., Hoyle, C.H.V. and Burnstock, G. (1995). Transmission: Nitric oxide. In: *The Autonomic Nervous System, Vol. 1* (reprinted): *Autonomic Neuroeffector Mechanisms.* (Eds.) Burnstock, G. and Hoyle, C.H.V. pp. 509–539. Harwood Academic Publishers: The Netherlands.

Loewi, O. (1921). Über humorale Übertrangbarkeit der Herznervenwirkung. XI. Mitteilung. *Pflügers Arch. Gesamte Physiol.*, **189**, 239–242.

Loewy, A.D. and Spyer, K.M. (1990). *Central Regulations of Autonomic Functions.* Oxford University Press: New York.

Maggi, C.A. (1991). The pharmacology of the efferent function on sensory nerves. *J. Auton. Pharmacol.*, **11**, 173–208.

Marrazzi, A.S. (1939). Electrical studies on the pharmacology of autonomic synapses. II. The action of a sympathomimetic drug (epinephrine) on sympathetic ganglia. *J. Pharmacol. Exp. Ther.*, **65**, 395–404.

Milner, P. and Burnstock, G. (1995). Neurotransmitters in the autonomic nervous system. In: *Handbook of Autonomic Nervous Dysfunction.* (Ed.) Korczyn, A.D. pp. 5–32. Marcel Dekker: New York.

Moncada, S., Palmer, R.M.J. and Higgs, E.A. (1991). Nitric oxide: Physiology, pathophysiology, and pharmacology. *Pharmacol. Rev.*, **43**, 109–142.

Palmer, R.M.J., Rees, D.D., Ashton, D.S. and Moncada, S. (1988). Arginine is the physiological precursor for the formation of nitric oxide in endothelium-dependent relaxation. *Biochem. Biophys. Res. Commun.*, **153**, 1251–1256.

Rand, M.J. (1992). Nitrergic transmission: nitric oxide as a mediator of non-adrenergic, non-cholinergic neuro-effector transmission. *Clin. Exp. Pharmacol. Physiol.*, **19**, 147–169.

Rubanyi, G.M. and Polokoff, M.A. (1994). Endothelins: Molecular biology, biochemistry, pharmacology, physiology, and pathophysiology. *Pharmacol. Rev.*, **46**, 328–415.

Saffrey, M.J. and Burnstock, G. (1994). Growth factors and the development and plasticity of the enteric nervous system. *J. Auton. Nerv. Syst.*, **49**, 183–196.

Saffrey, M.J., Hassall, C.J.S., Allen, T.G.J. and Burnstock, G. (1992). Ganglia within the gut, heart, urinary bladder and airways: studies in tissue culture. *Int. Rev. Cytol.*, **136**, 93–144.

Vizi, E.S. (1979). Prejunctional modulation of neurochemical transmission. *Prog. Neurobiol.*, **12**, 181–290.

von Euler, U.S. and Gaddum, J.H. (1931). An unidentified depressor substance in certain tissue extracts. *J. Physiol.*, **72**, 74–87.

Yanagisawa, M., Kurihara, H., Kimura, S., Tomobe, Y., Kobayashi, M., Mitsui, Y., Yazaki, Y., Goto, K. and Masaki, T. (1988). A novel potent vasoconstrictor peptide produced by vascular endothelial cells. *Nature*, **332**, 411–415.

Zhang, J. and Snyder, S.H. (1995). Nitric oxide in the nervous system. *Annu. Rev. Pharmacol. Toxicol.*, **35**, 213–233.

Preface

The eye is our window to the world and the loss of its function has a most profound effect on our ability to interact with the world around us. Despite this there is a tendency to underestimate the range, complexity and subtlety of the processes that sustain its function. In this volume we present a view that attempts to draw together various aspects of our understanding of the autonomic neural control of the eye and linked aspects of its pathophysiology.

On one level the contributions highlight the role of the autonomic nervous system in sustaining the balance and normal functions of ocular tissues such as the cornea and on another they highlight its role in the specific visual functions centering around the control of accommodation and pupillary diameter. Whilst the regulation of accommodation is inextricably embedded in the processes of vision, it is intriguing that the preganglionic neurons in the Edinger-Westphal nuclei associated with accommodation and the pupillary light reflex also seem to influence choroidal blood flow and possibly intraocular pressure in the cat. Picking up the latter point regarding intraocular pressure as an example, the issue of the control of intraocular pressure is of enormous clinical significance for the major blinding disease glaucoma. The evidence for the role and pharmacology of the autonomic neural influences in the control of aqueous humour formation thus provides crucial insight into this matter.

Clearly there are many levels of interaction between the systems sustaining the function of the eye and its autonomic innervation and our objective has been to provide in each of the areas covered an insight into the relative significance of the different components. In the retinal vessels for example the autonomic nervous system seems to have little influence, with autoregulatory processes providing the main control of retinal vascular tone and blood flow. While the dual control of vessels by perivascular autonomic and sensory-motor nerves at the advential-medial border and by endothelial vasodilator and vasoconstrictor mechanisms from the intimal side is now recognised (see Burnstock and Ralevic, 1994), there are many special features of the eye that require a level of detail beyond the scope of the present volume such as the afore-mentioned control of intraocular pressure, the blood-aqueous barrier and the origin in the CNS of some nerves controlling blood flow in the eye. Indeed, in some cases these features of the eye lack the research necessary to provide an adequate account.

We have also decided not to devote a separate chapter to diseases of the eye concerned with abnormalities in autonomic function, but rather to refer to disease conditions within

Adam M. Sillito

the separate chapters. However for fuller coverage of this topic the Editors recommend the excellent review by Collins and O'Brien (1990). Finally it is our particular hope that this volume on the Nervous Control of the Eye will act as a stimulus to further work in this important field.

Geoffrey Burnstock and Adam Sillito

REFERENCES

Burnstock, G. and Ralevic, V. (1994). New insights into the local regulation of blood flow by perivascular nerves and endothelium. *British Journal of Plastic Surgery*, **47**, 527–543.
Collins, B.K. and O'Brien, D. (1990). Autonomic dysfunction of the eye. *Seminars in Veterinary Medicine and Surgery (Small Animal)*, **5**, 24–36.

Contributors

Burnstock, Geoffrey
Autonomic Neuroscience Institute
Royal Free Hospital School of Medicine
Rowland Hill Street
London, NW3 2PF
UK

Dart, John K.G.
Corneal and External Disease Service
Moorfields Eye Hospital
City Road
London, EC1V 2PD
UK

Elder, Mark J.
Department of Ophthalmology
Christchurch Hospital
Private Bag 4710
Christchurch
New Zealand

Flitcroft, D. Ian
Institute of Opthalmology
60 Eccles Street
Dublin 7
Ireland

Gamlin, Paul D.R.
Vision Science Research Center
626 Worrell Building
University of Alabama at Birmingham
Birmingham, AL 35294-4390
USA

Greenwood, John
Department of Clinical Ophthalmology
Institute of Ophthalmology
University College London
Bath Street
London, EC1V 9EL
UK

Judge, Stuart J.
University Laboratory of Physiology
Parks Road
Oxford OX1 3PT
UK

Jumblatt, James E.
Department of Ophthalmology and
    Visual Sciences
School of Medicine
Kentucky Lions Eye Research Institute
University of Louisville
Louisville, KY 40292
USA

Klassen, Henry
Professorial Unit
Moorfields Eye Hospital
City Road
London, EC1V 2PD
UK

Lund, Raymond D.
Institute of Ophthalmology
Neural Transplant Program
11-43 Bath Street
London, EC1V 9EL
UK

Marfurt, Carl F.
Northwest Center for Medical Education
School of Medicine
Indiana University
3400 Broadway
Gary, IN 46408
USA

Penfold, Philip L.
Department of Clinical Ophthalmology
University of Sydney
Sydney Eye Hospital
Sydney, NSW 2001
Australia

Pintor, Jesús
Department of Biochemistry and
   Molecular Biology IV
University School of Optics and
   Optometry
University Complutense of Madrid
Arcos de Jalón s/n
28037 Madrid
Spain

Provis, Jan M.
Departments of Anatomy, and Histology
   and Clinical Opthalmology
University of Sydney
Sydney, NSW 2006
Australia

Sillito, Adam M.
Institute of Ophthalmology
University College London
Bath Street
London, EC1V 9EL
UK

Young, Michael J.
The Schepens Eye Research Institute
Harvard Medical School
20 Stamford Street
Boston, MA 02114
USA

1 Innervation and Pharmacology of the Iris and Ciliary Body

James E. Jumblatt

Department of Ophthalmology and Visual Sciences, University of Louisville, School of Medicine, Louisville, Kentucky 40292, USA

In this chapter, the physiological and pharmacological basis for autonomic neuroeffector transmission in the iris and ciliary body is reviewed. The anterior uvea of the eye is richly supplied with peripheral nerve fibres from sympathetic, parasympathetic and sensory sources. Recent studies show that ocular autonomic nerves contain a variety of peptide and non-peptide cotransmitters in addition to the 'classical' neurotransmitters noradrenaline and acetylcholine. Although functional receptors for many of these putative transmitters have been identified in ocular tissues, the physiological significance of cotransmission or modulation in the iris and ciliary body is only beginning to come to light. This chapter begins with a detailed description of the peripheral innervation of the iris and ciliary body and the histological distribution of various neurotransmitters to specific target structures. The subsequent section describes the physiology and pharmacology of autonomic neuroeffector transmission in the iris sphincter and dilator muscles and discusses recent evidence for dual, reciprocal autonomic regulation of these muscles. The final section describes the anatomy and transport physiology of the ciliary epithelium and reviews current evidence for autonomic neural influences on aqueous humour formation. The central role of the adenylyl cyclase system in control of aqueous humour secretion is discussed. Due to species variability in the anatomy, innervation and neurotransmitter content of the anterior uvea, emphasis is placed on human data wherever possible.

KEY WORDS: innervation; autonomic; iris; ciliary body; ciliary epithelium; aqueous humour.

INTRODUCTION

The iris and ciliary body comprise the anterior portion of the uvea — the densely vascularized layer of the eye that is located between the outer scleral coat and the inner neuroepithelial layer. The iris, which contains the opposing sphincter and dilator pupillae muscles, controls the pupillary aperture and the amount of light entering to the back of the eye. The ciliary muscle, which comprises the bulk of the ciliary body, controls the shape of the lens to facilitate accommodation for near vision. The ciliary processes, which line the posterior surface of the ciliary body, are responsible for secretion of aqueous humour. While it is generally accepted that the autonomic nervous system plays a regulatory or modulatory role in each of these tissues, it is becoming increasingly apparent that the mediators and mechanisms underlying autonomic neuroeffector transmission vary considerably from one target tissue to another. The aim of this chapter is to review the pharmacology and physiology of autonomic neuroeffector transmission in the iris and ciliary processes, with particular emphasis on the biochemical basis for cellular responses and the role(s) of recently discovered, non-classical neurotransmitters and cotransmitters in these tissues. As in many areas of science, the rapid proliferation of descriptive information on the expression of various neurotransmitters and receptors in the anterior uvea has greatly outpaced the elucidation of their physiological function. By comparing and contrasting the patterns of neuroregulation in two different target structures — smooth muscle and secretory epithelium — the present chapter will hopefully provide useful insights into the mechanisms by which neuroeffector transmission can be fine-tuned by individual cotransmitters or modulators and suggest possible new areas for investigation.

INNERVATION OF THE ANTERIOR UVEA

The peripheral nerve supply to the anterior uvea includes sympathetic, parasympathetic and sensory components (Figure 1.1). The sympathetic nerves are derived from the superior cervical ganglion, while the parasympathetic nerves originate from the ciliary, accessory ciliary and pterygopalatine ganglia. Postganglionic sympathetic and parasympathetic nerve fibres are distributed in varying proportions to the iris sphincter and dilator muscles, the ciliary muscle, uveal blood vessels and ciliary processes. Sensory nerve fibres, which derive from the trigeminal ganglion, are found throughout the anterior uvea. These nerves convey afferent responses to mechanical, thermal and chemical stimuli and also play an efferent, modulatory role in ocular inflammation or injury. The anatomy of sympathetic, parasympathetic and sensory innervations to the iris and ciliary body is presented in the following section, which includes descriptions of the various neurotransmitters and cotransmitters identified in each type of nerve.

SENSORY INNERVATION

The trigeminal nerve (Figure 1.1A) emerges in the midlateral margin of the pons and extends to the trigeminal ganglion, which contains the cell bodies of its sensory root (Dutton, 1994; ten Tusscher et al., 1994). The trigeminal ganglion gives rise to the ophthalmic (V1), maxillary (V2) and mandibular (V3) branches. The ophthalmic branch divides into the lacrimal, frontal and nasociliary nerves. The nasociliary nerve gives rise to a major branch that travels to the eye with the long ciliary nerves and a minor branch that passes through the ciliary ganglion and enters the eye with the short posterior ciliary nerves. In addition to the major nasociliary component of sensory innervation, a small maxillary sensory projection to the eye has been described in monkeys (Ruskell, 1974).

The intraocular distribution of sensory nerve fibres has been investigated by several methods, including analysis of nerve fibre degeneration following ophthalmic neurotomy

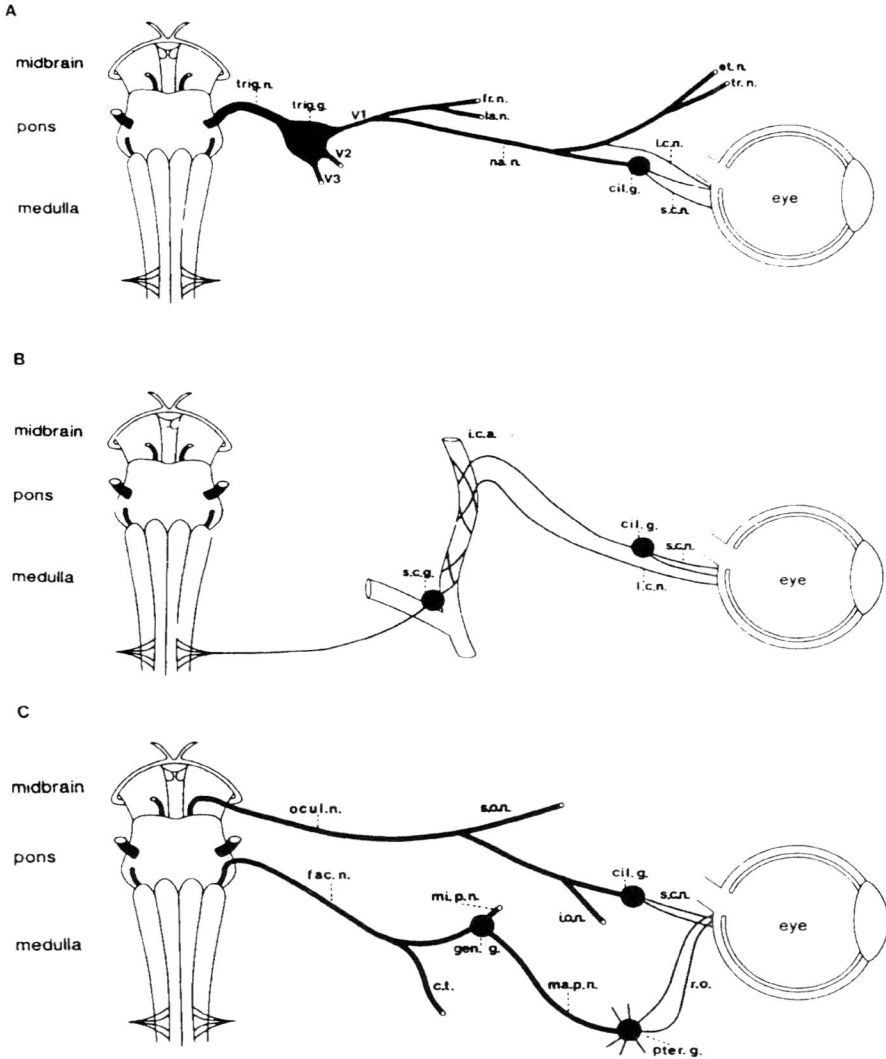

Figure 1.1   Peripheral innervation of the anterior uvea. (A) The sensory innervation. Abbreviations: trigeminal nerve (trig. n.), trigeminal ganglion (trig. g.), ophthalmic nerve (V1), maxillary nerve (V2), mandibular nerve (V3), frontal nerve (fr. n.), lacrimal nerve (la. n.), nasociliary nerve (na. n.), posterior ethmoid nerve (et. n.), infratrochlear nerve (tr. n.), ciliary ganglion (cil. g.), long ciliary nerve (l. c. n.) and short ciliary nerves (s. c. n.). (B) Sympathetic innervation. Abbreviations: superior cervical ganglion (s. c. g.), internal carotid artery (i. c. a.), ciliary ganglion (cil. g.), long ciliary nerve (l. c. n.) and short ciliary nerves (s. c. n.). (C) Parasympathetic innervation. Abbreviations: oculomotor nerve (ocul. n.), superior oculomotor nerve (s. o. n.), inferior oculomotor nerve (i. o. n.), ciliary ganglion (cil. g.), short ciliary nerves (s. c. n.), facial nerve (fac. n.), chorda tympani (c. t.), geniculate ganglion (gen. g.), minor petrosal nerve (mi. p. n.), major petrosal nerve (ma. p. n.), pterygopalatine ganglion (pter. g.), rami oculares (r. o.). See text for description of pathways.
    (Reproduced from ten Tusscher *et al.*, 1994, with permission of Kluwer Academic Publishers).

(Bergmanson, 1977; Ruskell, 1994) and intra-axonal transport of peroxidase injected into the trigeminal ganglion (Lehtosalo, Uusitalo and Palkama, 1984). In the iris, sensory fibres are distributed to the sphincter muscle, large blood vessels and anterior stromal melanocytes. In the ciliary body, sensory fibres innervate the ciliary processes, ciliary body blood vessels and portions of the longitudinal ciliary muscle near the scleral spur (Stone, Kuwayama and Laties, 1987; Ruskell, 1994). Sensory receptors in the anterior uvea consist mainly of mechanoreceptors and polymodal nociceptors (Mintenig *et al.*, 1995). When viewed by electron microscopy, sensory terminals appear as thin (15–50 μm in diameter), myelinated or unmyelinated axons with periodic varicosities containing mitochondria, small clear vesicles and large granular vesicles (Lehtosalo, Uusitalo and Palkama, 1984; Beckers *et al.*, 1992; Ruskell, 1994).

Sensory Neurotransmitters

Substance P   Substance P (SP) is a member of the tachykinin peptide family that also includes neurokinins A, K and B (Maggio, 1988). SP was the first peptide neurotransmitter to be localized in ocular sensory nerves (Tervo *et al.*, 1981, 1982a; Stone, Laties and Brecha, 1982). SP immunoreactivity is found in approximately 20% of rat trigeminal ganglion neurons, mostly in the smaller cells (Lee *et al.*, 1985; Skofitsch and Jacobowitz, 1985). Colocalization studies have shown that many SP-immunoreactive neurons in the anterior uvea also contain calcitonin gene-related peptide (CGRP), another ocular sensory neurotransmitter (Gibbins *et al.*, 1985; Matsuyama *et al.*, 1986). In the human anterior uvea, SP-immunoreactive fibres are found in proximity to the iris sphincter muscle and blood vessels of the iris stroma, ciliary processes and ciliary muscle (Tervo *et al.*, 1982b; Stone and Kuwayama, 1985). The presence of SP in the human iris and ciliary body has been confirmed by direct biochemical methods (Geppetti *et al.*, 1992). In rabbits, SP-immunoreactive nerves degenerate after combined ophthalmic and maxillary neurotomy, demonstrating their sensory origin (Tervo *et al.*, 1982a). SP is released locally in ocular tissues in response to antidromic trigeminal nerve stimulation and various mediators (capsaicin, bradykinin, histamine, nicotine, prostaglandins) that excite intraocular sensory terminals (Mandahl and Bill, 1984; Zhang, Butler and Cole, 1984; Wahlestedt, Bynke and Håkanson, 1985; Bill, 1991). Electrically-evoked release of SP from ocular sensory nerves is subject to prejunctional inhibition by α_2-adrenergic and κ-opioid agonists and facilitation by nicotinic and A_2 purinergic agonists (see Fuder, 1994). Three related tachykinins — neurokinin A, neurokinin B and neuropeptide K — are also present in the rabbit iris-ciliary body (Taniguchi *et al.*, 1986; Beding-Barnekow and Brodin, 1989). Their exact neural locations, however, have not been established.

Calcitonin gene-related peptide   CGRP is a 37-amino acid peptide that is expressed in about 40% of trigeminal ganglion neurons (Lee *et al.*, 1985), a portion of which also express SP (Gibbins *et al.*, 1985; Lee *et al.*, 1985; Terenghi *et al.*, 1985; Matsuyama *et al.*, 1986). In human eyes, the distribution of CGRP-immunoreactive fibres in the iris and ciliary body generally parallels that of SP, except that few CGRP fibres occur in the ciliary processes (Stone and McGlinn, 1988). CGRP is released from ocular sensory nerves by antidromic nerve stimulation or capsaicin treatment (Wahlestedt *et al.*, 1986; Geppetti *et al.*, 1992).

Cholecystokinin Cholecystokinin (CCK), a member of the cholecystokinin-gastrin family of neuropeptides, has been localized in ocular sensory nerves in rat and guinea-pig eyes (Stone *et al.*, 1984; Bjorklund *et al.*, 1985a; Kuwayama *et al.*, 1987a). CCK-immunoreactivity is expressed in approximately 2% of trigeminal ganglion neurons, some of which are also immunoreactive for SP (Kuwayama *et al.*, 1987a). CCK-immunoreactive nerve fibres are localized in the iris stroma and in association with blood vessels of the ciliary body and ciliary processes (Stone *et al.*, 1984; Bjorklund *et al.*, 1985a). To date, the presence of CCK in human ocular tissues has not been confirmed.

Galanin Galanin, a 29-amino acid peptide, has been identified in ocular sensory nerves of rat and porcine eyes (Stromberg *et al.*, 1987; Stone, McGlinn and Kuwayama, 1988). In the rat iris, galanin-immunoreactive nerve fibres are found in proximity to the iris dilator and sphincter muscles and stromal blood vessels (Stromberg *et al.*, 1987). In the porcine eye, galanin-immunoreactive fibres are also detected within the ciliary processes and in apposition to blood vessels of the ciliary body (Stone, McGlinn and Kuwayama, 1988). Galanin immunoreactivity disappears from the rat iris following trigeminal ganglionectomy, indicating its sensory origin in this species (Stromberg *et al.*, 1987). In cat eyes, galanin-immunoreactive fibres have a more extensive ocular distribution and appear to derive from autonomic rather than sensory sources (Grimes *et al.*, 1994).

Vasopressin Vasopressin (VP) or antidiuretic hormone is a 9 amino-acid peptide with a wide range of biological activities (Hays, 1985). The presence of VP-immunoreactive peptides has been detected in the anterior uvea of rabbits and rats (Too *et al.*, 1989; Palm, Keil and Severs, 1994). In rabbit eyes, the highest concentration of VP occurs in the iris sphincter muscle which contracts in response to low concentrations of $[Arg^8]$-vasopressin and related peptides (Too *et al.*, 1989). The VP immunoreactivity largely disappears from the rabbit sphincter muscle after trigeminal nerve denervation, suggesting a sensory derivation (Too *et al.*, 1989). The histological distribution of VP-containing nerve fibres in the anterior uveal has not been reported.

SYMPATHETIC INNERVATION

The sympathetic efferent innervation to the eye (Figure 1.1B) originates in the intermediolateral cell column of spinal segments C_8–T_3 (ciliary spinal center of Budge), in which the cell bodies of the preganglionic neurons are located (Dutton, 1994; ten Tusscher *et al.*, 1994). Efferent axons leave the spinal column with the ventral roots of thoracic nerves I–III, then pass through the white communicating rami and upward through the cervical sympathetic chain to the superior cervical ganglia (SCG). Within the SCG, single preganglionic fibres make synaptic contacts with several postganglionic neurons, providing a basis for amplification and divergence of efferent signals (Strack *et al.*, 1988). Postganglionic axons exit the SCG and form a plexus along the internal carotid artery, following the artery through the carotid canal and foramen lacerum to the region of the trigeminal ganglion and cavernous sinus. One group of efferent fibres passes through the trigeminal ganglion, joins the nasociliary branch of the ophthalmic nerve and enters the globe via the long posterior ciliary nerves. Inside the eye, these fibres form a plexus within the suprachoroidal space from which fibres project to the iris smooth muscles and blood vessels. Another group of fibres from the cavernous plexus travels along the internal carotid artery to the ciliary ganglion, passes through the ganglion without relay and enters the globe with the short ciliary nerves to innervate the ciliary body and uveal blood vessels.

The ocular distribution of adrenergic nerve fibres has been determined by catecholamine histofluorescence microscopy (Laties and Jacobowitz, 1964, 1966; Ehinger, 1966; Ehinger and Falck, 1970) and, in more recent studies, confirmed by immunocytochemical detection of the enzyme tyrosine hydroxylase (Hirai et al., 1994; Tamm et al., 1995). In the iris, dense networks of adrenergic fibres occur in proximity to iris dilator muscle, uveal blood vessels and stromal melanocytes. The ciliary processes also contain numerous adrenergic fibres that surround blood vessels and form a plexus beneath the ciliary epithelium. In rabbit ciliary epithelium, adrenergic fibres appear to penetrate the epithelium and form contacts with nonpigmented epithelial cells (Yamada, 1988), whereas in human ciliary epithelium the nerve fibres do not extend beyond the pigmented cell layer (Ehinger, 1966). The iris sphincter and ciliary muscles receive only a sparse adrenergic innervation. Adrenergic nerve fibres disappear from all regions of the anterior uvea following ipsilateral superior cervical ganglionectomy, confirming the sympathetic origin of these neurons (Ehinger, Falck and Rosengren, 1969). When examined by electron microscopy, adrenergic terminals appear as ramified axons with ovoid varicosities (0.3–0.9 μm in diameter) containing numerous mitochondria, small (300–500 Å diameter), dense-core synaptic vesicles and occasional large (600–800 Å) granular vesicles (Richardson, 1964). The small vesicles contain mainly catecholamines, whereas the larger vesicles contain both catecholamines and neuropeptides (Neuman et al., 1984).

Sympathetic Neurotransmitters

Noradrenaline Noradrenaline (NA), the principal neurotransmitter in ocular sympathetic nerves (Sears, 1975), is synthesized in a stepwise fashion from the amino acid tyrosine (Blaschko, 1939). In the initial rate-limiting step, tyrosine is converted to dihydroxyphenylalanine (DOPA) in a reaction catalyzed by tyrosine hydroxylase (TH). TH activity and DOPA synthesis are increased during neural firing and are subject to feedback regulation by α_2-adrenoceptors located on sympathetic terminals (Rittenhouse and Zigmond, 1991; Jumblatt, 1994). The next step in NA synthesis is the decarboxylation of DOPA by the enzyme DOPA decarboxylase to form dopamine (DA). DA then enters synaptic vesicles where it is converted to NA by dopamine β-hydroxylase. NA is present at high concentrations (>20 nmol/g tissue) in iris tissues (Hiromatsu, Araie and Fujimori, 1994) and at lower levels (0.3–0.8 ng/ml) in aqueous humour (Cooper, Constable and Davidson, 1984; Liu, 1992). In monkey eyes, iridectomy causes the aqueous humour concentration of NA to decline by more than 80%, suggesting that the NA derives mainly from sympathetic nerve terminals of the iris (Cooper, Constable and Davidson, 1984). NA is released from isolated iris-ciliary body preparations in response to electrical field stimulation or tyramine (Farnebo and Hamberger, 1970; Neufeld and Page, 1975). In the rabbit iris-ciliary body, the neural release of NA is inhibited by activation of prejunctional α_2-adrenergic, muscarinic, DA_2-dopaminergic, prostaglandin E and neuropeptide Y receptors and enhanced by angiotensin II (Jumblatt, Liu and North, 1987; Jumblatt and North, 1988; Jumblatt and Hackmiller, 1990; Ohia and Jumblatt, 1990a,b; Ogidigben, Chu and Potter, 1993; Fuder, 1994). Once released NA is transported back into nerve terminals by a cocaine-sensitive re-uptake mechanism. The efficient re-uptake of NA may account for the low levels of NA found in aqueous humour (Sears, 1975).

Dopamine   DA, a precursor in NA biosynthesis, is co-stored and co-released with NA from some peripheral sympathetic nerves (Bell, 1988). DA is present at low levels in anterior uveal tissues and aqueous humour (Cooper, Constable and Davidson, 1984; Hiromatsu, Araie and Fujimori, 1994), although its source is unclear. In monkey eyes, the aqueous humour concentration of DA (0.4 ng/ml) does not change following total iridectomy, suggesting that the DA derives from a source other than iris sympathetic nerves, possibly from the ciliary body or retina (Cooper, Constable and Davidson, 1984). Although there is little evidence to support a neurotransmitter role for DA in the anterior uvea, dopaminergic drugs are of considerable pharmacological interest as potential antiglaucoma medications (Mittag, 1996).

Serotonin   Serotonin (5-hydroxytryptamine; 5-HT) is an indoleamine neurotransmitter synthesized from the amino acid tryptophan. Using histochemical methods, 5-HT-containing nerve fibres have been identified by in the ciliary body and cornea of several species (Osborne, 1983; Osborne and Tobin, 1987; Tobin, Unger and Osborne, 1988). Low levels of 5-HT are also present in human aqueous humour (Martin, Brennan and Lichter, 1988). The source of the serotonergic nerve fibres in these tissues is unclear, but in the rabbit they are thought to be of sympathetic origin (Neufeld, Ledgard and Yoza, 1983; Tobin, Unger and Osborne, 1988). Serotonin receptors have been identified in the iris and ciliary body by biochemical and radioligand binding assays (Mallorga and Sugrue, 1987; Barnett and Osborne, 1993). It is unclear, however, whether these receptors receive any functional serotonergic innervation.

Purines   Adenosine 5'-triphosphate (ATP) functions as a cotransmitter with NA in many peripheral sympathetic nerves innervating blood vessels and smooth muscles (Burnstock, 1990). ATP is stored in both large and small noradrenergic storage granules at concentrations ranging from 2–20% of the NA concentration (Fried *et al.* 1984). ATP also occurs as a cotransmitter in some peripheral parasympathetic and sensory nerves (Hoyle, 1992). Although histochemical methods are available for visualization of ATP in peripheral nerves (Richards and Da Prada, 1977), such studies have not yet been performed in the iris or ciliary body. Upon release, ATP is rapidly metabolized by extracellular nucleotidases to produce adenosine 5'-diphosphate, adenosine 5'-monophosphate and adenosine, all of which may also serve as neurotransmitters. Little information exists on the neurotransmitter role(s) of purines in anterior uveal tissues (see also Chapter 6). Recent evidence that topically applied adenosine A_1 purinergic agonists are effective in lowering intraocular pressure in rabbit eyes (Crosson, 1995) should spur further investigation in this area.

Neuropeptide Y.   Neuropeptide Y (NPY) is a 36-amino acid peptide of the pancreatic polypeptide family (Emson and De Quidt, 1984). In the SCG, more than 50% of nerve cell bodies exhibit strong immunoreactivity for NPY (Gibbins, 1991). NPY has been localized in ocular sympathetic nerve fibres of several mammalian species, including humans (Terenghi *et al.*, 1983; Stone, Laties and Emson, 1986; Stone, 1986a). The distribution of NPY in the anterior uvea generally parallels that of noradrenergic fibres and differs little among mammalian species (Stone, Kuwayama and Laties, 1987). Numerous NPY-immunoreactive fibres are present in the iris dilator muscle and ciliary processes, whereas the iris sphincter and ciliary muscles are sparsely innervated. Superior cervical ganglionectomy in rats or guinea-pigs results in the disappearance of nearly all NPY-immunoreactive fibres from the ipsilateral anterior uvea, demonstrating the sympathetic origin of these fibres (Terenghi *et al.*, 1983; Zhang *et al.*, 1984). NPY is stored in large synaptic vesicles and is preferentially released at high (>10 Hz) nerve firing frequencies (Lundberg *et al.*, 1986; Granstam and Nilsson, 1990). Functional NPY receptors coupled to adenylyl cyclase inhibition have been detected in the rabbit ciliary process epithelium, suggesting a possible role of NPY in aqueous humour regulation (Bausher and Horio, 1990; Cepelik and Hynie, 1990; Jumblatt and Gooch, 1990).

Opioid neuropeptides. The prodynorphin-derived peptides dynorphin $A_{(1-8)}$, dynorphin $A_{(1-17)}$ and α-neo-endorphin have been colocalized with NPY in postganglionic sympathetic nerves projecting to the guinea-pig iris, where they are distributed to both the iris sphincter and dilator muscles (Gibbins and Morris, 1987). While another opioid peptide, leu-enkephalin, has been identified by immunohistochemistry in the rat iris (Bjorklund *et al.*, 1984), it is believed to be of parasympathetic rather than sympathetic origin (Stone *et al.*, 1988; Stone, 1996). Although there is some pharmacological evidence for opiate receptors in the iris sphincter muscle (Drago *et al.*, 1980; Fanciullaci *et al.*, 1980), little is known about the function of endogenous opioid peptides in this tissue.

PARASYMPATHETIC INNERVATION

The parasympathetic innervation to the anterior uvea (Figure 1.1C) derives from the ciliary, accessory ciliary and pterygopalatine ganglia (Dutton, 1994; ten Tusscher *et al.*, 1994). The parasympathetic nerve supply to the iris and ciliary muscles originates with preganglionic neurons located in the Edinger-Westphal nucleus in the midbrain. Efferent axons leave the midbrain and travel via the oculomotor nerve and inferior oculomotor branch to the ciliary ganglion (located in the muscle cone of the orbit) where they synapse with postganglionic neurons. In the ciliary ganglia, the ratio of preganglionic to postganglionic neurons is 1:1 or greater (Hulme and Purves, 1983; May and Warren, 1993). The cell bodies of the postganglionic neurons destined for the iris and ciliary body are located in the periphery of the ganglion (ten Tusscher, Klooster and Vrensen, 1988). Postganglionic fibres enter the eyeball via several branches from the short posterior ciliary nerves. Inside the eye, the fibres travel anteriorly within the suprachoroidal space to innervate the iris and ciliary body. Most (>95%) of the postganglionic fibres project to the ciliary muscle and the remainder are distributed to the iris sphincter muscle, blood vessels and other uveal targets (Warwick, 1954; Laties and Jacobowitz, 1966).

A portion of the ocular parasympathetic innervation derives from the accessory ciliary ganglia, which have been identified in many mammalian species (Grimes and von Sallman, 1960; Kuchiiwa, Kuchiiwa and Suzuki, 1989). In humans, the accessory ciliary ganglia appear as small clusters of neurons located on the short ciliary nerves 0.5–2.5 mm distal to the main ganglion. Depending on the species, accessory ganglia may also be located in the episclera, scleral canals, or even in the uveal tract (Kuchiiwa, Kuchiiwa and Suzuki, 1989). In the cat, the accessory ciliary ganglia receive their preganglionic inputs from oculomotor parasympathetic fibres that bypass the main ciliary ganglion by way of the trigeminal nerve (Kuchiiwa *et al.*, 1993). The ocular distribution and function of postganglionic nerve fibres from the accessory ciliary ganglia are not well defined, but there is some evidence for their involvement in the pupillary miotic response to near vision (Nathan and Turner, 1942).

Another parasympathetic pathway to the eye by way of the facial nerve and pterygopalatine ganglia (Figure 1.1C) has been described in rats, rabbits, cats, monkeys and humans (Ruskell, 1965, 1970a,b; Kuwayama *et al.*, 1987b; Lin, Grimes and Stone, 1988). The preganglionic parasympathetic fibres arise in the superior salivatory nucleus (lacrimal nucleus) in the pons and travel via the facial nerve and major petrosal nerve to relay in the pterygopalatine (sphenopalatine) ganglia. In primates, the unmyelinated postganglionic fibres leave the ganglion caudally and travel to the eyeball by way of the rami orbitales, the retro-orbital plexus and the rami oculares (Ruskell, 1970b). The postganglionic pterygopalatine nerve fibres are distributed mainly to the choroidal vasculature (Ruskell, 1985).

Cholinergic nerves in the anterior uvea have been detected histochemically using the thiocholine technique for acetylcholinesterase (Koelle and Friedenwald, 1950; Laties and Jacobowitz, 1964, 1966). High densities of cholinergic fibres are found in the iris sphincter and ciliary muscles. The iris dilator muscle contains moderate numbers of cholinergic fibres, some of which appear to make axo-axonic contacts with sympathetic or sensory fibres (Nomura and Smelser, 1974). When examined by electron microscopy, the parasympathetic terminals appear as unmyelinated, varicose axons, often running parallel to the innervated blood vessels or muscle fibres (Richardson, 1964). The varicosities contain clusters of small (<450 Å in diameter), agranular synaptic vesicles and occasional large (>800 Å) granular vesicles that may represent storage sites for vasoactive intestinal polypeptide (VIP) or other neuropeptides (Butler *et al.*, 1984).

Parasympathetic neurotransmitters

Acetylcholine Acetylcholine (ACh) is the principal neurotransmitter of postganglionic parasympathetic nerves in the eye (Sastry, 1985). The synthesis of acetylcholine from choline and acetylcoenzyme A is catalyzed by the enzyme choline acetyltransferase (ChAT) (Tucek, 1988). ChAT activity has been measured in ocular tissues from several species, including humans (De Roetth, 1950; Mindel and Mittag, 1976; Erickson-Lamy *et al.*, 1990). The distribution of ChAT activity in the iris and ciliary body generally mirrors that of cholinergic nerve fibres as visualized by the thiocholine technique (Laties and Jacobowitz, 1964), with highest enzyme activity in the sphincter and ciliary muscles (Mindel and Mittag, 1976; Erickson-Lamy *et al.*, 1990). In cat and rat irides, ACh levels measured by direct biochemical methods range from 70–220 ng ACh/mg tissue (Consolo *et al.*, 1972; Takayanagi, Shiraishi and Satoh, 1992). ACh released from ocular cholinergic terminals is rapidly hydrolyzed by acetylcholinesterase (AChE), accounting for the acute miotic and accommodative effects of AChE inhibitors (e.g., physostigmine,

TABLE 1.1
Distribution of autonomic and sensory neurotransmitters in the anterior uvea.

Nerve/Ganglion	Neurotransmitter	Target	Species	Reference
Sympathetic Superior cervical	Noradrenaline	Iris dilator, sphincter, uveal melanocytes, blood vessels, ciliary processes	Rat Rabbit, cat Monkey Monkey, human	Hirai et al., 1994 Laties and Jacobowitz, 1964 Laties and Jacobowitz, 1966 Ehinger, 1966
Sympathetic (?)	Dopamine	Iris	Rat Monkey	Hiromatsu, Araie and Fujimori, 1994 Cooper, Constable and Davidson, 1984
Sympathetic (?)	Serotonin	Iris, ciliary processes	Rabbit	Tobin, Unger and Osborne, 1988
Sympathetic (?)	Purines (ATP, adenosine)	Iris dilator	Rat Rabbit	Fuder and Muth, 1993 Muramatsu, Kigoshi and Oda, 1994
Sympathetic Superior cervical	Neuropeptide Y	Iris dilator, uveal melanocytes, blood vessels, ciliary processes	Guinea-pig, rat Guinea-pig, rat, cat, monkey Human	Terenghi et al., 1983 Stone, Laties and Emson, 1986 Stone, 1986a
Sympathetic Superior cervical	Dynorphin	Iris sphincter, dilator	Guinea-pig	Gibbins and Morris, 1987
Parasympathetic Ciliary Pterygopalatine	Acetylcholine	Iris sphincter, dilator, ciliary muscle, ciliary processes	Rabbit Cat, monkey	Laties and Jacobowitz, 1964 Laties and Jacobowitz, 1966
Parasympathetic Pterygopalatine	Vasoactive intestinal peptide	Choroid, uveal blood vessels, ciliary muscle, ciliary processes	Rat Rabbit Cat Human	Bjorklund et al., 1985a Butler et al., 1984 Uddman et al., 1980 Miller et al., 1983; Stone, 1986b
Parasympathetic Ciliary	Enkephalin	Iris sphincter, dilator	Rat	Bjorklund et al., 1984; Stone et al., 1988
Sensory Trigeminal	Substance P	Iris sphincter, uveal blood vessels, melanocytes, ciliary muscle, ciliary processes	Rat Rabbit Human Rabbit, cat, monkey	Hirai et al., 1994 Tervo et al., 1981; Tervo et al., 1982b; Stone and Kuwayama, 1985 Stone Laties and Brecha, 1982
Sensory Trigeminal	Calcitonin gene-related peptide	Iris sphincter, uveal blood vessels, melanocytes, ciliary muscle	Rat Guinea-pig, rat, cat Monkey, human	Hirai et al., 1994 Terenghi et al., 1985 Stone and McGlinn, 1988
Sensory Trigeminal	Cholecystokinin	Iris sphincter, blood vessels, ciliary processes	Guinea-pig Rat	Stone et al., 1984; Kuwayama et al., 1987a Bjorklund et al., 1985a
Sensory Trigeminal	Galanin	Iris dilator, sphincter, blood vessels, ciliary processes	Rat Porcine	Stromberg et al., 1987 Stone, McGlinn and Kuwayama, 1988
Sensory Trigeminal	Vasopressin	Iris sphincter	Rabbit Rat	Too et al., 1989 Palm, Keil and Severs, 1994

echothiophate) applied topically to the eye (Lund-Karlsen and Fonnum, 1976; Kaufman, 1984). Release of ACh from parasympathetic nerves in rabbit or bovine irides is subject to prejunctional modulation by muscarinic, α_2-adrenergic, A_1-purinergic, prostaglandin, opioid and nicotinic agonists (Fuder, 1994). In skeletal muscle, release of ACh is blocked by botulinum-A toxin (Simpson, 1981), a potent bacterial neurotoxin that is used clinically to treat imbalances of the extraocular muscles.

Vasoactive intestinal polypeptide   Vasoactive intestinal polypeptide (VIP), a 28-residue neuropeptide, is expressed predominantly in ocular parasympathetic neurons of pterygopalatine origin (Uddman *et al.*, 1980; Butler *et al.*, 1984). In the human eye, VIP-immunoreactive fibres are found mainly in association with uveal blood vessels, especially in the choroid. In addition, moderate numbers of VIP-immunoreactive fibres are present in the iris stroma, anterior ciliary body and ciliary processes (Miller *et al.*, 1983; Stone, 1986b). At present, little information is available on the biosynthesis, release or inactivation of VIP in ocular tissues. VIP receptors mediating activation of adenylyl cyclase have been identified in ciliary process epithelium from rabbit and human eyes (Mittag, Tormay and Podos, 1987; Bausher, Gregory and Sears, 1989; Crook *et al.*, 1994), suggesting that VIP may play a role in the regulation of aqueous humour formation (Bill, 1991).

The histological distribution of autonomic and sensory neurotransmitters in the anterior uvea is summarized in Table 1.1. The large array of neuropeptides and their complex patterns of colocalization suggest numerous possibilities for functional interactions, as discussed in the following section.

REGULATION OF THE IRIS

The pupil functions as a diaphragm to regulate the passage of light from the anterior to the posterior segments of the eye. The pupillary aperture is controlled by two opposing smooth muscles, the sphincter and dilator pupillae, that are regulated by the Edinger-Westphal nucleus of the midbrain. It is well established that light-induced pupillary miosis is mediated primarily by parasympathetic contraction of the sphincter muscle, whereas mydriasis in response to darkness or emotions is mediated by sympathetic contraction of the dilator muscle (Duke-Elder and Scott, 1971; Davson, 1990). Considerable physiological and pharmacological evidence suggests, in addition, that both iris muscles are dually innervated and reciprocally regulated by inhibitory as well as excitatory autonomic inputs. The iris sphincter muscle is also subject to regulation by sensory neuropeptides released in response to ocular inflammation or injury. The physiological and pharmacological basis for neural regulation of the iris muscles is described below. Although the sphincter and dilator muscles are discussed separately, the reader should keep in mind that these muscles function as a coordinated unit to regulate pupil size.

ANATOMY OF THE IRIS MUSCLES

The iris sphincter and dilator muscles are unusual among smooth muscles of the body in that they are both derived from neuroectoderm rather than mesoderm (Davson, 1990).

The sphincter muscle consists of a ring of smooth muscle fibres, ~1 mm in width, that surrounds the inner pupillary margin. The sphincter muscle is tightly associated with layers of connective tissue on its anterior and posterior surfaces. Thus, pupillary constriction can occur even after the muscle has been transected. By electron microscopy, the sphincter muscle consists of bundles of smooth muscle fibres surrounded by basement membrane and separated by collagenous tissue containing blood vessels and nerve fibres (Rodrigues, Hackett and Donohoo, 1991). Individual cells within these fibre bundles are connected by gap junctions. The dilator pupillae muscle consists of a sheet of muscle fibres oriented radially in the posterior layers of the iris between the stroma and posterior pigmented epithelium. The dilator muscle is composed of thin, poorly differentiated smooth muscle fibres termed myoepithelial cells, which are joined by gap junctions similar to those found in sphincter muscle. The dilator muscle has its origin in the ciliary body and insertion in the iris stroma just posterior to the sphincter muscle (Rodrigues, Hackett and Donohoo, 1991).

IRIS SPHINCTER MUSCLE — PARASYMPATHETIC REGULATION

Early studies showed that stimulation of the IIIrd nerve or exposure to light causes miosis that can be blocked by the muscarinic antagonist atropine (Lowenstein and Loewenfeld, 1969). Atropine also antagonizes neurally-evoked contractions of isolated iris sphincter muscles from rat, rabbit, dog, bovine and human eyes (Banno, Imaizumi and Watanabe, 1985; Wahlestedt *et al.*, 1985; Yoshitomi and Ito, 1986; Suzuki, Yoshino and Kobayashi, 1987; Yoshitomi, Ito and Inomata, 1988). The iris sphincter muscle contracts in response to a variety of cholinomimetic agents (e.g., pilocarpine, acetylcholine, carbamylcholine and AChE inhibitors), both *in vitro* (Banno, Imaizumi and Watanabe, 1985; Howe *et al.*, 1986; Suzuki and Kobayashi, 1988) and *in vivo* (Lowenstein and Loewenfeld, 1953; Kaufman, 1984). Pharmacological and biochemical studies show that excitatory muscarinic receptors in the iris sphincter muscle are predominantly of the M_3 type (Bognar, Wesner and Fuder, 1990; Honkanen, Howard and Abdel-Latif, 1990; Woldemussie, Feldmann and Chen, 1993) and are coupled to phosphoinositide (PI) hydrolysis and formation of inositol 1,4,5-triphosphate (IP_3) and diacylglycerol (DAG) (Abdel-Latif, 1989). Bovine and rabbit iris sphincter muscle also contain a minor population of M_2 type muscarinic receptors, possibly coupled to adenylyl cyclase inhibition (Honkanen, Howard and Abdel-Latif, 1990; Abdel-Latif and Zhang, 1991). Although their function is not yet established, the M_2 receptors may contribute to sphincter muscle contraction by reducing the intracellular levels of cyclic adenosine 3′,5′-monophospate (cAMP), a smooth muscle relaxant (Tachado, Akhtar and Abdel-Latif, 1989).

Both electrically evoked and muscarinic agonist-induced contractions of iris sphincter muscle are dependent on the presence of external calcium (Narita and Watanabe, 1982; Howe *et al.*, 1986; Suzuki and Kobayashi, 1986), but are resistant to organic calcium channel blockers, suggesting that L-type calcium channels are probably not involved (Narita and Watanabe, 1982). The biochemical mechanisms that underlie cholinergic activation of the sphincter muscle have been investigated extensively by Abdel-Latif (reviewed in Abdel-Latif, 1989). In the bovine sphincter, carbamylcholine induces an initial phasic contraction followed by a smaller tonic component (Akhtar and Abdel-Latif,

1991). The phasic response, which persists in the absence of external calcium, is mediated by IP_3-triggered release of calcium from intracellular stores, whereas the tonic component is due to the combined, synergistic effects of DAG/protein kinase C-mediated protein phosphorylation and calcium entry via membrane calcium channels. Kinetic studies show an excellent correlation between muscarinic receptor occupancy, IP_3 production, myosin light chain phosphorylation and sphincter muscle contraction (Abdel-Latif, 1989; Akhtar and Abdel-Latif, 1991).

Somewhat paradoxically, high concentrations of cholinergic agonists cause relaxation of canine iris sphincter muscle, whereas low agonist concentrations produce vigorous contractions as described above (Abdel-Latif *et al.*, 1992). Biochemical evidence suggests that this relaxation results from M_3 muscarinic receptor-mediated stimulation of cAMP synthesis via activation of a calcium-dependent form of adenylyl cyclase (Tachado *et al.*, 1994). To date, muscarinic relaxation has been documented only in canine sphincter muscle and its physiological significance is uncertain. It is tempting to speculate, however, that relaxation in response to high agonist concentrations may represent a braking mechanism to prevent cholinergic over-stimulation of the muscle.

As discussed previously, VIP is a cotransmitter in a subpopulation of ocular parasympathetic nerves of pterygopalatine origin (Table 1.1). Exogenously administered VIP causes relaxation of cholinergically-contracted rabbit and bovine sphincter muscles, an effect attributed to stimulation of cAMP biosynthesis (Hayashi and Masuda, 1982). At higher concentrations, VIP inhibits electrically evoked contractions of the bovine sphincter muscle. The mechanism for this inhibition is unclear, but may be due in part to prejunctional inhibition of ACh release from parasympathetic terminals, as recently demonstrated in the rabbit iris-ciliary body (Vittitow and Jumblatt, 1996). Further studies are needed to clarify the neurotransmitter and/or modulator functions of VIP in iris sphincter muscle. An obstacle to such studies, however, has been the unavailability of suitable VIP antagonists.

IRIS SPHINCTER MUSCLE — SYMPATHETIC REGULATION

Adrenergic innervation to the iris sphincter muscle was first suggested from histological evidence (Laties and Jacobowitz, 1964; Nomura and Smelser, 1974). Subsequent physiological and pharmacological studies of diverse species reveal both inhibitory and excitatory adrenergic influences on sphincter muscle tone. β-Adrenergic agonists (e.g., adrenaline, isoprenaline) cause relaxation of isolated iris sphincter muscles from cat, bovine, canine and human eyes (van Alphen, Kern and Robinette, 1965; van Alphen, 1976; Yoshitomi and Ito, 1986; Yoshitomi, Ito and Inomata, 1988; Tachado, Akhtar and Abdel-Latif, 1989; Patil and Weber, 1991). These effects are attributed to β-adrenergic stimulation of cAMP formation (Tachado, Akhtar and Abdel-Latif, 1989). In addition to β-adrenoceptors mediating relaxation, α-adrenoceptors mediating contraction have been reported in canine, rabbit and human iris sphincter muscles (Wahlestedt *et al.*, 1985; Yoshitomi and Ito, 1986; Yoshitomi, Ito and Inomata, 1988). It is unclear from these studies, however, whether the excitatory α-adrenoceptors in sphincter muscle are sites of innervation or are merely pharmacological targets. The physiological significance of inhibitory β-adrenergic input to sphincter muscle is also questionable. Although β-adrenergic relaxation of sphincter muscle plays a significant role in pupillary dilation in amphibian eyes (Morris, 1976), such a role seems unlikely in humans in view of evidence that β-adrenergic antagonists affect neither baseline pupil size nor the pupillary light reflex (Johnson, Brubaker and Trautman, 1978; Namba, Utsumi and Nakajima, 1980).

IRIS SPHINCTER MUSCLE — SENSORY REGULATION

In an early study, Perkins (1957) reported that antidromic stimulation of the trigeminal nerve in rabbits induces a long-lasting miosis. Subsequent studies showed that the miotic response to sensory nerve stimulation is resistant to antagonism by atropine, but can be blocked by intracameral injection of selective SP antagonists (Stjernschantz, Geijer and Bill, 1979; Mandahl and Bill, 1983; Mandahl and Bill, 1984). In isolated rabbit iris sphincter muscle, transmural electrical stimulation induces a prolonged, atropine-resistant contraction that is mimicked by exogenous SP and capsaicin, an agent that induces release of endogenous SP from sensory terminals (Ueda *et al.*, 1981, 1982). Both the transmurally-stimulated and capsaicin-induced contractions are abolished by trigeminal sensory denervation (Ueda *et al.*, 1982). Taken together, these results point to SP as the principal mediator of sensory neurogenic miosis in the rabbit eye (Bill, 1991).

The contractile effect SP on isolated iris sphincter muscle exhibits considerable species variability. SP strongly contracts sphincter muscle from rabbit, bovine and porcine eyes, but has little or no effect on rat, cat, monkey or human sphincters (Zhang, Butler and Cole, 1984; Tachado *et al.*, 1991; Yoshitomi *et al.*, 1995). A possible biochemical explanation for these differences was provided by Tachado *et al.* (1991) who observed that SP stimulates PI hydrolysis and IP_3 formation in sphincter muscles (rabbit, porcine and bovine) that contract in response to SP, but induces cAMP synthesis and relaxation of the non-contracting (canine, feline and human) sphincter muscles. On the basis of these findings, the investigators propose that two different SP receptors may exist in iris sphincter muscle, one coupled to PI hydrolysis and the other to adenylyl cyclase (Tachado *et al.*, 1991). In support of this concept, the rabbit iris sphincter muscle was recently shown to contain two tachykinin receptor subtypes (NK_1 and NK_3), one of which (NK_1) is primarily responsible for mediating the miotic responses to SP and trigeminal nerve stimulation (Hall, Mitchell and Morton, 1991; Hall, Mitchell and Morton, 1993).

Other sensory neuropeptides reported to induce sphincter muscle contraction include CGRP, CCK, VP and galanin. CGRP contracts the porcine iris sphincter (Geppetti *et al.*, 1990), but has no effect on rabbit or monkey sphincter muscles (Muramatsu, Nakanishi and Fujiwara, 1987; Bill, 1991). CCK strongly contracts both monkey and human sphincter muscles at nanomolar concentrations (Bill, 1991). The effects of CCK are blocked by the selective antagonists lorglumide and loxiglumide, suggesting the involvement of CCK-A receptors (Almegard, Stjernschantz and Bill, 1992). The remarkable contractile potency of CCK compared with SP or CGRP in primate sphincter muscle suggests that CCK is an important biological mediator of sensory neurogenic miosis in humans (Bill, 1991; Almegard, Stjernschantz and Bill, 1992). Other reports indicate that galanin and VP induce contractions of isolated rabbit iris sphincter muscle (Ekblad *et al.*, 1985; Too *et al.*, 1989). However, the role of these neuropeptides in sensory neurogenic responses in the rabbit eye has not been assessed.

IRIS DILATOR MUSCLE — SYMPATHETIC REGULATION

The involvement of α_1-adrenoceptors in mediating sympathetic neural contraction of the dilator muscle has been demonstrated in several species, both *in vivo* and *in vitro* (van Alphen, Kern and Robinette, 1965; Duke-Elder and Scott, 1971; Koss and Gherezghiher, 1988; Koss, Logan and Gherezghiher, 1988). In the cat, electrical stimulation of the preganglionic cervical sympathetic nerve induces pupillary dilation that is potently inhibited by the α_1-adrenergic antagonists phenoxybenzamine and WB-4101 (Koss and Gherezghiher, 1988; Koss, Logan and Gherezghiher, 1988). Similarly, phentolamine and other α_1-antagonists block transmurally-stimulated contraction of isolated dilator muscles from rat, canine and human eyes (Narita and Watanabe, 1982; Yoshitomi, Ito and Inomata, 1985; Yoshitomi and Ito, 1986). The α_1-adrenoceptors mediating noradrenergic contractions of the rabbit iris dilator muscle have recently been classified as α_{1B}-type based on their high affinity for chloroethylclonidine, moderate affinities for WB-4101 and 5-methylurapidil, and low affinity for prazosin (Takayanagi, Shiraishi and Kokubu, 1992).

The cellular mechanisms of α_1-adrenergic contraction of the dilator muscle have been studied by electrophysiological and biochemical methods. Using intracellular recording techniques, Hill *et al.* (1993) found that sympathetic nerve stimulation induces excitatory junction potentials (EJPs) in rat dilator myoepithelial cells that persist for several seconds. The EJPs are blocked by chloroethylclonidine and prazosin, showing the responses to be mediated by α_1-adrenoceptors. The organic calcium antagonist nifedipine abolishes the neurally-evoked EJPs, but has no effect on the contractile response, suggesting that α_1-adrenergic contractions are triggered by release of calcium from internal stores. Consistent with this conclusion, a further study has shown that stimulation of α-adrenoceptors in rabbit dilator muscle cells induces transient increases in intracellular calcium that are unaffected by calcium removal or organic calcium channel blockers (Kokubu, Satoh and Takayanagi, 1993). The biochemical responses to α_1-adrenoceptor stimulation (increased PI hydrolysis, formation of IP$_3$ and DAG, myosin light chain phosphorylation) are essentially the same as those described previously for muscarinic activation of the sphincter muscle (Abdel-Latif, Smith and Akhtar, 1985; Abdel-Latif, 1989). In rabbit dilator muscle, the PI response to α-adrenoceptor agonists is enhanced by sympathectomy (denervation supersensitivity) and diminished by chronic receptor activation (Abdel-Latif, Green and Smith, 1979; Akhtar and Abdel-Latif, 1986; Yousufzai and Abdel-Latif, 1987), with no change in α_1-adrenoceptor density as measured by radioligand binding (Page and Neufeld, 1978). Hence, α_1-adrenoceptor responsiveness in dilator muscle is thought to be regulated at the level of receptor-effector coupling (Abdel-Latif, 1989).

In addition to α_1-adrenoceptors mediating contraction, β-adrenoceptors mediating relaxation have been described in dilator muscle preparations from cat, rabbit and monkey eyes (van Alphen, Kern and Robinette, 1965). The presence of β-adrenoceptors in rabbit dilator muscle has also been confirmed by radioligand binding using [^{125}I]-cyanopindolol (Kahle, Kaulen and Wollensak, 1990). β-Adrenergic relaxation of the dilator muscle is presumably mediated by stimulation of cAMP biosynthesis, as has been demonstrated in sphincter muscle (Tachado, Akhtar and Abdel-Latif, 1989). Although the biological function of β-adrenoceptors in dilator muscle is unknown, these receptors may have a modulatory role in limiting the contractile responses to α_1-adrenoceptor activation.

The involvement of purines in sympathetic neural regulation of the dilator muscle is suggested by recent experiments of Muramatsu, Kigoshi and Oda (1994). Transmural electrical stimulation of the isolated rabbit dilator elicits phasic contractions consisting of both adrenergic and non-adrenergic components. The non-adrenergic contractions are mimicked by ATP and abolished by guanethidine or prolonged treatment with 2-methylthio-ATP, suggesting that they are mediated by purines released from the sympathetic nerves. Both the electrically-evoked and ATP-induced contractions were potentiated by α,β-methylene ATP and unaffected by suramin, indicating the involvement of an unusual type of P2 purinoceptor. Other evidence indicates that purines may act prejunctionally to modulate sympathetic neurotransmitter release in the iris (Fuder, 1994). In isolated rat irides, exogenous ATP and adenosine strongly inhibit electrically evoked release of NA via activation of prejunctional A_1-adenosine and P2Y-like purinoceptors (Fuder et al., 1992; Fuder and Muth, 1993). Purinergic antagonists significantly enhance evoked NA release in this preparation, suggesting that the prejunctional purine receptors undergo tonic activation by endogenously released purines (Fuder, 1994). Clearly, further studies are needed to elucidate the physiological role of purinergic neurotransmission in iris dilator muscle.

As discussed previously, NPY is present in many of the sympathetic nerve fibres that innervate the iris. The influence of NPY on sympathetic neural contraction of rabbit dilator muscle has been examined by Piccone et al. (1988). NPY has no effect on resting tension of the dilator muscle or on noradrenergic contractions elicited by transmural electrical stimulation. However, NPY potentiates the contractile responses to low concentrations of phenylephrine, an α_1-agonist. The mechanism for this potentiation is unclear, but may be due to NPY-mediated inhibition of adenylyl cyclase, as has been demonstrated in the rabbit iris-ciliary body (Bausher and Horio, 1990; Jumblatt and Gooch, 1990). NPY also mediates prejunctional inhibition of NA release in both rabbit and human irides (Ohia and Jumblatt, 1990b; Jumblatt, Ohia and Hackmiller, 1993). Taken together, these findings suggest that NPY functions as a modulator of sympathetic neurotransmission in the iris dilator, providing feedback regulation of NA secretion and postjunctional enhancement of NA-induced muscle contraction.

IRIS DILATOR MUSCLE — PARASYMPATHETIC REGULATION

Parasympathetic cholinergic innervation to the dilator muscle has been documented for several species (Laties and Jacobowitz, 1964; Nishida and Sears, 1969; Nomura and Smelser, 1974), including humans (Hutchins and Hollyfield, 1984). Considerable physiological evidence indicates that parasympathetic input to the iris dilator has an inhibitory influence on muscle contraction. In isolated human dilator muscle, transmural electrical stimulation in the presence of phentolamine induces a pronounced relaxation that is antagonized by atropine (Yoshitomi, Ito and Inomata, 1985). The electrically-evoked relaxation is blocked by tetrodotoxin (TTX), mimicked by carbamylcholine and potentiated by the AChE inhibitor neostigmine, showing the response to be mediated by neurally-released ACh. Similar findings have been reported in rat, bovine, dog and porcine dilator muscle preparations (Narita and Watanabe, 1982; Suzuki, Oso and Kobayashi, 1983; Yoshitomi and Ito, 1986; Ryang *et al.*, 1990). In the rat dilator muscle, carbamylcholine has a biphasic effect on muscle tone, causing relaxation at low concentrations and contraction at higher concentrations (Narita and Watanabe, 1982). Pharmacological evidence suggests that both responses are mediated by a single population of muscarinic receptors of the M_3 subtype (Shiraishi and Takayanagi, 1993; Masuda *et al.*, 1995). The biochemical mechanism for muscarinic relaxation of the dilator muscle is unknown, but possibly involves the activation of a calcium-dependent form of adenylyl cyclase, as previously suggested for cholinergic relaxation of canine iris sphincter muscle (Tachado *et al.*, 1994). Other studies have demonstrated that sympathetic terminals in rat, rabbit and human irides are endowed with prejunctional muscarinic receptors of the M_2 type that mediate inhibition of NA release (Jumblatt and North, 1988; Jumblatt and Hackmiller, 1994). These receptors are tonically activated by endogenously released ACh under *in vitro* conditions (Jumblatt and North, 1988), and may contribute to parasympathetic inhibition of the dilator muscle *in vivo*. Such inhibition is thought to play an important physiological role in humans during the miotic and redilation phases of the pupillary light reflex (Heller *et al.*, 1990).

The effects of VIP on motor functions of dilator muscle have been investigated in rabbit and bovine dilator muscle preparations (Hayashi, Mochizuki and Masuda, 1983; Suzuki and Kobayashi, 1983). In bovine dilator muscle, VIP induces a relaxation of resting muscle tension that is smaller in magnitude but longer in duration than that obtained with cholinergic agonists (Suzuki and Kobayashi, 1983). High concentrations of VIP also inhibit neurally-mediated contractions of this muscle but have no effect on NA-induced muscle contraction or carbachol-induced relaxation, suggesting that VIP may act at both pre- and postjunctional sites to inhibit contraction. In the isolated rabbit dilator muscle, VIP-induced muscle relaxations are attributed to stimulation of cAMP accumulation (Hayashi, Mochizuki and Masuda, 1983). The physiological role of VIP in parasympathetic neural regulation of the iris dilator and sphincter muscles remains to be demonstrated.

PROSTAGLANDIN EFFECTS

Prostaglandins (PGs) of various types are synthesized in the iris in response to adrenergic and cholinergic stimulation (Engstrom and Dunham, 1982; Yousufzai and Abdel-Latif, 1984) and may contribute indirectly to the responses described above. Several types of prostaglandins (PGE_1, PGE_2, $PGF_{2\alpha}$) induce sphincter muscle contraction and/or miosis in rabbit, feline, canine and bovine eyes (Gustafsson, Hedqvist and Lagercrantz, 1975; Gustafsson, Hedqvist and Lundgren, 1980; Mandahl and Bill, 1983; Yoshitomi and Ito, 1986; Yousufzai, Chen and Abdel-Latif, 1988; Woodward *et al.*, 1989; Suzuki, Yoshino and Kobayashi, 1989), but have little or no miotic effect in monkey or humans (Stern and Bito, 1982; Bito, 1984; Camras and Miranda, 1989). This absence of pupillary side effects in humans has been a significant factor in the development of $PGF_{2\alpha}$ derivatives for treatment of glaucoma (for discussion see Bito, 1984).

Pharmacological evidence for the involvement of PGs in cholinergic neuromuscular transmission has been obtained in studies of isolated bovine sphincter muscle (Gustafsson, Hedqvist and Lagercrantz, 1975; Gustafsson, Hedqvist and Lundgren, 1980). Cholinergic contractions induced by transmural electrical stimulation were shown to be enhanced by several PGs and partially inhibited by SC19220, a prostaglandin EP_1-receptor antagonist (Gustafsson, Hedqvist and Lagercrantz, 1975). In a subsequent study, cyclooxygenase inhibitors (indomethacin, meclofenamic acid and eicosatetraynoic acid) were found to inhibit both electrically-evoked and acetylcholine-induced contractions of the sphincter muscle, suggesting that endogenous PGs may function as intermediates or modulators of the cholinergic response (Gustafsson, Hedqvist and Lundgren, 1980).

The potentiating effect of PGs on cholinergic neurotransmission in bovine sphincter muscle is attributed, in part, to prejunctional facilitation of ACh release from parasympathetic nerves (Gustafsson, Hedqvist and Lundgren, 1980; Suzuki, Yoshino and Kobayashi, 1989). PGs have also been demonstrated to mediate prejunctional inhibition of NA release from ocular sympathetic nerves (Ohia and Jumblatt, 1990a), providing a possible complementary mechanism for enhancement of miosis. In the latter study, inhibitors of PG synthesis had no effect on basal or stimulus-evoked NA release, indicating that endogenously generated eicosanoids do not influence adrenergic neurosecretion under the experimental conditions employed (i.e., rapid superfusion). It should be noted, however, that PGs are not inactivated metabolically by ocular tissues, and their removal from the eye *in vivo* occurs by the slow (~2 µl/min) process of aqueous humour outflow (Bito, 1984). Consequently, endogenous PGs may have a greater modulatory influence on autonomic neuromuscular transmission in the iris *in vivo* than *in vitro*. Such a role may be especially important under conditions of ocular inflammation, irritation or trauma.

REGULATION OF AQUEOUS HUMOUR FORMATION

Aqueous humour is continuously formed and secreted into the posterior chamber of the eye by the ciliary processes. Before discussing autonomic regulation of aqueous humour formation, it is necessary to give some background on the anatomy and physiology of the ciliary processes and secretory epithelium.

ANATOMY OF CILIARY PROCESSES AND CILIARY EPITHELIUM

The ciliary processes consist of 60–70 radially arranged ridges or outfoldings of the uveal coat that project from the posterior border of the ciliary body into the posterior chamber (Davson, 1990; Morrison and Freddo, 1996). Each process is lined on its external surface with a double layer of epithelial cells — the nonpigmented and pigmented ciliary epithelium (Figure 1.2). The nonpigmented ciliary epithelium (NPE), which faces the posterior chamber, is derived embryologically from the neuroepithelium that also gives rise to the neural retina. The pigmented ciliary epithelium (PE) represents a forward extension of the retinal pigmented epithelium. The PE and NPE cell layers face each other apex-to-apex, i.e., with their basal surfaces facing the ciliary stroma and posterior chamber, respectively. This is an atypical organization for secretory epithelium, which is usually arranged so that fluids are transported in a basal-to-apical direction. The NPE cells are connected on their lateral surfaces by tight junctions that provide a barrier to diffusion of large solutes (e.g., proteins) into the aqueous humour, but permit limited passage of

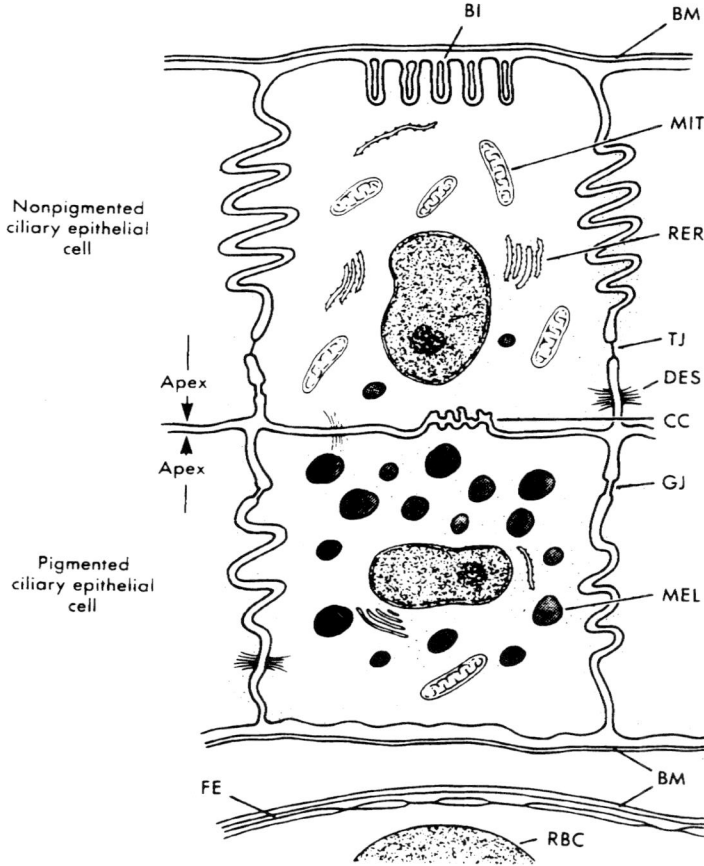

Figure 1.2   Schematic diagram of the ciliary epithelial bilayer. Abbreviations: basal infoldings facing aqueous humour (BI), basement membrane (BM), mitochondrion (MIT), rough endoplasmic reticulum (RER), tight junction (TJ), desmosome (DES), ciliary channels (CC), gap junctions (GJ), melanosome (MEL), fenestrated vascular endothelium (FE), red blood cell (RBC). See text for details. (Reproduced from Caprioli, 1992, with permission of Mosby Year Book Inc).

small ions. In addition, cells within each layer are joined by gap junctions as are cells between the two layers, making the bilayer a functional syncytium. The NPE cells contain numerous outpocketings and interdigitations on their basolateral surfaces that maximize the effective surface area for aqueous humour secretion. The stroma of ciliary processes is highly vascularized and contains dense networks of capillaries located in close proximity to the epithelium. The capillaries are fenestrated, allowing passage of plasma solutes and small proteins from the blood into the ciliary process stroma. The tight junctions between NPE cells thus represent the major diffusional barrier between the uveal blood supply and the aqueous humour (Morrison and Freddo, 1996).

PHYSIOLOGY OF AQUEOUS HUMOUR FORMATION

Aqueous humour formation involves the net movement of fluids from the ciliary stroma, across the ciliary epithelium and into the posterior chamber of the eye. This process occurs by the combined mechanisms of passive diffusion, ultrafiltration and active secretion (Davson, 1990). Although the relative contribution of each of these processes to aqueous humour formation has long been debated, the currently accepted view is that active secretion, rather than passive diffusion or ultrafiltration, is of primary importance (Davson, 1990; Krupin and Civan, 1996). This conclusion is supported by the following observations: (i) experimentally-induced changes in systemic blood pressure have little effect on aqueous humour formation (Bill, 1973); (ii) the concentrations of ascorbic acid and several other solutes in aqueous humour are substantially higher than in plasma, suggesting that aqueous humour is not merely an ultrafiltrate of plasma (Davson, 1990); (iii) the osmotic fluid permeability of the intact ciliary epithelium is too low to account for the physiological rate of aqueous humour formation (Brodwall and Fischbarg, 1982); (iv) protein leakage from the capillaries into the ciliary stroma creates an oncotic pressure that favors reabsorption rather than ultrafiltration of aqueous fluid (Bill, 1975). The key role of active transport in aqueous humour formation is underscored by the ability of ouabain, an inhibitor of Na^+/K^+ ATPase, to decrease aqueous formation by >70% (Bonting and Becker, 1964).

Current models of aqueous humour secretion are based on the transport of solutes (primarily Na^+) from the ciliary stroma to the posterior chamber. This transport creates an osmotic gradient that drives the passive movement of water across the ciliary epithelium and into the aqueous humour (Cole, 1977; Davson, 1990). The ionic transport mechanisms that underlie this process are complex. In addition to the Na^+/K^+ ATPase, numerous ion transport mechanisms (symports and antiports) and selective ion channels have been detected in intact ciliary epithelium and isolated NPE or PE cells (reviewed in Krupin and Civan, 1996). A simplified model of aqueous humour secretion, as proposed by Coca-Prados et al. (1995) is depicted in Figure 1.3. According to this model, Na^+, K^+ and Cl^- ions are taken up from the stroma by a $Na^+/K^+/2Cl^-$ symport system located on the basal surfaces of PE cells and diffuse via gap junctions into the NPE cells. The Na^+ ions are then pumped into the posterior chamber by Na^+/K^+ ATPase, which is concentrated on the basolateral membranes of the NPE cells. The Na^+ pump exchanges three Na^+ ions for two K^+ ions, creating a local hyperosmotic environment in the extracellular clefts that separate the NPE cells and inducing the passive flow of water into these clefts and, ultimately, to the posterior chamber. As Na^+ ions are being actively extruded from NPE cells, K^+ and Cl^- ions flow down their electrochemical gradients into the aqueous humour via selective K^+ and Cl^- channels. Some Na^+ secretion also occurs by paracellular diffusion across 'leaky' junctions between the NPE cells (Figure 1.3). The driving force for diffusional movement of Na^+ is a transepithelial electric potential, with the aqueous humour side negative with respect to the ciliary stroma (Kishida et al., 1981; Krupin et al., 1984).

Carbonic anhydrase (CA), which catalyzes hydration and dehydration of carbon dioxide, also plays a key role in aqueous humour formation (Maren, 1994). In 1954, Becker reported that acetazolamide, a CA inhibitor, lowers intraocular pressure (IOP) when

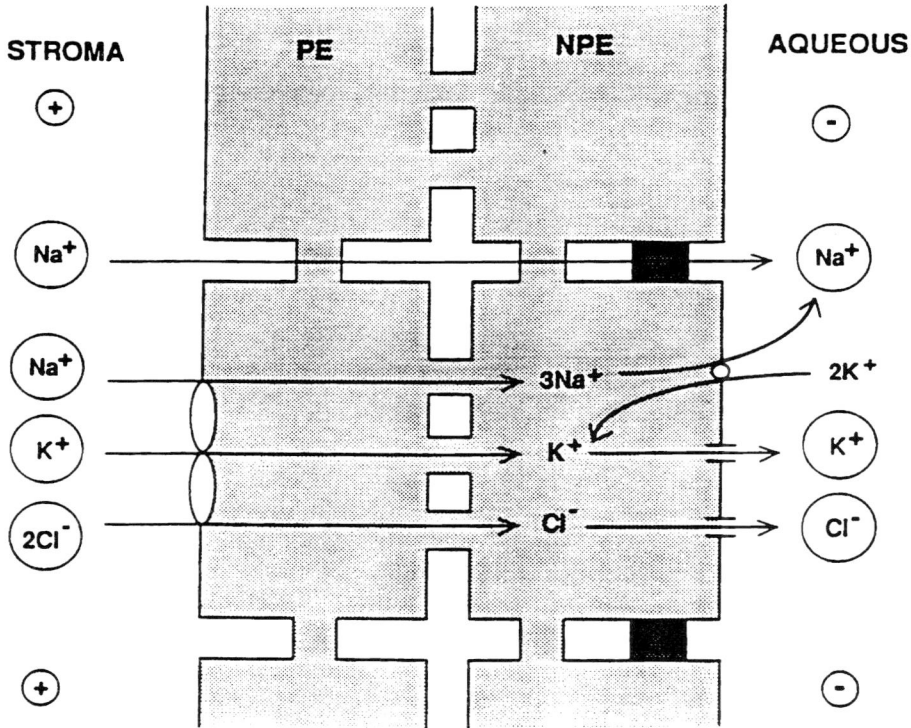

Figure 1.3   Hypothetical model of the major ion transport pathways in ciliary epithelium Na^+, Cl^- and K^+ ions are taken up by pigmented ciliary epithelial cells (PE) via the $Na^+/K^+/2Cl^-$ symport system located on the basal (stroma-facing) membranes and diffuse to nonpigmented ciliary epithelial cells (NPE) via gap junctions connecting the two cell layers. Na^+ is actively extruded into the aqueous system by Na^+/K^+ ATPase, whereas K^+ and Cl^- ions passively diffuse into the aqueous system via selective ion channels located on the basolateral surfaces of the NPE cells. In addition, a standing transepithelial potential (stroma positive with respect to aqueous) generates the driving force for paracellular diffusion of Na^+ ions into the aqueous via the 'leaky' gap junctions between adjacent NPE cells. (Reproduced with permission from Coca-Prados *et al.*, 1995).

administered systemically in humans. The IOP-lowering effect of acetazolamide was later shown to be due to a decrease in aqueous humour production. *In vivo* studies in the dog and monkey reveal that acetazolamide inhibits the influx of Na^+, HCO_3^- and Cl^- ions into the aqueous humour (Zimmerman *et al.*, 1976; Maren, 1977). Despite numerous investigations in this area, the precise role CA and HCO_3^- in aqueous humour secretion has not been resolved (Maren, 1994; Krupin and Civan, 1996). Some of the proposed functions of HCO_3^- include facilitation of Cl^- uptake via a HCO_3^-/Cl^- exchange mechanism located on the basolateral surfaces of PE cells (Helbig *et al.*, 1989), cellular hyperpolarization and potentiation of Cl^- channels in NPE cells (Carre *et al.*, 1992), and modulation of the transepithelial potential due to upregulation of K^+ channels (Krupin and Civan, 1996). Several distinct isoenzymes of CA have been described in ocular tissues (Maren, 1994). CA-II is the predominant isoenzyme found in the cytosol of both NPE and PE cells (Wistrand, Schenholm and Lonnerholm, 1986). In addition, a membrane-associated form of CA (possibly CA-IV) has recently been localized to the basolateral surfaces of NPE cells (Ridderstrale, Wistrand and Brechue, 1994). Both CA-II and CA-IV isoenzymes are inhibited by acetazolamide, however, and their relative contributions to aqueous humour secretion have not been established (Maren, 1994).

AUTONOMIC REGULATION AND PHARMACOLOGY

It will be recalled from the first part of this chapter that the ciliary processes are richly innervated by both sympathetic and parasympathetic nerve fibres (Table 1.1). Although important species differences exist, it is generally accepted that the autonomic nerves play an active role in aqueous humour regulation, as underscored by the longstanding clinical use of autonomic drugs for treatment of glaucoma. Aqueous humour formation, as measured by fluorophotometry and other techniques, is not a static process, but increases and decreases in a diurnal rhythm with corresponding changes in IOP (Brubaker, 1991). In humans, aqueous humour inflow is highest during the day (~2.5 µl/min), but falls to ~1.5 µl/min during sleep (Topper and Brubaker, 1985). In rabbits, the diurnal pattern of aqueous inflow is reversed, with low rates of inflow during daytime and increased rates at night (Smith and Gregory, 1989). Although the physiological mechanisms for circadian regulation of aqueous humour inflow are incompletely understood, the sympatho-adrenal system is strongly implicated (Braslow and Gregory, 1987; Brubaker, 1991). The physiological and pharmacological evidence for autonomic regulation of aqueous humour formation is reviewed in the following section, concluding with a discussion of the intracellular signalling pathways thought to underlie neurotransmitter-mediated effects on ciliary epithelial secretion.

Sympathetic regulation

Sympathetic neural influences on aqueous humour dynamics have been investigated in rabbits, cats, dogs and primates (Stone, 1996). In general, these studies have employed one of two methods: electrical stimulation of the cervical sympathetic nerves, or surgical denervation. Before discussing the results, it would be helpful to consider some of the limitations of these methods. Stimulation of the cervical nerve trunk simultaneously activates all sympathetic nerves to the eye. Due to the multiplicity of innervated tissues (blood vessels, ciliary epithelium, intraocular and extraocular muscles, trabecular meshwork), the responses obtained are often complex and difficult to interpret. Postganglionic sympathetic denervation, on the other hand, causes intraocular release of catecholamines and other neurotransmitters from degenerating nerve terminals. This may result in inflammatory responses or other secondary effects that have no direct relationship to the cessation of neural activity (Sears, 1984). Furthermore, chronic sympathetic denervation may produce long-term, adaptive changes in the eye, such as neural remodeling or altered neuroreceptor sensitivity (Langham and Weinstein, 1967; Bjorklund et al., 1985b; Brubaker, 1991). Despite these reservations, both experimental approaches have yielded useful insights into the role of sympathetic nerves in aqueous humour regulation.

The effects of sympathetic nerve stimulation on aqueous humour dynamics have been investigated extensively in rabbits. In anesthetized animals, high frequency (>10 Hz) electrical stimulation of the preganglionic sympathetic nerve trunk or the superior cervical ganglion elicits an acute decrease in IOP that is attributed to uveal vasoconstriction and a reduction in blood volume, with a secondary decrease in aqueous humour formation (Greaves and Perkins, 1952; Langham and Rosenthal, 1966). In conscious rabbits, prolonged (3 h) stimulation of the ocular sympathetic nerves at 9–12 Hz causes a similar acute reduction of IOP that gradually returns to baseline (Belmonte *et al.*, 1987). The IOP decrease is blocked by phentolamine but is unaffected by the β-adrenergic antagonist timolol, suggesting the response is primarily due to α-adrenergic vasoconstriction. In contrast to the effects obtained with high stimulation frequencies, low frequency (< 5 Hz) sympathetic nerve stimulation in conscious rabbits induces an increase in IOP (Gallar and Liu, 1993). This response is accompanied by release of NA, but not NPY, into the aqueous humour. Because NPY is preferentially secreted at high stimulation frequencies, it has been suggested that differential release of NPY *vs* NA may underlie the frequency-dependent effects of sympathetic nerve stimulation on aqueous humour formation (Gallar and Liu, 1993). To test this hypothesis, however, would require topically effective NPY antagonists which are currently unavailable.

The effects of sympathetic denervation on aqueous humour dynamics are variable, depending on the species and mode of denervation (Stone, 1996). In rabbits, section of the preganglionic sympathetic nerves has little acute or long term effect on aqueous humour inflow or IOP (Langham and Taylor, 1960). Preganglionic nerve section or superior cervical ganglionectomy, however, abolishes the circadian increases in aqueous humour inflow, IOP and release of NA into aqueous humour during the dark phase of the light:dark cycle (Braslow and Gregory, 1987; Liu and Dacus, 1991; Yoshitomi and Gregory, 1991; Yoshitomi, Horio and Gregory, 1991), indicating that these circadian changes are dependent on intact sympathetic innervation. Superior cervical ganglionectomy in rabbits also causes a transient (3–4 h) decrease in IOP that is attributed to increased outflow of aqueous humour mediated by the discharge of catecholamines from degenerating nerve terminals (Langham and Taylor, 1960; Sears and Barany, 1960). In monkeys, superior cervical ganglionectomy has no significant long-term effects on aqueous humour formation or IOP (Gabelt *et al.*, 1995). Similarly, human patients with Horner's syndrome (i.e., a sympathetic neural deficit in one or both eyes) exhibit normal rates and circadian patterns of aqueous humour formation in the denervated eyes (Wentworth and Brubaker, 1981). Hence, the ocular sympathetic nerves are thought to play a relatively minor role in regulation of aqueous humour inflow in the primate eye.

An extensive literature exists concerning the pharmacological effects of exogenously administered adrenergic agents on aqueous humour formation (Sears, 1984; Sugrue, 1989). Much recent work has focused on β-adrenergic agents, which have a major influence on aqueous humour secretion in humans and are consequently used widely for treatment of glaucoma (Brubaker, 1991; Kaufman and Mittag, 1994). Adrenaline, a mixed α- and β-adrenergic agonist, stimulates aqueous humour inflow in normal human subjects (Townsend and Brubaker, 1980), an effect that is blocked by the β-adrenergic antagonist timolol (Higgins and Brubaker, 1980). Isoprenaline, a selective β-adrenergic agonist, also stimulates aqueous humour inflow in humans, but only during sleep when the basal rate of aqueous formation is low (Larson and Brubaker, 1988). Conversely, timolol and other β-antagonists decrease the formation of aqueous humour during the daytime, but are ineffective during sleep (Topper and Brubaker, 1985; Brubaker, 1991). In rabbits, timolol partially inhibits the circadian increase in aqueous humour inflow during the dark phase, but has no effect in sympathetically denervated eyes (Gregory, 1990). β$_2$-Adrenoceptors positively coupled to adenylyl cyclase have been identified in ciliary process epithelium from human, rabbit and bovine eyes (Nathanson, 1980, 1981; Cepelik and Cernohorsky, 1981; Elena *et al.*, 1984). Taken together, these findings support the hypothesis that activation of β$_2$-adrenoceptors and subsequent stimulation of cAMP synthesis lead to increased aqueous humour formation by the ciliary epithelium (Nathanson, 1980).

α$_2$-Adrenergic mechanisms are also implicated in regulation of aqueous humour formation. In humans and experimental animals, α$_2$-adrenoceptor agonists (e.g., clonidine, apraclonide, brimonidine, UK-14,304-18) markedly inhibit aqueous humour formation (Bill and Heilmann, 1975; Lee, Topper and Brubaker, 1984; Burke and Potter, 1986; Gabelt *et al.*, 1994). The presence of α$_2$-adrenoceptors negatively coupled to adenylyl cyclase has been documented in rabbit and human ciliary processes and ciliary epithelium (Mittag and Tormay, 1985; Bausher, Gregory and Sears, 1987; Bausher and Horio, 1995). Activation of α$_2$-adrenoceptors in these tissues does not modify basal cAMP synthesis, but inhibits the increased cAMP formation induced by β-adrenoceptor agonists and other agents (Bausher, Gregory and Sears, 1987; Bausher and Horio, 1995). These findings support the postulated stimulatory influence of cAMP on aqueous humour formation and are consistent with the observation that α$_2$-adrenergic agonists and β-blockers cause a similar (30–40%) decrease in aqueous humour inflow in humans (Gharagozloo, Relf and Brubaker, 1988; Brubaker, 1991).

Interaction of α$_2$-adrenergic agonists with prejunctional α$_2$-adrenoceptors to inhibit the release of NA from ocular sympathetic nerves may also occur (Farnebo and Hamberger, 1971; Jumblatt, Liu and North, 1987; Jumblatt, Ohia and Hackmiller, 1993). Although a reduction of NA release would theoretically diminish β-adrenergic tone in the ciliary epithelium, the extent to which prejunctional inhibition contributes to the α$_2$-agonist-mediated decrease in aqueous humour formation *in vivo* remains unclear. In rabbits and cats, the ocular hypotensive response to the α$_2$-agonist UK-14,304-18 is attenuated by sympathetic denervation, supporting a possible prejunctional site of action (Burke and Potter, 1986). In cynomolgus monkeys, however, sympathetic denervation has no effect on the suppression of aqueous humour inflow by apraclonidine or brimonidine (Gabelt *et al.*, 1994), indicating that the responses to these agonists are mediated primarily by postjunctional α$_2$-adrenoceptors. In support of this finding, apraclonidine is also effective in lowering IOP in the sympathetically denervated eyes of humans with unilateral Horner's syndrome (Morales, Ho and Crosson, 1992).

As described earlier, NPY exists as a cotransmitter in sympathetic nerves of the ciliary processes (Table 1.1). Several lines of evidence suggest a possible role for NPY in regulation of aqueous humour formation. When administered intravenously in rabbits, NPY markedly constricts the uveal blood vessels and reduces ocular blood flow (Nilsson, 1991), including blood flow to the ciliary processes (Funk, Wagner and Wild, 1992). NPY receptors coupled to inhibition of adenylyl cyclase have been identified in both rabbit ciliary processes and ciliary epithelium (Bausher and Horio, 1990; Cepelik and Hynie, 1990; Jumblatt and Gooch, 1990). These NPY receptors are functionally analogous to α_2-adrenoceptors in that they have no effect on basal adenylyl cyclase activity but inhibit β-adrenergic stimulation of cAMP formation (Bausher and Horio, 1990; Jumblatt and Gooch, 1990). Other studies show that NPY mediates prejunctional inhibition of NA release from sympathetic nerves in isolated rabbit and human iris-ciliary body preparations (Ohia and Jumblatt, 1990b; Jumblatt, Ohia and Hackmiller, 1993). Based on the above observations, NPY is postulated to have an inhibitory influence on aqueous humour formation (Nilsson and Bill, 1994).

Recent pharmacological evidence suggests that purinergic mechanisms may also contribute to regulation of aqueous humour formation (Crosson and Gray, 1994; Crosson, 1995). In rabbits, topical administration of the relatively selective adenosine A_1 agonists N^6-cyclohexyladenosine (CHA) and R-(–)N^6-(2-phenylisopropyl)adenosine (R-PIA) cause a dose-dependent reduction of IOP and an associated decrease in aqueous humour formation. These responses are unaltered by sympathetic denervation and are correlated with A_1-purinoceptor-mediated inhibition of cAMP formation in the ciliary processes (Crosson and Gray, 1994; Crosson, 1995). *In vitro* studies show that CHA and R-PIA have no significant prejunctional effects on NA release in the rabbit iris-ciliary body, confirming that the inhibitory effect on aqueous humour inflow is mediated primarily by activation of postjunctional A_1 purine receptors (Crosson, 1995). Biochemical studies of cultured human NPE cells and bovine PE cells reveal that both cell types contain A_1 receptors coupled to adenylyl cyclase inhibition, A_2 receptors coupled to adenylyl cyclase activation and P2 receptors linked to phospholipase C and PI hydrolysis (Wax *et al.*, 1993), suggesting that the effects of endogenous or exogenous purines on aqueous humour formation are likely to be complex. Nevertheless, these findings should prompt further investigation of the involvement of purinergic mechanisms in aqueous humour regulation.

Parasympathetic regulation

Cholinomimetic drugs (e.g., pilocarpine, physostigmine) have been used for more than a century to treat glaucoma. It is well established that cholinomimetic drugs lower intraocular pressure by facilitating outflow of aqueous humour via the trabecular meshwork and Schlemm's canal (Kaufman, 1984). Less is known, however, about parasympathetic or cholinergic influences on aqueous humour formation. The effects of parasympathetic nerve stimulation on aqueous humour dynamics have been examined in several species (reviewed in Stone, 1996). In rabbits, preganglionic stimulation of the oculomotor nerve increases aqueous humour inflow and IOP (Uusitalo, 1972; Stjernschantz, 1976). The rise in IOP can be partially blocked by curare, suggesting that contraction of the extraocular muscles contributes to this response (Stjernschantz, 1976). In the enucleated arterially-perfused cat eye, electrical stimulation of the ciliary ganglion induces a sustained increase in aqueous humour formation (Macri and Cevario, 1975). The pressure-dependence of this response, however, suggests that increased ultrafiltration rather than secretion of aqueous humour may be the underlying mechanism.

The involvement of the facial nerve in aqueous humour regulation has been examined in rabbit, feline and monkey eyes (Gloster, 1961; Nillson, Linder and Bill, 1985). In all three species, electrical stimulation of the facial nerve causes an increase in uveal blood flow and IOP. Conversely, preganglionic denervation of the facial nerve lowers IOP in monkey eyes (Ruskell, 1970a). The rise in IOP induced by facial nerve stimulation is attributed to uveal vasodilation and increased episcleral venous pressure. This vasodilation is resistant to muscarinic antagonists and may be mediated by the parasympathetic cotransmitter VIP (Nillson, Linder and Bill, 1985). Other evidence suggests, however, that VIP may have a direct influence on aqueous humour formation (Bill, 1991). In cynomolgus monkeys, intracameral administration of VIP markedly stimulates aqueous humour inflow (Nilsson, Sperber and Bill, 1986; Nilsson *et al.*, 1990). The presence of VIP receptors coupled to adenylyl cyclase activation has been documented in rabbit and human NPE cells (Mittag, Tormay and Podos, 1987; Crook *et al.*, 1994). In rabbit ciliary epithelium, both VIP- and β-adrenoceptor-stimulated cAMP formation can be inhibited by α_2-agonists, muscarinic agonists and NPY (Bausher, Gregory and Sears, 1989; Jumblatt and Gooch, 1990; Jumblatt, North and Hackmiller, 1990), showing that these receptors co-regulate a common adenylyl cyclase system (Nilsson and Bill, 1994).

Pharmacological studies of the effects of cholinomimetic drugs on aqueous humour formation have yielded contradictory results (Kaufman, 1984). In rabbits, for example, cholinergic agonists are reported to increase (Becker, 1962; Miichi and Nagataki, 1982), decrease (Berggren, 1965, 1970), or have no effect (Green and Padgett, 1979) on aqueous humour formation. Biochemical evidence suggests that ciliary epithelium contains at least two types of muscarinic receptors, one of which (possibly M_2 or M_4) mediates adenylyl cyclase inhibition and the other (M_3) activation of phospholipase C and PI hydrolysis (Tobin and Osborne, 1988; Mallorga *et al.*, 1989; Wax and Coca-Prados, 1989; Jumblatt, North and Hackmiller, 1990; Crook and Polansky, 1992). It is tempting to speculate that simultaneous muscarinic stimulation of PI hydrolysis and inhibition of cAMP synthesis might have mutually opposing actions on ciliary epithelial secretion (Krupin and Civan, 1996). Such functional antagonism might explain the inconsistent effects of cholinomimetic agents on aqueous humour inflow, as cited above.

SIGNALLING PATHWAYS AND AQUEOUS HUMOUR SECRETION

An emergent theme from the preceding discussion is the central importance of the adenylyl cyclase system in regulation of aqueous humour formation. Thus, we have seen that agents which activate adenylyl cyclase in ciliary epithelium (β-adrenoceptor agonists, VIP) stimulate aqueous humour formation, while agents that inhibit adenylyl cyclase (e.g., α_2-adrenoceptor agonists, A_1 purinergic agonists, NPY) reduce aqueous formation. The term 'formation' is used broadly here to mean the net influx of fluid into the eye by the combined processes of secretion, ultrafiltration and transepithelial diffusion. Current methods used for *in vivo* measurement of aqueous humour inflow (e.g., fluorophotometry) do not readily distinguish between these mechanisms. Because active secretion is the predominant mechanism for aqueous humour formation, however, one may assume that the adenylyl cyclase system has a direct influence on ciliary epithelial fluid transport. To gain a better understanding of this signalling pathway, we now turn our attention to the role of cAMP and other intracellular messengers in regulation of ion transport in non-pigmented and pigmented ciliary epithelial cells.

The observation that ouabain strongly inhibits aqueous humour formation (Bonting and Becker, 1964) has prompted biochemical studies of Na^+/K^+ ATPase in ciliary epithelium as a potential target for regulation by neurotransmitters and second messenger systems (Krupin and Civan, 1996). Na^+/K^+ ATPase activity is regulated primarily by the intracellular concentration of Na^+, which is rate limiting in intact cells. However, there is increasing evidence that at least some forms of Na^+/K^+ ATPase can be acutely modulated by intracellular signals (cAMP, Ca^{2+}, protein kinase C), presumably via protein phosphorylation (Bertorello and Katz, 1995; Ewart and Klip, 1995). Treatment of isolated rabbit ciliary epithelial membranes with cAMP-dependent protein kinase (PKA) or intact tissues with cAMP activators (forskolin, dibutyryl-cAMP, cAMP phosphodiesterase inhibitors) inhibits Na^+/K^+ ATPase activity (Delamere, Socci and King, 1990; Delamere and King, 1992). Consistent with these observations, NPE cells contain the cAMP-regulated phosphoprotein DARPP-32 (Stone et al., 1986), a protein phosphatase inhibitor that has been implicated in regulation of Na^+/K^+ ATPase activity in renal tubule cells (Aperia et al., 1991). Other results show that Na^+/K^+ ATPase in cultured human NPE cells is stimulated by activation of protein kinase C, suggesting a second signalling pathway by which Na^+ pump activity may be acutely regulated (Mito and Delamere, 1993). The inhibitory effects of cAMP on Na^+/K^+ ATPase in ciliary epithelium, however, are difficult to reconcile with the positive influence of adenylyl cyclase activation on aqueous humour formation in vivo. Thus, we must consider other ion transport systems as potential targets for regulation by cAMP.

As indicated in Figure 1.3, Na^+, K^+ and Cl^- are transported into ciliary epithelium by a $Na^+/K^+/2Cl^-$ cotransport system located on the basal (stroma-facing) membranes of PE cells. A recent study by Crook and Polansky (1994) shows that $Na^+/K^+/2Cl^-$ cotransport in fetal human ciliary epithelial cells can be stimulated ~1.5–2-fold by forskolin or cell-permeable cAMP analogs. Furthermore, the stimulatory effects of these agents are blocked by the protein kinase A inhibitor H-89, showing the responses to be due to cAMP-mediated protein phosphorylation. Other results indicate that $Na^+/K^+/2Cl^-$ cotransport in these cells is down-regulated by activation of protein kinase C (Crook, von Brauchitsch and Polansky, 1992). As in the kidney, the $Na^+/K^+/2Cl^-$ cotransporter in ciliary epithelial cells can be blocked by the 'high ceiling' diuretics furosemide or bumetanide. In theory, stimulation of $Na^+/K^+/2Cl^-$ uptake by cAMP may contribute to the secretory activity of these cells. However, furosemide and similar agents have little effect on aqueous humour formation in vivo, making it unlikely that $Na^+/K^+/2Cl^-$ transport has a rate-limiting role in aqueous humour secretion (Maren, 1994; Krupin and Civan, 1996).

As discussed previously, K^+ and Cl^- are released from NPE cells into the aqueous humour via selective K^+ and Cl^- channels located on the basolateral membranes (Figure 1.3). Recent evidence indicates that both of these ion channels are subject to regulation by intracellular signalling pathways. When exposed to hypoosmotic medium, NPE cells display an initial swelling followed by a regulatory volume decrease (RVD) that is mediated by activation of Cl^- channels and, to a lesser extent, K^+ channels (Farahbakhsh and Fain, 1988; Civan et al., 1992; Adorante and Cala, 1995). Pharmacological analysis of the RVD response shows that K^+ channels, which under normal conditions are largely in an 'open' state, are further activated by intracellular Ca^{2+} and cAMP (Civan et al., 1992; Civan, Coca-Prados and Peterson-Yantorno, 1994; Adorante and Cala, 1995). Cl^- channels, which have a rate-limiting role in the RVD response, are little affected by intracellular Ca^{2+} (Adorante and Cala, 1995) but are down-regulated by activation of protein kinase C (Civan, Coca-Prados and Peterson-Yantorno, 1994). Electrophysiological studies of isolated bovine and canine NPE cells show that Cl^- channels are acutely activated by β-adrenergic agonists and/or dibutyryl-cAMP (Chen *et al.*, 1994; Edelman, Loo and Sachs, 1995). These findings suggest that cAMP-dependent Cl^- channels in NPE cells may contribute to β-adrenergic or VIP-mediated stimulation of aqueous humour formation. Much more work will be required, however, to elucidate the biochemical regulation of these channels and to establish the physiological and pharmacological links between ion transport in ciliary epithelium and aqueous humour formation *in vivo*.

REFERENCES

Abdel-Latif, A.A. (1989). Calcium-mobilizing receptors, polyphosphoinositides, generation of second messengers and contraction in the mammalian iris smooth muscle: historical perspectives and current status. *Life Sciences*, **45**, 757–786.

Abdel-Latif, A.A. and Zhang, Y. (1991). Effects of surgical sympathetic denervation on myo-inositol trisphosphate production and contraction in the dilator and sphincter smooth muscles of the rabbit iris: evidence for interaction between the cyclic AMP and calcium signaling systems. *Journal of Neurochemistry*, **57**, 447–457.

Abdel-Latif, A.A., Green, K. and Smith, J.P. (1979). Sympathetic denervation and the triphosphoinositide effect in the iris smooth muscle: a biochemical method for the determination of α-adrenergic receptor denervation supersensitivity. *Journal of Neurochemistry*, **32**, 225–228.

Abdel-Latif, A.A., Smith, J.P. and Akhtar, R.A. (1985). Polyphosphoinositides and muscarinic cholinergic and a_1-adrenergic receptors in the iris smooth muscle. In *Inositol and Phosphoinositides: Metabolism and Regulation*, edited by J.E. Bleasdale, J. Eichberg and G. Hauser, pp. 275–298. Clifton: Humana Press.

Abdel-Latif, A.A., Yousufzai, S.Y., De, S. and Tachado, S.E. (1992). Carbachol stimulates adenylate cyclase and phospholipase C and muscle contraction-relaxation in a reciprocal manner in dog iris sphincter smooth muscle. *European Journal of Pharmacology*, **226**, 351–361.

Adorante, J.S. and Cala, P.M. (1995). Mechanisms of regulatory volume decrease in nonpigmented human ciliary epithelial cells. *American Journal of Physiology*, **268**, C721–C731.

Akhtar, R.A. and Abdel-Latif, A.A. (1986). Surgical sympathetic denervation increases α_1-adrenoceptor-mediated accumulation of myo-inositol trisphosphate and muscle contraction in rabbit iris dilator smooth muscle. *Journal of Neurochemistry*, **46**, 96–104.

Akhtar, R.A. and Abdel-Latif, A.A. (1991). The effect of M&B 22948 on carbachol-induced inositol trisphosphate accumulation and contraction in iris sphincter smooth muscle. *European Journal of Pharmacology*, **206**, 291–295.

Almegard, B., Stjernschantz, J. and Bill, A. (1992). Cholecystokinin contracts isolated human and monkey iris sphincters; a study with CCK receptor antagonists. *European Journal of Pharmacology*, **211**, 183–187.

Aperia, A., Fryckstedt, J., Svensson, L., Hemmings, H.C.Jr., Nairn, A.C. and Greengard, P. (1991). Phosphorylated Mr 32,000 dopamine- and cAMP-regulated phosphoprotein inhibitor of Na^+, K^+-ATPase activity in renal tubule cells. *Proceedings of the National Academy of Sciences USA*, **88**, 2798–2801.

Banno, H., Imaizumi, Y. and Watanabe, M. (1985). Pharmaco-mechanical coupling in the response to acetylcholine and substance P in the smooth muscle of the rat iris sphincter. *British Journal of Pharmacology*, **85**, 905–911.

Barnett, N.L. and Osborne, N.N. (1993). The presence of serotonin (5-HT$_1$) receptors negatively coupled to adenylate cyclase in rabbit and human iris-ciliary processes. *Experimental Eye Research*, **52**, 209–216.

Bausher, L.P. and Horio, B. (1990). Neuropeptide Y and somatostatin inhibit stimulated cyclic AMP production in rabbit ciliary processes. *Current Eye Research*, **9**, 371–377.

Bausher, L.P. and Horio, B. (1995). Regulation of cyclic AMP production in adult human ciliary processes. *Experimental Eye Research*, **60**, 43–48.

Bausher, L.P., Gregory, D.S. and Sears, M.L. (1987). Interaction between α_2- and β_2-adrenergic receptors in rabbit ciliary processes. *Current Eye Research*, **6**, 497–505.

Bausher, L.P., Gregory, D.S. and Sears, M.L. (1989). α_2-Adrenergic and VIP receptors in responses in rabbit ciliary processes interact. *Current Eye Research*, **8**, 47–54.

Becker, B. (1954). Decrease in intraocular pressure in man by a carbonic anhydrase inhibitor, diamox. *American Journal of Ophthalmology*, **37**, 13–14.

Becker, B. (1962). The measurement of rate of aqueous flow with iodide. *Investigative Ophthalmology*, **1**, 52–58.

Beckers, H.J., Klooster, J., Vrensen, G.F. and Lamers, N.P. (1992). Ultrastructural identification of trigeminal nerve endings in the rat cornea and iris. *Investigative Ophthalmology and Visual Science*, **33**, 1979–1986.

Beding-Barnekow, B. and Brodin, E. (1989). Neurokinin A, neurokinin B and neuropeptide K in the rabbit iris: a study comparing different extraction methods. *Regulatory Peptides*, **25**, 199–206.

Bell, C. (1988). Dopamine release from sympathetic nerve terminals. *Progress in Neurobiology*, **30**, 193–208.

Belmonte, C., Bartels, S.P., Liu, J.H.K and Neufeld, A.H. (1987). Effects of stimulation of the ocular sympathetic nerves on IOP and aqueous humour flow. *Investigative Ophthalmology and Visual Science*, **28**, 1649–1654.

Berggren, L. (1965). Effect of parasympathomimetic and sympathomimetic drugs on secretion *in vitro* by the ciliary processes of the rabbit eye. *Investigative Ophthalmology*, **4**, 91–97.

Berggren, L. (1970). Further studies on the effect of autonomic drugs on *in vitro* secretory activity of the rabbit eye ciliary processes. *Acta Ophthalmologica*, **48**, 293–302.

Bergmanson, J.P.G. (1977). The ophthalmic innervation of the uvea in monkeys. *Experimental Eye Research*, **24**, 225–240

Bertorello, A.M. and Katz, A.I. (1995). Regulation of Na^+-K^+ pump activity: pathways between receptors and effectors. *News in Physiological Sciences*, **10**, 253–259.

Bill, A. (1973). The role of ciliary body blood flow and ultrafiltration in aqueous humour formation. *Experimental Eye Research*, **16**, 287–293.

Bill, A. (1975). Blood circulation and fluid dynamics in the eye. *Physiological Reviews*, **55**, 383–417.

Bill, A. (1991). Effects of some neuropeptides on the uvea. *Experimental Eye Research*, **53**, 3–11.

Bill, A. and Heilmann, K. (1975). Ocular effects of clonidine in cats and monkeys (*Macaca irus*). *Experimental Eye Research*, **21**, 481–488.

Bito, L.Z. (1984). Prostaglandins, other eicosanoids, and their derivatives as potential antiglaucoma agents. In *Applied Pharmacology in the Medical Treatment of Glaucomas*, edited by S.M. Drance, pp. 477–505. Orlando: Grune and Stratton.

Bjorklund, H., Hoffer, B., Olson, L., Palmer, M. and Seiger, A. (1984). Enkephalin immunoreactivity in iris nerves: distribution in normal and grafted irides, persistence and enhanced fluorescence after denervations. *Histochemistry*, **80**, 1–7.

Bjorklund, H., Fahrenkrug, J., Sieger, A., Vanderhaeghen, J.J. and Olson, L. (1985a). On the origin and distribution of vasoactive intestinal polypeptide, peptide HI-, and cholecystokinin-like immunoreactive nerve fibres in the rat iris. *Cell and Tissue Research*, **242**, 1–7.

Bjorklund, H., Hökfelt, T., Goldstein, M., Terenius, L. and Olson, T. (1985b). Appearance of the noradrenergic markers tyrosine hydroxylase and neuropeptide Y in cholinergic nerves of the iris following sympathectomy. *Journal of Neuroscience*, **5**, 1633–1640.

Blaschko, H. (1939). The specific action of l-DOPA-decarboxylase. *Journal of Physiology*, **96**, 50–51P.

Bognar, I.T., Wesner, M.T. and Fuder, H. (1990). Muscarine receptor types mediating autoinhibition of acetylcholine release and sphincter contraction in the guinea-pig iris. *Naunyn-Schmiedeberg's Archives of Pharmacology*, **341**, 22–29.

Bonting, S.L. and Becker, B. (1964). Studies on sodium-potassium activated adenosinetriphosphatase. *Investigative Ophthalmology*, **3**, 523–533.

Braslow, R.A. and Gregory, D.S. (1987). Adrenergic decentralization modifies the circadian rhythm of intraocular pressure. *Investigative Ophthalmology and Visual Science*, **28**, 1730–1732.

Brodwall, J. and Fischbarg, J. (1982). The hydraulic conductivity of rabbit ciliary epithelium *in vitro*. *Experimental Eye Research*, **34**, 121–129.

Brubaker, R.F. (1991). Flow of aqueous humour in humans. *Investigative Ophthalmology and Visual Science*, **32**, 3145–3166.

Burke, J.A. and Potter, D.E. (1986). Ocular effects of a relatively selective a_2- agonist (UK-14,304–18) in cats, rabbits and monkeys. *Current Eye Research*, **5**, 665–676.

Burnstock, G. (1990). Noradrenaline and ATP as cotransmitters in sympathetic nerves. *Neurochemistry International*, **17**, 357–368.

Butler, J.M., Ruskell, G.L., Cole, D.F., Unger, W.G., Zhang, S.Q., Blank, M.A., McGregor, G.P. and Bloom, S.R. (1984). Effects of VIIth (facial) nerve degeneration on vasoactive intestinal polypeptide and substance P levels in ocular and orbital tissues of the rabbit. *Experimental Eye Research*, **39**, 523–532.

Camras, C.B. and Miranda, O.C. (1989). The putative role of prostaglandins in surgical miosis. *Progress in Clinical and Biological Research*, **312**, 197–210.

Caprioli, J. (1992). The ciliary epithelia and aqueous humor. In *Alder's Physiology of the Eye*, edited by W.M. Hart, pp. 228–241. St. Louis: Mosby.

Carre, D.A., Tang, C.S.R., Krupin, T. and Civan, M.M. (1992). Effect of bicarbonate on intracellular potential of rabbit ciliary epithelium *Current Eye Research*, **11**, 609–624.

Cepelik, J. and Cernohorsky, M. (1981). The effects of adrenergic agonists and antagonists on the adenylate cyclase in albino rabbit ciliary processes. *Experimental Eye Research*, **32**, 291–299.

Cepelik, J. and Hynie, S. (1990). Inhibitory effects of neuropeptide Y on adenylate cyclase of rabbit ciliary processes. *Current Eye Research*, **2**, 121–128.

Chen, S., Inoue, R., Inomata, H. and Ito, Y. (1994). Role of cyclic AMP-induced Cl^- conductance in aqueous humour formation by the dog ciliary epithelium. *British Journal of Pharmacology*, **112**, 1137–1145.

Civan, M.M., Peterson-Yantorno, K., Coca-Prados, M. and Yantorno, R.E. (1992). Regulatory volume decrease by cultured non-pigmented ciliary epithelial cells. *Experimental Eye Research*, **54**, 181–191.

Civan, M.M, Coca-Prados, M. and Peterson-Yantorno, K. (1994). Pathways signaling the regulatory volume decrease in cultured nonpigmented ciliary epithelial cells. *Investigative Ophthalmology and Visual Science*, **35**, 2876–2886.

Coca-Prados, M., Anguita, J., Chalfant, M.L. and Civan, M. (1995). PKC-sensitive Cl^- channels associated with ciliary epithelial homologue of pI_{Cln}. *American Journal of Physiology*, **268**, C572–C579.

Cole, D.F. (1977). Secretion of the aqueous humour. *Experimental Eye Research*, **25**, 161–176.

Consolo, S., Farattini, S., Ladinsky, H. and Thoenen, H. (1972). Effects of chemical sympathectomy on the content of acetylcholine, choline and choline acetyltransferase activity in the cat spleen and iris. *Journal of Physiology*, **220**, 639–646.

Cooper, R.L., Constable, I.J. and Davidson, L. (1984). Aqueous humour catecholamines. *Current Eye Research*, **3**, 809–813.

Crook, R.B. and Polansky, J.R. (1992). Neurotransmitters and neuropeptides stimulate inositol phosphates and intracellular calcium in cultured human nonpigmented ciliary epithelium. *Investigative Ophthalmology and Visual Science*, **33**, 1706–1716.

Crook, R.B. and Polansky, J.R. (1994). Stimulation of Na^+, K^+, Cl^- cotransport by forskolin-activated adenylyl cyclase in fetal human nonpigmented epithelial cells. *Investigative Ophthalmology and Visual Science*, **35**, 3374–3383.

Crook, R.B., von Brauchitsch, D.K. and Polansky, J.R. (1992). Potassium transport in nonpigmented epithelial cells of human ocular ciliary body: inhibition of a Na^+,K^+, Cl^- cotransporter by protein kinase C. *Journal of Cellular Physiology*, **153**, 214–220.

Crook, R.B., Lui, G.M., Alvarado, J.A., Fauss, D.J. and Polansky, J.R. (1994). High affinity vasoactive intestinal peptide receptors on fetal human nonpigmented ciliary epithelial cells. *Current Eye Research*, **13**, 271–279.

Crosson, C.E. (1995). Adenosine receptor activation modulates intraocular pressure in rabbits. *Journal of Pharmacology and Experimental Therapeutics*, **273**, 320–326.

Crosson, C.E. and Gray, T. (1994). Modulation of intraocular pressure by adenosine agonists. *Journal of Ocular Pharmacology*, **10**, 379–383

Davson, H. (1990). The pupil. In *Davson's Physiology of the Eye*, 5th edition, pp. 754–758. New York: Pergamon Press.

De Roetth, A. (1950). Choline acetylase activity in ocular tissues. *Archives of Ophthalmology*, **43**, 849–855.

Delamere, N.A. and King, K.L. (1992). The influence of cyclic AMP upon Na, K-ATPase activity in rabbit ciliary epithelium. *Investigative Ophthalmology and Visual Science*, **33**, 430–435.

Delamere, N.A., Socci, R.R. and King, K.L. (1990). Alteration of sodium, potassium-adenosine triphosphatase activity in rabbit ciliary processes by cyclic adenosine monophosphate-dependent protein kinase. *Investigative Ophthalmology and Visual Science*, **31**, 2164–2170.

Drago, F., Gorgone, G., Spina, F., Panissidi, G., Dal Bello, A., Moro, F. and Scapagnini, U. (1980). Opiate receptors in the rabbit iris. *Naunyn-Schmiedeberg's Archives of Pharmacology*, **315**, 1–4.

Duke-Elder, S. and Scott, G.I. (1971). The pupillary reflexes. In *System of Ophthalmology*, volume XII, edited by S. Duke-Elder, pp. 640–691. St. Louis: Mosby.

Dutton, J.J. (1994). *Atlas of Clinical and Surgical Orbital Anatomy*. Philadelphia: W.B. Saunders Co.

Edelman, J.L., Loo, D.D.F. and Sachs, G. (1995). Characterization of potassium and chloride channels in the basolateral membrane of bovine nonpigmented ciliary epithelial cells. *Investigative Ophthalmology and Visual Science*, **36**, 2706–2716.

Ehinger, B. (1966). Adrenergic nerves to the eye and to related structures in man and in the cynomolgus monkey (*Macaca irus*). *Investigative Ophthalmology and Visual Science*, **5**, 42–52.

Ehinger, B. and Falck, B. (1970). Innervation of iridic melanophores. *Zeitschrift für Zellforschung und Mikroskopische Anatomie*, **105**, 538–542.

Ehinger, B., Falck, B., and Rosengren, E. (1969). Adrenergic denervation of the eye by unilateral cervical sympathectomy. *Graefes Archive for Clinical and Experimental Ophthalmology*, **177**, 206–211

Ekblad, E., Håkanson, R., Sundler, F. and Wahlestedt, C. (1985). Galanin: neuromodulatory and direct contractile effects on smooth muscle preparations. *British Journal of Pharmacology*, **86**, 241–247.

Elena, P.P., Fredi-Reygrobellet, D., Moulin, G. and Lapalus, P. (1984). Pharmacological characteristics of b-adrenergic sensitive adenylate cyclase in nonpigmented and in pigmented cells of bovine ciliary process. *Current Eye Research*, **3**, 1383–1389.

Emson, P.C. and De Quidt, M.E. (1984). NPY – a new member of the pancreatic polypeptide family. *Trends in Neuroscience*, **7**, 31–35.

Engstrom, P and Dunham, E.W. (1982). Alpha-adrenergic stimulation of prostaglandin release from rabbit iris-ciliary body *in vitro*. *Investigative Ophthalmology and Visual Science*, **22**, 757–767.

Erickson-Lamy, K.A., Johnson, C.D., True-Gabelt, B. and Kaufman, P. (1990). Ciliary muscle choline acetyltransferase and acetylcholinesterase after ciliary ganglionectomy. *Experimental Eye Research*, **51**, 295–299.

Ewart, H.S. and Klip, A. (1995). Hormonal regulation of the Na^+-K^+-ATPase: mechanisms underlying rapid and sustained changes in pump activity. *American Journal of Physiology*, **269**, C295–C311.

Fanciullacci, M., Boccuni, M., Pietrini, U. and Sicuteri, F. (1980). Search for opiate receptors in human pupil. *International Journal of Clinical and Pharmacological Research*, **1**, 109–113.

Farahbakhsh, N.A. and Fain, G.L. (1988). Volume regulation of nonpigmented cells from ciliary epithelium. *Investigative Ophthalmology and Visual Science*, **28**, 934–944.

Farnebo, L.O. and Hamberger, B. (1970). Release of norepinephrine from isolated rat iris by field stimulation. *Journal of Pharmacology and Experimental Therapeutics*, **172**, 332–341.

Farnebo, L.O. and Hamberger, B. (1971). Drug-induced changes in the release of [^{3}H]-noradrenaline from field stimulated rat iris. *British Journal of Pharmacology*, **43**, 97–106.

Fried, G., Lagercrantz, H., Klein, R. and Thureson-Klein, A. (1984). Large and small noradrenergic vesicles — origin, contents, and functional significance. In *Catecholamines: Basic and Peripheral Mechanisms*, edited by E. Usdin, pp.45–53. New York: Alan R. Liss

Fuder, H. (1994). Functional consequences of prejunctional receptor activation or blockade in the iris. *Journal of Ocular Pharmacology*, **10**, 109–123.

Fuder, H. and Muth, U. (1993). ATP and endogenous agonists inhibit evoked [^{3}H]-noradrenaline release in rat iris via A_1 and P_{2Y}-like purinoceptors. *Naunyn-Schmiedeberg's Archives of Pharmacology*, **348**, 352–357.

Fuder, H., Brink, A., Meincke, M. and Tauber, U. (1992). Purinoceptor-mediated modulation by endogenous and exogenous agonists of stimulation-evoked [^{3}H]-noradrenaline release on rat iris. *Naunyn-Schmiedeberg's Archives of Pharmacology*, **345**, 417–423.

Funk, R., Wagner, W. and Wild, J. (1992). Microendoscopic observations of the hemodynamics in the ciliary process vasculature of the rabbit. *Current Eye Research*, **6**, 543–551.

Gabelt, B.T., Robinson, J.C., Hubbard, W.C., Peterson, C.M., Debink, N., Wadhwa, A. and Kaufman, P.L. (1994). Apraclonidine and brimonidine effects on anterior ocular and cardiovascular physiology in normal and sympathectomized monkeys. *Experimental Eye Research*, **59**, 633–644.

Gabelt, B.T., Robinson, J.C., Gange, S.J. and Kaufman, P.L. (1995). Superior cervical ganglionectomy in monkeys: aqueous humour dynamics and their responses to drugs. *Experimental Eye Research*, **60**, 575–584.

Gallar, J. and Liu, J.H.K. (1993). Stimulation of the cervical sympathetic nerves increases intraocular pressure. *Investigative Ophthalmology and Visual Science*, **34**, 596–605.

Geppetti, P., Patacchini, R., Cecconi, R., Tramontana, M., Meini, S., Romani, A., Nardi, M. and Maggi, C.A. (1990). Effects of capsaicin, tachykinins, calcitonin gene-related peptide and bradykinin in the pig iris sphincter muscle. *Naunyn-Schmiedeberg's Archives of Pharmacology*, **341**, 301–307.

Geppetti, P., Del Bianco, E., Cecconi, R., Tramontana, M., Romani, A. and Theodorsson, E. (1992). Capsaicin releases calcitonin gene-related peptide from the human iris and ciliary body *in vitro*. *Regulatory Peptides*, **41**, 83–92.

Gharagozloo, N.Z., Relf, S.J. and Brubaker, R.F. (1988). Aqueous flow is reduced by the alpha-adrenergic agonist, apraclonidine hydrochloride (ALO 2145). *Ophthalmology*, **95**, 1217–1220.

Gibbins, I.L. (1991). Vasomotor, pilomotor and secretomotor neurons distinguished by size and neuropeptide content in superior cervical ganglia of mice. *Journal of the Autonomic Nervous System*, **34**, 171–184.

Gibbins, I.L. and Morris, J.L. (1987). Co-existence of neuropeptides in sympathetic, cranial autonomic and sensory neurons innervating the iris of the guinea-pig. *Journal of the Autonomic Nervous System*, **21**, 67–82.

Gibbins, I.L., Furness, J.B., Costa, M., MacIntyre, I., Hillyard, C.J. and Girgis, S. (1985). Co-localization of calcitonin gene-related peptide-like immunoreactivity with substance P in cutaneous, vascular and visceral sensory neurons of guinea pigs. *Neuroscience Letters*, **57**, 125–130.

Gloster, J. (1961). Influence of facial nerve on intraocular pressure. *British Journal of Ophthalmology*, **45**, 259–278.

Granstam, E. and Nilsson, S.F.E. (1990). Non-adrenergic sympathetic vasoconstriction in the eye and some other facial tissues of the rabbit. *European Journal of Pharmacology*, **175**, 175–186.

Greaves, D.P. and Perkins, E.S. (1952). Influence of the sympathetic nervous system on the intraocular pressure and vascular circulation of the eye. *British Journal of Ophthalmology*, **36**, 258–264.

Green, K. and Padgett, D. (1979). Effect of various drugs on pseudofacility and aqueous humor formation in the rabbit eye. *Experimental Eye Research*, **28**, 239–246.

Gregory, D.S. (1990). Timolol reduces IOP in normal NZW rabbits during the dark only. *Investigative Ophthalmology and Visual Science*, **31**, 715–721.

Grimes, P. and von Sallman, L. (1960). Comparative anatomy of the ciliary nerves. *Archives of Ophthalmology*, **64**, 111–121.

Grimes, P.A., McGlinn, A.M., Koeberlein, B. and Stone, R.A. (1994). Galanin immunoreactivity in autonomic innervation of the cat eye. *Journal of Comparative Neurology*, **348**, 234–243.

Gustafsson, L., Hedqvist, P. and Lagercrantz, H. (1975). Potentiation by prostaglandins E_1, E_2 and $F_{2\alpha}$ of the contraction response to transmural stimulation in the bovine iris sphincter muscle. *Acta Physiologica Scandinavica*, **95**, 26–33.

Gustafsson, L., Hedqvist, P. and Lundgren, G. (1980). Pre- and postjunctional effects of prostaglandin E_2, prostaglandin synthetase inhibitors and atropine on cholinergic neurotransmission in guinea pig ileum and bovine iris. *Acta Physiologica Scandinavica*, **110**, 401–411.

Hall, J.M., Mitchell, D. and Morton, I.K. (1991). Neurokinin receptors in the rabbit iris sphincter characterized by novel agonist ligands. *European Journal of Pharmacology*, **199**, 9–14.

Hall, J.M., Mitchell, D. and Morton, I.K. (1993). Tachykinin receptors mediating responses to sensory nerve stimulation and exogenous tachykinins and analogues in the rabbit isolated iris sphincter. *British Journal of Pharmacology*, **109**, 1008–1013.

Hayashi, K. and Masuda, K. (1982). Effects of vasoactive intestinal polypeptide (VIP) and cyclic-AMP on the isolated sphincter pupillae muscles of the albino rabbit. *Japanese Journal of Ophthalmology*, **26**, 437–442.

Hayashi, K., Mochizuki, M. and Masuda, K. (1983). Effects of vasoactive intestinal polypeptide (VIP) and cyclic AMP on isolated dilator pupillae muscle of albino rabbit eye. *Japanese Journal of Ophthalmology*, **27**, 647–654.

Hays, R.M. (1985). Agents affecting the renal conservation of water. In *Goodman and Gilman's The Pharmacological Basis of Therapeutics*, 7th edition, edited by A.G. Gilman, L.S. Goodman, T.W. Rall and F. Murad, pp. 908–919. New York: Macmillan.

Helbig, H., Korbmacher, C., Wohlfarth, J., Coca-Prados, M. and Wiederholt, M. (1989). Intracellular voltage recordings in bovine non-pigmented ciliary epithelial cells in primary culture. *Current Eye Research*, **8**, 793–800.

Heller, P.H., Perry, F., Jewett, D.L. and Levine, J.D. (1990). Autonomic components of the human pupillary light reflex. *Investigative Ophthalmology and Visual Science*, **31**, 156–62.

Higgins, R.G. and Brubaker, R.F. (1980). Acute effect of epinephrine on aqueous humour formation in the timolol-treated normal eye as measured by fluorophotometry. *Investigative Ophthalmology and Visual Science*, **19**, 420–423.

Hill, C.E., Klemm, M., Edwards, F.R. and Hirst, G.D. (1993). Sympathetic transmission to the dilator muscle of the rat iris. *Journal of the Autonomic Nervous System*, **45**, 107–123.

Hirai, R., Tamamaki, N., Hukami, K. and Nojyo, Y. (1994). Ultrastructural analysis of tyrosine hydroxylase-, substance P-, and calcitonin gene-related peptide-immunoreactive nerve fibres in the rat iris. *Ophthalmic Research*, **26**, 169–180.

Hiromatsu, S., Araie, M. and Fujimori, K. (1994). Endogenous catecholamine concentrations in rat iris-ciliary body. *Japanese Journal of Ophthalmology*, **38**, 123–128.

Honkanen, R.E., Howard, E. and Abdel-Latif, A.A. (1990). M_3-muscarinic receptor subtype predominates in the bovine iris sphincter smooth muscle and ciliary processes. *Investigative Ophthalmology and Visual Science*, **31**, 590–594.

Howe, P.H., Akhtar, R.A., Naderi, S. and Abdel-Latif, A. (1986). Correlative studies on the effect of carbachol on myo-inositol trisphosphate accumulation, myosin light chain phosphorylation and contraction in the sphincter smooth muscle of rabbit iris. *Journal of Pharmacology and Experimental Therapeutics*, **239**, 574–583.

Hoyle, C.H.V. (1992). Transmission: Purines. In *Autonomic Neuroeffector Mechanisms*, edited by G. Burnstock and C.H.V. Hoyle, pp. 367–407. Chur, Switzerland: Harwood Academic Publishers.

Hulme, R.I. and Purves, D. (1983). Apportionment of the terminals from single preganglionic axons to target neurons in the rabbit ciliary ganglion. *Journal of Physiology*, **383**, 259–275.

Hutchins, J.B. and Hollyfield, J.G. (1984). Autoradiographic identification of muscarinic receptors in human iris smooth muscle. *Experimental Eye Research*, **38**, 515–520.

Johnson, S.H., Brubaker, R.F. and Trautman, J.C. (1978). Absence of an effect of timolol on the pupil. *Investigative Ophthalmology and Visual Science*, **17**, 924–928.

Jumblatt, J.E. (1994). Prejunctional α_2-adrenoceptors and adenylyl cyclase regulation in the rabbit iris-ciliary body. *Journal of Ocular Pharmacology*, **10**, 617–621.

Jumblatt, J.E. and Gooch, J.M. (1990). Neuropeptide Y modulates adenylate cyclase in rabbit iris, ciliary body and ciliary epithelium. *Experimental Eye Research*, **51**, 229–231.

Jumblatt, J.E. and Hackmiller, R.C. (1990). Potentiation of norepinephrine secretion by angiotensin II in the isolated rabbit iris-ciliary body. *Current Eye Research*, **9**, 169–176.

Jumblatt, J.E. and Hackmiller, R.C. (1994). M_2-type muscarinic receptors mediate prejunctional inhibition of norepinephrine release in the human iris-ciliary body. *Experimental Eye Research*, **58**, 175–180.

Jumblatt, J.E. and North, G.T. (1988). Cholinergic inhibition of adrenergic neurosecretion in the rabbit iris-ciliary body. *Current Eye Research*, **29**, 615–620.

Jumblatt, J.E., Liu, J.G.H and North, G.T. (1987). Alpha-2 adrenergic modulation of norepinephrine secretion in the perfused rabbit iris-ciliary body. *Current Eye Research*, **6**, 767–777.

Jumblatt, J.E., North, G.T. and Hackmiller, R.C. (1990). Muscarinic cholinergic inhibition of adenylate cyclase in the rabbit iris-ciliary body and ciliary epithelium. *Investigative Ophthalmology and Visual Science*, **31**, 1103–1108.

Jumblatt, J.E., Ohia, S.E. and Hackmiller, R.C. (1993). Prejunctional modulation of norepinephrine release in the human iris-ciliary body. *Investigative Ophthalmology and Visual Science*, **34**, 2790–2793.

Kahle, G., Kaulen, P. and Wollensak, J. (1990). Quantitative autoradiography of β-adrenergic receptors in rabbit eyes. *Experimental Eye Research*, **51**, 503–507.

Kaufman, P.L. (1984). Mechanisms of actions of the cholinergic drugs in the eye. In *Glaucoma*, edited by S.M. Drance and A.H. Neufeld, pp. 395–427. Orlando: Grune and Stratton.

Kaufman, P.L. and Mittag, T.W. (1994). Medical therapy of glaucoma. In *Glaucoma*, edited by P.L. Kaufman and T.W. Mittag, pp. 7–30. London: Mosby.

Kishida, K., Sasabe, T., Manabe, R. and Otosi, T. (1981). Electrical characteristics of the isolated rabbit ciliary body. *Japanese Journal of Ophthalmology*, **25**, 407–416.

Koelle, G.B and Friedenwald, J.S. (1950). The histochemical localization of cholinesterase in ocular tissues. *American Journal of Ophthalmology*, **33**, 253–262.

Kokubu, N., Satoh, M. and Takayanagi, I. (1993). Contractile responses and calcium movements induced by α_1-adrenoceptor stimulant, norepinephrine, in rabbit iris dilator muscle. *General Pharmacology*, **24**, 1541–1545.

Koss, M.C. and Gherezghiher, T. (1988). Pharmacological characterization of alpha-adrenoceptors involved in nictitating membrane and pupillary responses to sympathetic nerve stimulation in cats. *Naunyn-Schmiedeberg's Archives of Pharmacology*, **337**, 18–23.

Koss, M.C., Logan, L.G. and Gherezghiher, T. (1988). Alpha-adrenoceptor activation of nictitating membrane and iris in cats. *Naunyn-Schmiedeberg's Archives of Pharmacology*, **337**, 519–524.

Krupin, T. and Civan, M.M. (1996). Physiological basis of aqueous humour formation. In *The Glaucomas*, edited by R. Ritch, M.B. Shields and T. Krupin, volume 1, pp. 357–383. St. Louis: Mosby.

Krupin, T., Reinach, P.S., Candia, O.A. and Podos, S.M. (1984). Transepithelial electrical measurements on the isolated rabbit iris-ciliary body. *Experimental Eye Research*, **38**, 115–123.

Kuchiiwa, S., Kuchiiwa, T. and Suzuki, T. (1989). Comparative anatomy of the accessory ciliary ganglion in mammals. *Anatomy and Embryology*, **180**, 199–205.

Kuchiiwa, S., Kuchiiwa, T., Nakagawa, S. and Ushikai, M. (1993). Oculomotor parasympathetic pathway to the accessory ciliary ganglion bypassing the main ciliary ganglion by way of the trigeminal nerve. *Neuroscience Research*, **18**, 79–82.

Kuwayama, Y., Terenghi, G., Polak, J.M., Trojanowski, J.Q. and Stone, R.A. (1987a). A quantitative correlation of substance P-, calcitonin gene-related peptide- and cholecystokinin-like immunoreactivity with retrogradely labeled trigeminal ganglion cells innervating the eye. *Brain Research*, **405**, 220–226.

Kuwayama, Y., Grimes, P.A., Ponte, B. and Stone, R.A. (1987b). Autonomic neurons supplying the rat eye and the intraorbital distribution of vasoactive intestinal polypeptide (VIP)-like immunoreactivity. *Experimental Eye Research*, **44**, 907–922.

Langham, M.E. and Rosenthal, A.R. (1966). Role of cervical sympathetic nerve in regulating intraocular pressure and circulation. *American Journal of Physiology*, **210**, 786–794.

Langham, M.E. and Taylor, C.B. (1960). The influence of superior cervical ganglionectomy on the intraocular pressure. *Journal of Physiology*, 152, 447–458.

Langham, M.E. and Weinstein, G.W. (1967). Horner's syndrome: ocular supersensitivity to adrenergic amines. *Archives of Ophthalmology*, **78**, 462–469.

Larson, R.S. and Brubaker, R.F. (1988). Isoproterenol stimulates aqueous flow in humans with Horner's syndrome. *Investigative Ophthalmology and Visual Science*, **29**, 621–625.

Laties, A. and Jacobowitz, D. (1964). A histochemical study of the adrenergic and cholinergic innervation of the anterior segment of the rabbit eye. *Investigative Ophthalmology*, **3**, 592–600.

Laties, A. and Jacobowitz, D. (1966). A comparative study of the autonomic innervation of the eye in monkey, cat and rabbit. *Anatomical Record*, **156**, 383–396.

Lee, D.A., Topper, J.E. and Brubaker, R.F. (1984). Effect of clonidine on aqueous humour flow in normal human eyes. *Experimental Eye Research*, **38**, 239–246.

Lee, Y., Kawai, Y., Shiosaka, S., Takami, K., Kiyama, H., Hillyard, C.J., Girgis, S., MacIntyre, I., Emson, P.C. and Tohyama, M. (1985). Coexistence of calcitonin gene-related peptide and substance P-like peptide in single cells of the trigeminal ganglion of the rat: immunohistochemical analysis. *Brain Research*, **330**, 194–196.

Lehtosalo, J., Uusitalo, H. and Palkama, A. (1984). Sensory supply of the anterior uvea: a light and electron microscope study. *Experimental Brain Research*, **55**, 562–569.

Lin, T. Grimes, P.A and Stone, R.A. (1988). Nerve pathways between the pterygopalatine ganglion and eye in cats. *Anatomical Record*, **222**, 95–102.

Liu, J.H. (1992). Aqueous humour messengers in the transient decrease of intraocular pressure after ganglionectomy. *Investigative Ophthalmology and Visual Science*, **33**, 3181–3185.

Liu, J.H.K. and Dacus, A.C. (1991). Endogenous hormonal changes and circadian elevation of intraocular pressure. *Investigative Ophthalmology and Visual Science*, **32**, 496–500.

Lowenstein, O. and Loewenfeld, I.E. (1953). Effect of physostigmine and pilocarpine on iris sphincter of normal man. *Archives of Ophthalmology*, **50**, 311–319.

Lowenstein, O. and Loewenfeld, I.E. (1969). The pupil. In *The Eye*, Volume 3, edited by H Davson, pp. 255–337. New York: Academic Press.

Lund-Karlsen, R. and Fonnum, F. (1976). The effect of locally applied cholinesterase inhibitors and oximes on the acetylcholinesterase activity in different parts of the guinea pig eye. *Acta Pharmacologica Toxicologica*, **38**, 299–307.

Lundberg, J.M., Rudehill, A., Sollevi, A., Theodorsson-Norheim, E. and Hamberger, B. (1986). Frequency and reserpine-dependent chemical coding of sympathetic transmission: differential release of noradrenaline and neuropeptide Y from pig spleen. *Neuroscience Letters*, **63**, 96–100.

Macri, F.J. and Cevario, S.J. (1975). Ciliary ganglion stimulation. I. Effects on aqueous humour inflow and outflow. *Investigative Ophthalmology*, **14**, 28–33.

Maggio, J.E. (1988). Tachykinins. *Annual Review of Neuroscience*, **11**, 13–28.

Mallorga, P. and Sugrue, M.F. (1987). Characteristics of serotonin receptors in the iris and ciliary body of the albino rabbit. *Current Eye Research*, **6**, 527–533.

Mallorga, P., Babilon, R.W., Buisson, S. and Sugrue, M.F. (1989). Muscarinic receptors of the albino rabbit ciliary process. *Experimental Eye Research*, **48**, 509–522.

Mandahl, A. and Bill, A. (1983). In the eye (D-Arg[1], D-Pro[2], D-Trp[7,9])-SP is a substance P agonist, which modifies the responses to substance P, prostaglandin E[1] and antidromic trigeminal nerve stimulation. *Acta Physiologica Scandinavica*, **117**, 139–144.

Mandahl, A. and Bill, A. (1984). Effects of the substance P antagonist, (D-Arg[1], D-Pro[2], D-Trp[7,9], Leu[11])-SP on the miotic response to substance P, antidromic trigeminal nerve stimulation, capsaicin, prostaglandin E[1], compound 48/80 and histamine. *Acta Physiologica Scandinavica*, **120**, 27–35.

Maren, T.H. (1977). Ion secretion into the posterior aqueous humour of dogs and monkeys. *Experimental Eye Research*, **25**, 245–247.

Maren, T.H. (1994). Biochemistry of aqueous humour inflow. In *Glaucoma*, edited by P.L. Kaufman and T.W. Mittag, pp. 35–46. London: Mosby.

Martin, X.D., Brennan, M.C. and Lichter, P.R. (1988). Serotonin in human aqueous humor. *Ophthalmology*, **95**, 1221–1226.

Masuda, Y., Yamahara, N.S., Tanaka, M., Ryang, S., Kawai, T., Imaizumi, Y. and Watanabe, M. (1995). Characterization of muscarinic receptors mediating relaxation and contraction in the rat iris dilator muscle. *British Journal of Pharmacology*, **114**, 769–776.

Matsuyama, T., Wanaka, A., Yoneda, S., Kimura, K., Kamada, T., Girgis, S., MacIntyre, I., Emson, P.C. and Tohyama, M. (1986). Two distinct calcitonin gene-related peptide-containing peripheral nervous systems: distribution and quantitative differences between the iris and cerebral artery with special reference to substance P. *Brain Research*, **373**, 205–212.

May, P.J. and Warren, S. (1993). Ultrastructure of the macaque ciliary ganglion. *Journal of Neurocytology*, **22**, 1073–1095.

Miichi, H. and Nagataki, S. (1982). Effects of cholinergic drugs and adrenergic drugs on aqueous humour formation in the rabbit eye. *Japanese Journal of Ophthalmology*, **26**, 425–431.

Miller, A.S., Coster, D.J., Costa, M. and Furness, J.B. (1983). Vasoactive intestinal polypeptide immunoreactive nerve fibres in the human eye. *Australian Journal of Ophthalmology*, **11**, 185–193.

Mindel, J.S. and Mittag, T.W. (1976). Choline acetyltransferase in ocular tissues of rabbits, cats, cattle, and man. *Investigative Ophthalmology*, **15**, 808–814.

Mintenig, G.M., Sanchez-Vives, M.V., Martin, C., Gual, A. and Belmonte, C. (1995). Sensory receptors in the anterior uvea of the cat's eye: an *in vitro* study. *Investigative Ophthalmology and Visual Science*, **36**, 1615–1624.

Mito, T. and Delamere, N.A. (1993). Alteration of active Na-K transport on protein kinase C activation in cultured ciliary epithelium. *Investigative Ophthalmology and Visual Science*, **34**, 539–546.

Mittag, T.W. (1996). Adrenergic and dopaminergic drugs in glaucoma. In *The Glaucomas*, edited by R. Ritch, M.B. Schields and T. Krupin, volume III, pp. 1409–1424. St. Louis: Mosby.

Mittag, T.W. and Tormay, A. (1985). Drug responses of adenylate-cyclase in iris-ciliary body membranes: adenine labelling. *Investigative Ophthalmology and Visual Science*, **26**, 396–399.

Mittag, T.W., Tormay, A. and Podos, S.M. (1987). Vasoactive intestinal peptide and intraocular pressure: adenylate cyclase activation and binding sites for vasoactive intestinal peptide in membranes of ocular ciliary processes. *Journal of Pharmacology and Experimental Therapeutics*, **241**, 230–235.

Morales, J., Ho, P. and Crosson, C.E. (1992). Effect of apraclonidine on intraocular pressure and pupil size in patients with unilateral Horner's syndrome. *Investigative Ophthalmology and Visual Science*, **34** (suppl.), 929.

Morris, J.L. (1976). Motor innervation of the toad iris (*Bufo marinus*). *American Journal of Physiology*, **231**, 1272–1278.

Morrison, J.C. and Freddo, T.F. (1996). Anatomy, microcirculation, and ultrastructure of the ciliary body. In *The Glaucomas,* Volume I, edited by R. Ritch, M.B. Shields and T. Krupin, pp. 125–138. St. Louis: Mosby.

Muramatsu, I., Nakanishi, S. and Fujiwara, M. (1987). Comparison of the responses to the sensory neuropeptides, substance P, neurokinin A, neurokinin B and calcitonin gene-related peptide and to trigeminal nerve stimulation in the iris sphincter muscle of the rabbit. *Japanese Journal of Pharmacology*, **44**, 85–92.

Muramatsu, I., Kigoshi, S. and Oda, Y. (1994). Evidence for sympathetic, purinergic transmission in the iris dilator muscle of the rabbit. *Japanese Journal of Pharmacology*, **66**, 191–193.

Namba, K., Utsumi, T. and Nakajima, M. (1980). Effect of timolol on the pupillary dynamics under open-loop photic stimulus. *Folia Ophthalmologica (Japan)*, **31**, 118–124.

Narita, S. and Watanabe, M. (1982). Effects of calcium blockers and Mn^{2+} on the response of isolated rat iris sphincter and dilator muscles to agonists and Ca^{2+}. *Journal of Pharmacobiodynamics*, **5**, 285–294.

Nathan, P.W. and Turner, J.W.A. (1942). The efferent pathway for pupillary contraction. *Brain*, **65**, 343–351.

Nathanson, J.A. (1980). Adrenergic regulation of intraocular pressure: identification of β_2-adrenergic-stimulated adenylate cyclase in ciliary process epithelium. *Proceedings of the National Academy of Sciences USA*, **77**, 7420–7424.

Nathanson, J.A. (1981). Human ciliary process adrenergic receptor: pharmacological characterization. *Investigative Ophthalmology and Visual Science*, **21**, 798–804.

Neufeld, A.H. and Page, E.D. (1975). Regulation of adrenergic neuromuscular transmission in the rabbit iris. *Experimental Eye Research*, **20**, 549–561.

Neufeld, A.H., Ledgard, S.E. and Yoza, B.K. (1983). Changes in responsiveness of the β-adrenergic and serotonergic pathways of the rabbit corneal epithelium. *Investigative Ophthalmology and Visual Science*, **24**, 527–534.

Neuman, B., Wiederman, C.J., Fischer-Colbrie, R., Schober, M., Sperk, G. and Winkler, H. (1984). Biochemical and functional properties of large and small dense-core vesicles in sympathetic nerves of rat and ox vas deferens. *Neuroscience*, **13**, 921–931.

Nilsson, S.F.E. (1991). Neuropeptide Y (NPY): a vasoconstrictor in the eye, brain and other tissues in the rabbit. *Acta Physiological Scandinavica*, **141**, 455–467.

Nilsson, S.F.E. and Bill, A. (1994). Physiology and neurophysiology of aqueous humour inflow and outflow. In *Glaucoma*, edited by P.L.Kaufman and T.W. Mittag, pp. 17–34. London: Mosby.

Nilsson, S.F.E., Linder, J. and Bill, A. (1985). Characteristics of uveal vasodilation produced by facial nerve stimulation in monkeys, cats and rabbits. *Experimental Eye Research*, **40**, 841–852.

Nilsson, S.F.E., Sperber, G.O. and Bill, A. (1986). Effects of vasoactive intestinal polypeptide (VIP) on intraocular pressure, facility of outflow and formation of aqueous humour in the monkey. *Experimental Eye Research*, **43**, 849–857.

Nilsson, S.F.E., Maepea, O., Samuelsson, M. and Bill, A. (1990). Effects of timolol on terbutaline- and VIP-stimulated aqueous humour flow in the cynomolgus monkey. *Current Eye Research*, **9**, 863–872.

Nishida, S. and Sears, M. (1969). Fine structural innervation of the dilator muscle of the iris of the albino guinea-pig studied with permanganate fixation. *Experimental Eye Research*, **8**, 292–296.

Nomura, T. and Smelser, G.K. (1974). The identification of adrenergic and cholinergic nerve endings in the trabecular meshwork. *Investigative Ophthalmology*, **13**, 523–532.

Ogidigben, M., Chu, T-C. and Potter, D.E. (1993). Ocular hypotensive action of a dopaminergic (DA_2) agonist, 2,10,11–trihydroxy-N-propylnoraporphine. *Journal of Pharmacology and Experimental Therapeutics*, **267**, 822–827.

Ohia, S.E. and Jumblatt, J.E. (1990a). Prejunctional inhibitory effects of prostanoids on sympathetic neurotransmission in the rabbit iris-ciliary body. *Journal of Pharmacology and Experimental Therapeutics*, **255**, 11–16.

Ohia, S.E. and Jumblatt, J.E. (1990b). Inhibitory effects of neuropeptide Y on sympathetic neurotransmission in the rabbit iris-ciliary body. *Neurochemical Research*, **15**, 251–256.

Osborne, N.N. (1983). The occurrence of serotonergic nerves in bovine cornea. *Neuroscience Letters*, **35**, 15–18.

Osborne, N.N. and Tobin, A.B. (1987). Serotonin-accumulating cells in the iris-ciliary body and cornea of various species. *Experimental Eye Research*, **44**, 731–746.

Page, E.D. and Neufeld, A.H. (1978). Characterization of α- and β-adrenergic receptors in membranes prepared from the rabbit iris before and after development of supersensitivity. *Biochemical Pharmacology*, **27**, 953–958.

Palm, D.E., Keil, L.C. and Severs, W.B. (1994). Angiotensin, vasopressin, and atrial natriuretic peptide in the rat eye. *Proceedings of the Society for Experimental Biology and Medicine*, **206**, 392–395.

Patil, P.N and Weber, P.A. (1991). *In vivo* functional implications of isoproterenol-mediated relaxation of isolated human iris sphincter. *Journal of Ocular Pharmacology*, **7**, 297–300.

Perkins, E.S. (1957). Influence of the fifth cranial nerve on the intraocular pressure of the rabbit eye. *British Journal of Ophthalmology*, **41**, 257–300.

Piccone, M., Littzi, J., Krupin, T., Stone, R.A., Davis, M. and Wax, M.B. (1988). Effects of neuropeptide Y on the isolated rabbit iris dilator muscle. *Investigative Ophthalmology and Visual Science*, **29**, 330–332.

Richards, J.G. and Da Prada, M. (1977). Uranaffin reaction: a new cytochemical technique for the localization of adenine nucleotides in organelles storing biogenic amines. *Journal of Histochemistry and Cytochemistry*, **25**, 1322–1336.

Richardson, K.C. (1964). The fine structure of the albino rabbit iris with special reference to the identification of adrenergic and cholinergic nerves and nerve endings in its intrinsic muscles. *American Journal of Anatomy*, **114**, 173–205.

Ridderstrale, Y., Wistrand, P.J. and Brechue, W.F. (1994). Membrane-associated CA activity in the eye of the CA II-deficient mouse. *Investigative Ophthalmology and Visual Science*, **35**, 2577–2584.

Rittenhouse, A.R. and Zigmond, R.E. (1991). Omega-conotoxin inhibits the acute activation of tyrosine hydroxylase and the stimulation of norepinephrine release by potassium depolarization of sympathetic nerve endings. *Journal of Neurochemistry*, **56**, 615–622.

Rodrigues, M.M., Hackett, J. and Donohoo, P. (1991). Iris. In *Duane's Foundations of Clinical Ophthalmology*, Volume I, edited by W. Tasman and E.A. Jaeger, pp. 1–18. Philadelphia: J.B. Lippincott.

Ruskell, G.L. (1965). The orbital distribution of the sphenopalatine ganglion in the rabbit. In *The Structure of the Eye; Symposium of the Eighth International Congress of Anatomists, Wiesbaden*, edited by J. Rohen, pp. 355–368. Stuttgart: Schattauer.

Ruskell, G.L. (1970a). An ocular parasympathetic nerve pathway of facial nerve origin and its influence on intraocular pressure. *Experimental Eye Research*, **10**, 319–330.

Ruskell, G.L. (1970b). The orbital branches of the pterygopalatine ganglion and their relationship with internal carotid nerve branches in primates. *Journal of Anatomy*, **106**, 323–339.

Ruskell, G.L. (1974). Ocular fibres of the maxillary nerve in monkeys. *Journal of Anatomy*, **118**, 195–203.

Ruskell, G.L. (1985). Facial nerve distribution to the eye. *American Journal of Optometry and Physiological Optics*, **62**, 793–798.

Ruskell, G.L. (1994). Trigeminal innervation of the scleral spur in cynomolgus monkeys. *Journal of Anatomy*, **184**, 511–518.

Ryang, S., Takei, S., Kawai, t., Imaizumi, Y. and Watanabe, M. (1990). Atropine-resistant relaxation induced by high K^+ in iris dilator muscle of the rat and pig. *British Journal of Pharmacology*, **100**, 401–406.

Sastry, R. (1985). Cholinergic systems and multiple cholinergic receptors in ocular tissues. *Journal of Ocular Pharmacology*, **1**, 201–226.

Sears, M.L. (1975). Catecholamines in relation to the eye. In *Handbook of Physiology*, Volume VI, edited by R.O. Greep and A.B. Astwood, pp. 553–590. Washington, D.C.: American Physiological Society.

Sears, M.L. (1984). Autonomic nervous system adrenergic agents. In *Pharmacology of the Eye*, edited by M.L. Sears, pp. 193–248. Berlin: Springer.

Sears, M.L. and Barany, E.H. (1960). Outflow resistance and adrenergic mechanisms. *Archives of Ophthalmology*, **64**, 839–849.

Shiraishi, K. and Takayanagi, I. (1993). Subtype of muscarinic receptors mediating relaxation and contraction in the rat iris dilator smooth muscle. *General Pharmacology*, **24**, 139–142.

Simpson, L.L. (1981). The origin, structure and pharmacological activity of botulinum toxin. *Pharmacological Reviews*, **33**, 155–188.

Skofitsch, G. and Jacobowitz, D.M. (1985). Calcitonin gene-related peptide coexists with substance P in capsaicin sensitive neurons and sensory ganglia of the rat. *Peptides*, **6**, 747–754.

Smith, S.D. and Gregory, D.S. (1989). A circadian rhythm of aqueous flow underlies the circadian rhythm of IOP in NZW rabbits. *Investigative Ophthalmology and Visual Science*, **30**, 775–778.

Stern, F.A. and Bito, L.Z (1982). Comparison of the hypotensive and other ocular effects of prostaglandins E_2 and $F_{2\alpha}$ on cat and rhesus monkey eyes. *Investigative Ophthalmology and Visual Science*, **22**, 588–598.

Stjernschantz, J. (1976). Effect of parasympathetic stimulation on intraocular pressure, formation of the aqueous humour, and outflow facility in rabbits, *Experimental Eye Research*, **22**, 639–645.

Stjernschantz, J., Geijer, C. and Bill, A. (1979). Electrical stimulation of the fifth cranial nerve in rabbits — effects on ocular blood flow, extravascular albumin content and intraocular pressure. *Experimental Eye Research*, **28**, 229–238.

Stone, R.A. (1986a). Neuropeptide Y and the ocular innervation of the human eye. *Experimental Eye Research*, **42**, 340–355.

Stone, R.A. (1986b). Vasoactive intestinal polypeptide and the ocular innervation. *Investigative Ophthalmology and Visual Science*, **27**, 951–957.

Stone, R.A. (1996). Nervous system and intraocular pressure. In *The Glaucomas*, Volume I, edited by R. Ritch, M.B. Shields and T. Krupin, pp. 357–383. St. Louis: Mosby.

Stone, R.A. and Kuwayama, Y. (1985). Substance P-like immunoreactive nerves in the human eye. *Archives of Ophthalmology*, **103**, 1207–1211.

Stone, R.A. and McGlinn, A.M. (1988). Calcitonin gene-related peptide immunoreactive nerves in human and rhesus monkey eyes. *Investigative Ophthalmology and Visual Science*, **29**, 305–310.

Stone, R.A., Laties, A.M. and Brecha, N.C. (1982). Substance P-like immunoreactive nerves in the anterior segment of the rabbit, cat and monkey eye. *Neuroscience*, **7**, 2459–2468.

Stone, R.A., Kuwayama, Y., Laties, A.M., McGlinn, A.M. and Schmidt, M.L. (1984). Guinea-pig ocular nerves contain a peptide of the cholecystokinin/gastrin family. *Experimental Eye Research*, **39**, 387–391.

Stone, R.A., Laties, A.M. and Emson, P.C. (1986). Neuropeptide Y and the ocular innervation of rat, guinea-pig, cat and monkey. *Neuroscience*, **17**, 1207–1216.

Stone, R.A., Laties, A.M., Hemmings, H.C., Ouimet, C.C. and Greengard, P. (1986). DARPP-32 in the ciliary epithelium of the eye: a neurotransmitter-regulated phosphoprotein of brain localizes to secretory cells. *Journal of Histochemistry and Cytochemistry*, **34**, 1465–1468.

Stone, R.A., Kuwayama, Y. and Laties, A.M. (1987). Regulatory peptides in the eye. *Experientia*, **43**, 791–800.

Stone, R.A., McGlinn, A.M. and Kuwayama, Y. (1988). Galanin-like immunoreactive nerves in the porcine eye. *Experimental Eye Research*, **46**, 457–461.

Stone, R.A., McGlinn, A.M., Kuwayama, Y. and Grimes, P.A. (1988). Peptide immunoreactivity of the ciliary ganglion and its accessory cells in the rat. *Brain Research*, **475**, 389–392.

Strack, A.M., Sawyer, W.B., Marubio, L.M. and Loewy, A.D. (1988). Spinal origin of sympathetic preganglionic neurons in the rat. *Brain Research*, **455**, 187–191.

Stromberg, I., Bjorklund, H., Melander, T., Rokeus, A., Hökfelt, T. and Olson, L. (1987). Galanin immunoreactive nerves in the rat iris: alterations induced by denervations. *Cell and Tissue Research*, **250**, 267–275.

Sugrue, M.F. (1989). The pharmacology of antiglaucoma drugs. *Pharmacology and Therapeutics*, **43**, 91–138.

Suzuki, R. and Kobayashi, S. (1983). Different effects of substance P and vasoactive intestinal peptide on the motor function of bovine intraocular muscles. *Investigative Ophthalmology and Visual Science*, **24**, 1566–1571.

Suzuki, R. and Kobayashi, S. (1986). Effects of divalent cations on the spontaneous synchronization in mammalian iris sphincter muscle cells. *Experimental Eye Research*, **42**, 407–415.

Suzuki, R. and Kobayashi, S. (1988). Possible mechanisms related to contraction of the bovine iris sphincter in the presence of acetylcholine and carbachol. *Documenta Ophthalmologica*, **70**, 293–300.

Suzuki, R., Oso, T. and Kobayashi, S. (1983). Cholinergic inhibitory response in the bovine iris dilator muscle. *Investigative Ophthalmology and Visual Science*, **24**, 760–765.

Suzuki, R., Yoshino, H. and Kobayashi, S. (1987). Different time courses of bovine iris sphincter and dilator muscles after stimulation. *Ophthalmic Research*, **19**, 344–350.

Suzuki, R., Yoshino, H. and Kobayashi, S. (1989). The different contributions of prostaglandins (E_1, E_2, $F_{2\alpha}$, D_2) to the tone and neurogenic response of bovine iris sphincter muscle. *Documenta Ophthalmologica*, **72**, 129–139.

Tachado, S.D., Akhtar, R.A. and Abdel-Latif, A.A. (1989). Activation of beta-adrenergic receptors causes stimulation of cyclic AMP, inhibition of inositol trisphosphate, and relaxation of bovine iris sphincter smooth muscle. Biochemical and functional interactions between the cyclic AMP and calcium signalling systems. *Investigative Ophthalmology and Visual Science*, **30**, 2232–2239.

Tachado, S.D., Akhtar, R.A., Yousufzai, S.Y. and Abdel-Latif, A.A. (1991). Species differences in the effects of substance P on inositol trisphosphate accumulation and cyclic AMP formation, and on contraction in isolated iris sphincter of the mammalian eye: differences in receptor density. *Experimental Eye Research*, **53**, 729–739.

Tachado, S.D., Virdee, K., Akhtar, R.A. and Abdel-Latif, A.A. (1994). M_3 muscarinic receptors mediate an increase in both inositol trisphosphate production and cyclic AMP formation in dog iris sphincter smooth muscle. *Journal of Ocular Pharmacology*, **10**, 137–147.

Takayanagi, I., Shiraishi, K. and Kokubu, N. (1992). α_{1B}-adrenoceptor mechanisms in rabbit iris dilator. *Japanese Journal of Pharmacology*, **59**, 301–305.

Takayanagi, I., Shiraishi, K. and Satoh, M. (1992). Effects of ageing on responses of rabbit iris smooth muscles to agonists and field stimulation. *General Pharmacology*, **23**, 463–469.

Tamm, E.R., Flugel-Koch, C., Mayer, B. and Lutjen-Drecoll, E. (1995). Nerve cells in the human ciliary muscle: ultrastructural and immunocytochemical characterization. *Investigative Ophthalmology and Visual Science*, **36**, 414–426.

Taniguchi, t., Fujiwara, M., Masuo, Y. and Kanazawa, I. (1986). Levels of neurokinin A, neurokinin B and substance P in rabbit iris sphincter muscles. *Japanese Journal of Pharmacology*, **42**, 590–593.

ten Tusscher, M.P.M., Klooster, J. and Vrensen, G.F.J.M. (1988). The innervation of the rabbit's anterior eye segment: a retrograde tracing study. *Experimental Eye Research*, **46**, 717–730.

ten Tusscher, M.P.M., Beckers, H.J.M., Vrensen, G.F.J.M. and Klooster, J. (1994). Peripheral neural circuits regulating IOP? *Documenta Ophthalmologica*, **87**, 291–313.

Terenghi, G., Polak, J.M., Allen, J.M., Zhang, S.Q., Unger, W.G. and Bloom, S.R. (1983). Neuropeptide Y-immunoreactive nerves in the uvea of guinea-pig and rat. *Neuroscience Letters*, **42**, 33–38.

Terenghi, G., Polak, J.M., Ghatei, M.A., Mulderry, P.K., Butler, J.M., Unger, W.G. and Bloom, S.R. (1985). Distribution and origin of calcitonin gene-related peptide (CGRP) immunoreactivity in the sensory innervation of the mammalian eye. *Journal of Comparative Neurology*, **233**, 505–516.

Tervo, K., Tervo, T., Eranko, L., Eranko, O. and Cuello, A.C. (1981). Immunoreactivity for substance P in the Gasserian ganglion, ophthalmic nerve and anterior segment of the rabbit eye. *Histochemical Journal*, **13**, 435–443.

Tervo, K., Tervo, T., Eranko, L. and Eranko O. (1982a). Effect of sensory and sympathetic denervation on substance P immunoreactivity in nerve fibres of the rabbit eye. *Experimental Eye Research*, **34**, 577–585.

Tervo, K., Tervo, T., Eranko, L., Vannas, A., Cuello, A.C. and Eranko, O. (1982b). Substance P-immunoreactive nerves in the human cornea and iris. *Investigative Ophthalmology and Visual Science*, **23**, 671–674.

Tobin, A.B. and Osborne, N.N (1988). Evidence for the presence of cholinergic muscarinic receptors negatively linked to adenylate cyclase in the iris-ciliary body. *Neurochemistry International*, **4**, 517–522.

Tobin, A.B., Unger, W. and Osborne, N.N. (1988). Evidence for the presence of serotonergic nerves and receptors in the iris-ciliary body complex of the rabbit. *Journal of Neuroscience*, **8**, 3713–3721.

Too, H.P., Todd, K., Lightman, S.L., Horn, A., Unger, W.G. and Hanley, M.R. (1989). Presence and actions of vasopressin-like peptides in the rabbit anterior uvea. *Regulatory Peptides*, **25**, 259–266.

Topper, J.E. and Brubaker, R.F. (1985). Effects of timolol, epinephrine, and acetazolamide on aqueous flow during sleep. *Investigative Ophthalmology and Visual Science*, **26**, 1315–1319.

Townsend, D.J. and Brubaker, R.F. (1980). Immediate effect of epinephrine on aqueous formation in the normal human eye as measured by fluorophotometry. *Investigative Ophthalmology and Visual Science*, **19**, 256–266.

Tucek, S. (1988). Choline acetyltransferase and the synthesis of acetylcholine. In *Handbook of Experimental Pharmacology*, Volume 86, edited by V.P. Whittaker, pp. 125–165. Berlin and Heidelberg: Springer Verlag

Uddman, R., Alumets, J., Ehinger, B., Håkanson, R., Loren, I. and Sundler, F. (1980). Vasoactive intestinal peptide nerves in ocular and orbital structures of the cat. *Investigative Ophthalmology and Visual Science*, **19**, 878–885.

Ueda, N., Muramatsu, I., Sakakibara, Y. and Fujiwara, M. (1981). Noncholinergic, nonadrenergic contraction and substance P in rabbit iris sphincter muscle. *Japanese Journal of Pharmacology*, **31**, 1071–1079.

Ueda, N. Muramatsu, I., Hayashi, H. and Fujiwara, M. (1982). Trigeminal nerve: the possible origin of substance P-ergic response in isolated rabbit iris sphincter muscle. *Life Sciences*, **31**, 369–375.

Uusitalo, H. (1972). Effect of sympathetic and parasympathetic stimulation on the secretion and outflow of aqueous humour in the rabbit eye, *Acta Physiologica Scandinavica*, **86**, 315–326.

van Alphen, G.W.H.M. (1976). The adrenergic receptors of the intraocular muscles of the human eye. *Investigative Ophthalmology*, **15**, 502–511.

van Alphen, G.W.H.M., Kern, R. and Robinette, S. (1965). Adrenergic receptors of the intraocular muscles: comparison to cat, rabbit and monkey. *Archives of Ophthalmology*, **74**, 253–259.

Vittitow, J.L. and Jumblatt, J.E. (1996). Prejunctional modulators of cholinergic transmission in the iris-ciliary body. *Investigative Ophthalmology and Visual Science*, **37**, 835S.

Wahlestedt, C., Bynke, G. and Håkanson, R. (1985). Pupillary constriction by bradykinin and capsaicin: mode of action. *European Journal of Pharmacology*, **106**, 577–615.

Wahlestedt, C., Bynke, G., Beding, B., von Leithner, P. and Håkanson, R. (1985). Neurogenic mechanisms in control of the rabbit iris sphincter muscle. *European Journal of Pharmacology*, **117**, 303–309.

Wahlestedt, C., Beding, B., Ekman, R., Oksala, O., Stjernschantz, J. and Håkanson, R. (1986). Calcitonin gene-related peptide in the eye: release by sensory nerve stimulation and effects associated with neurogenic inflammation. *Regulatory Peptides*, **16**, 107–115.

Warwick, R. (1954). The ocular parasympathetic nerve supply and its mesencephalic sources. *Journal of Anatomy*, **88**, 71–93.

Wax, M.B. and Coca-Prados, M. (1989). Receptor-mediated phosphoinositide hydrolysis in human ocular ciliary epithelial cells. *Investigative Ophthalmology and Visual Science*, **30**, 1675–1679.

Wax, M., Sanghavi, D.M., Lee, C.H. and Kapadia, M. (1993). Purinergic receptors in ocular ciliary epithelial cells. *Experimental Eye Research*, **57**, 89–95.

Wentworth, W.O. and Brubaker, R.F. (1981). Aqueous humour dynamics in a series of patients with third neuron Horner's syndrome. *American Journal of Ophthalmology*, **92**, 407–415.

Wistrand, P.J., Schenholm, M. and Lonnerholm, G. (1986). Carbonic anhydrase isoenzymes CA-I and CA-II in the human eye. *Investigative Ophthalmology and Visual Science*, **27**, 418–428.

Woldemussie, E., Feldmann, B.J and Chen, J. (1993). Characterization of muscarinic receptors in cultured human iris sphincter and ciliary smooth muscle cells. *Experimental Eye Research*, **56**, 385–392.

Woodward, D.F., Burke, J.A., Williams, L.S., Palmer, B.P., Wheeler, L.A., Woldemussie, E., Ruiz, G. and Chen, J. (1989). Prostaglandin $F_{2\alpha}$ effects on intraocular pressure negatively correlate with FP-receptor stimulation. *Investigative Ophthalmology and Visual Science*, **30**, 1838–1842.

Yamada, E. (1988). Intraepithelial nerve fibres in rabbit ocular ciliary epithelium. *Archives of Histology and Cytology*, **51**, 43–51.

Yoshitomi, T. and Gregory, D.S. (1991). Ocular adrenergic nerves contribute to control of the circadian rhythm of aqueous flow in rabbits. *Investigative Ophthalmology and Visual Science*, **32**, 523–528.

Yoshitomi, T. and Ito, Y. (1986). Double reciprocal innervations in dog iris sphincter and dilator muscles. *Investigative Ophthalmology and Visual Science*, **27**, 83–91.

Yoshitomi, T., Ito, Y. and Inomata, H. (1985). Adrenergic excitatory and cholinergic inhibitory innervations in the human iris dilator. *Experimental Eye Research*, **40**, 453–459.

Yoshitomi, T., Ito, Y. and Inomata, H. (1988). Functional innervation and contractile properties of the human iris sphincter muscle. *Experimental Eye Research*, **46**, 979–986.

Yoshitomi, T., Horio, B. and Gregory, D.S. (1991). Changes in aqueous norepinephrine and cyclic adenosine monophosphate during the circadian cycle in rabbits. *Investigative Ophthalmology and Visual Science*, **32**, 1609–1613.

Yoshitomi, T., Ishikawa, H., Haruno, I. and Ishikawa, S. (1995). Effect of histamine and substance P on the rabbit and human iris sphincter muscle. *Graefes Archive for Clinical and Experimental Ophthalmology*, **233**, 181–185.

Yousufzai, S.Y.K. and Abdel-Latif, A.A. (1984). The effects of α_1-adrenergic and muscarinic cholinergic stimulation on prostaglandin release by rabbit iris. *Prostaglandins*, 28, 399–415.

Yousufzai, S.Y. and Abdel-Latif, A.A. (1987). α_1-Adrenergic receptor induced subsensitivity and supersensitivity in rabbit iris-ciliary body. Effects on myo-inositol trisphosphate accumulation, arachidonate release, and prostaglandin synthesis. *Investigative Ophthalmology and Visual Science*, **28**, 409–419.

Yousufzai, S.Y., Chen, A.L. and Abdel-Latif, A.A. (1988). Species differences in the effects of prostaglandins on inositol trisphosphate accumulation, phosphatidic acid formation, myosin light chain phosphorylation and contraction in iris sphincter of the mammalian eye: interaction with the cyclic AMP system. *Journal of Pharmacology and Experimental Therapeutics*, **247**, 1064–1072.

Zhang, S.Q., Butler, J.M. and Cole, D.F. (1984). Sensory neural mechanisms in contraction of the rabbit isolated iris sphincter pupillae: analysis of the response to capsaicin and electrical field stimulation. *Experimental Eye Research*, **38**, 153–163.

Zhang, S.Q., Terenghi, G., Unger, W.G., Ennis, K.W. and Polak, J. (1984). Changes in substance P- and neuropeptide Y-immunoreactive fibres in rat and guinea-pig irides following unilateral sympathectomy. *Experimental Eye Research*, **39**, 365–372.

Zimmerman, T.J., Garg, L.C., Vogh, B.P. and Maren, T.H. (1976). The effect of acetazolamide on the movement of ions into the posterior chamber of the dog eye. *Journal of Pharmacology and Experimental Therapeutics*, **196**, 510–516.

2 Nervous Control of the Cornea

Carl F. Marfurt

Northwest Center for Medical Education, Indiana University School of Medicine, 3400 Broadway, Gary, IN 46408, USA

The cornea is the most richly innervated surface tissue in the body. It receives a dense sensory innervation from the trigeminal ganglion and a modest sympathetic innervation from the superior cervical ganglion. A sparse parasympathetic innervation has also been demonstrated in some species. Ocular fibres enter the cornea in various planes from the corneoscleral limbus and give rise to an elaborate, highly branched stromal nerve network. The nerves eventually enter the corneal epithelium and, after additional branching, terminate as free nerve endings. The nerves are not static structures, but demonstrate continuous elongation and terminal rearrangement under normal physiological conditions. The corneal innervation is neurochemically complex and individual corneal nerves may contain one or more of a dozen different neuropeptides and neurotransmitters, as well as a variety of neuroenzymes, cytoskeletal proteins, and cytoplasmic markers. Corneal afferent fibres serve important sensory, reflex, and trophic functions. The predominant, if not exclusive, sensory perception elicited by corneal stimulation is pain. Corneal unimodal and polymodal nociceptors are classified according to their relative abilities to transduce mechanical, thermal, or chemical stimuli. Stimulation of these nerves results in transmission of sensory information to the brain, initiation of the protective blink reflex, and the intraocular release of neuropeptides by axon reflex (neurogenic inflammation). Interruption of the ocular sensory innervation by disease or trauma produces a degenerative corneal condition known as neurotrophic keratitis. The pathogenesis of neurotrophic keratitis is multifactorial, but is due in part to the loss of trophic factors (possibly neuropeptides) supplied by corneal sensory nerves. Corneal sympathetic nerves also exert important trophic effects on the corneal epithelium; the latter nerves regulate corneal epithelial ion transport processes, cell proliferation and mitogenesis, and cell migration during corneal wound healing. Most mammalian corneas contain high concentrations of acetylcholine, choline acetyltransferase, acetylcholinesterase, and cholinergic receptors. Only a part of the cholinergic system is associated with the corneal nerves; most of it is associated with the corneal epithelial cells. The functions of the corneal cholinergic system are uncertain; however, the system has been implicated in the regulation of epithelial cell growth and proliferation, ion transport and sensory transduction mechanisms. Development of the corneal innervation begins *in utero* and is completed, depending on the species, either pre- or postnatally. It has been postulated that the developing nerves stimulate perinatal epithelial differentiation and the acquisition of corneal transparency.

KEY WORDS: corneal innervation; corneal pain; neurotrophic keratitis; sympathetic nerves

Correspondence: Tel: +1 219-980-6666; Fax: +1 219-980-6566; E-mail: cmarfurt@meded.iun.indiana.edu

INTRODUCTION

The cornea constitutes the most anterior surface of the eye. Histologically, it comprises a stratified squamous epithelium facing the ocular tear film, a dense connective tissue stroma, and a simple cuboidal endothelium facing the anterior chamber. Although the morphology of the cornea is relatively simple, its three layers are highly specialized for the refraction and transmission of light, and for the prevention of intraocular infection.

One of the most distinguishing features of the cornea is its rich nerve supply. The cornea receives a dense sensory innervation from the trigeminal ganglion and a modest sympathetic innervation from the superior cervical ganglion. In addition to the well known sensory and reflex functions of the corneal afferent fibres, corneal sensory and sympathetic nerves exert various "trophic" or nutritive effects on the cornea. These effects include the maintenance of epithelial cellular integrity, modulation of cell proliferation and mitosis, stimulation of ion transport, and regulation of wound healing after corneal injuries. In light of their numerous sensory, reflex and trophic functions, damage to the corneal innervation by trauma or disease may have serious visual consequences. The present chapter summarizes current knowledge of the origins, distribution patterns, ultrastructure, neurochemistry, electrophysiology, functions and development of the corneal innervation.

ORIGINS OF THE CORNEAL INNERVATION

SENSORY NERVES

Corneal sensory nerves originate predominantly, if not exclusively, from cell bodies in the medial, or ophthalmic, region of the ipsilateral trigeminal ganglion (Arvidson, 1977; Morgan, Nadelhaft and DeGroat, 1978; Marfurt, 1981; Marfurt and DelToro, 1987; Morgan, Janetta and DeGroat, 1987; Marfurt and Echtenkamp, 1988; tenTusscher, Klooster and Vrensen, 1988; Marfurt, Kingsley and Echtenkamp, 1989; Keller et al., 1991; Lavail, Welkin and Spencer, 1993). The results of retrograde nerve tracing studies suggest that 50–450 neurons innervate each cornea, the actual number depending on the species. Corneal-innervating neurons are predominantly small or medium in size, averaging 20–23 µm in diameter in rodents and 31–33 µm in larger mammals (Nishimori et al., 1986; Sugimoto, Takemura and Wakisaka, 1988; Marfurt, Kingsley and Echtenkamp, 1989; Keller et al., 1991). The relatively small sizes of the neuronal somata are consistent with the fact that these cells give origin to unmyelinated or finely myelinated fibres that conduct in the C-fibre or A-δ range (see below).

The sensory nerves reach the eye mainly, if not exclusively, via the nasociliary branch of the ophthalmic nerve. In humans, the nasociliary nerve gives rise to two or three long ciliary nerves which course directly to the posterior pole of the eye, and a communicating branch carrying sensory fibres to the ciliary ganglion (Figure 2.1). Approximately six short ciliary nerves, carrying a mixture of sensory and autonomic fibres, emerge from the anterior pole of the ciliary ganglion and, together with the long ciliary nerves, penetrate the posterior aspect of the globe in close proximity to the optic nerve.

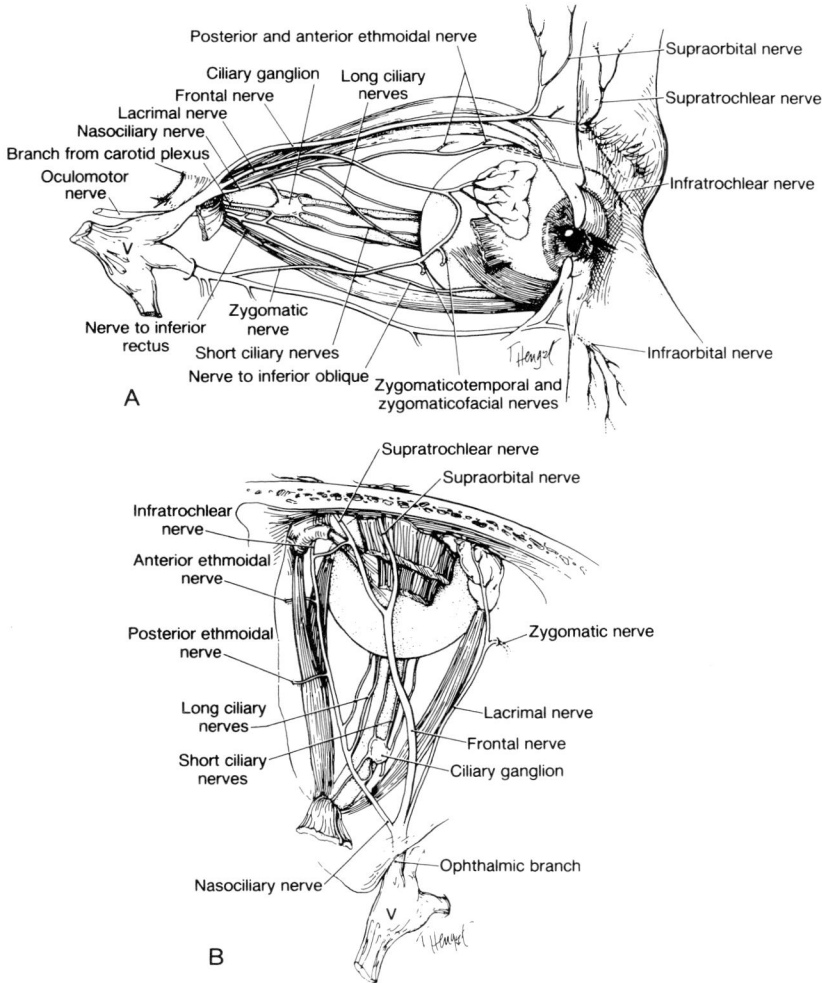

Figure 2.1   The branches of the ophthalmic division of the trigeminal nerve as seen from the lateral side (A) and from above (B). Sensory nerves to the eye travel mainly in the nasociliary nerve and its ocular branches, the long and short ciliary nerves. (Reprinted from Walsh and Hoyt's Clinical Neuro-ophthalmology Vol. 2., 1985, edited by N. R. Miller, with permission of Williams and Wilkins).

Whether additional sensory fibres reach the cornea via the maxillary division of the trigeminal nerve has been debated. Clinically, interruption of the maxillary nerve in the orbital floor (Vonderahe, 1928) and transection of the maxillary portion of the trigeminal sensory root (Karvounis and Frangos, 1972) have been reported to produce impaired sensibility or total anesthesia in the lower half of the cornea. In contrast, other reports indicate that transection of the maxillary nerve at the foramen rotundum (Rowbotham, 1939) and traumatic injury of the infraorbital nerve (Norn, 1975) are without effect on corneal sensibility. The results of orbital dissections and nerve tracing studies in monkeys and cats have revealed a minor maxillary input to the cornea in these species (Ruskell, 1974; Morgan, Nadelhaft and DeGroat, 1978; Marfurt and Echtenkamp, 1988); however, transection of the cat maxillary nerve or rabbit infraorbital nerve revealed an absence of degenerating corneal fibres (Zander and Weddell, 1951b; Rodger, 1953). It may be concluded from these studies that the maxillary nerve in most cases provides little or no innervation to the cornea and, if present, is unlikely to provide meaningful preservation of corneal sensibility and trophic support following ophthalmic nerve lesions.

AUTONOMIC NERVES

In addition to a rich sensory innervation, the cornea receives a sparse to modest sympathetic innervation from the ipsilateral superior cervical ganglion (SCG) (Morgan, DeGroat and Janetta, 1987; Marfurt, Kingsley and Echtenkamp, 1989). In humans (Watson and Vijayan, 1995), sympathetic postganglionic fibres leave the SCG in the internal carotid nerve and ascend in the internal carotid plexus before entering the carotid canal in the petrous portion of the temporal bone. At the foramen lacerum, most of the ocular sympathetic fibres move anteriorly, away from the artery, and advance inferiorly and medially to the trigeminal ganglion to enter into a retro-orbital autonomic plexus with parasympathetic fibres from the pterygopalatine ganglion (Ruskell, 1970). The plexus forms a meshwork about the abducens, trochlear and ophthalmic nerves. Ultimately, most sympathetic nerves continue into the nasociliary nerve, and then into the long and short ciliary nerves, to reach the eye. Other sympathetic nerves pass forward in the walls of arteries, or independently, to reach the globe.

Although it is has long been assumed that the cornea is not innervated by parasympathetic nerves, recent observations have demonstrated their existence in some species. Application of the retrograde nerve tracing compound, horseradish peroxidase, to the central corneas of cats and rats consistently labels small numbers of parasympathetic neurons in the ipsilateral ciliary ganglion (Morgan, DeGroat and Janetta, 1987; Marfurt, Jones and Thrasher, 1998). In the rat, extirpation of the main ciliary ganglion causes degeneration of small numbers of corneal axons (Tervo et al., 1979). Conversely, surgical transection of rat ocular sensory and sympathetic nerves eliminates most, but clearly not all, corneal nerves (Marfurt, Jones and Thrasher, 1998).

NERVE DISTRIBUTION PATTERN

The anatomy of the mammalian corneal innervation has been the subject of intense investigation since the earliest description of these nerves over one hundred and sixty years ago. Regrettably, it is not possible in these limited pages to give recognition to the substantial contributions made by early investigators. For an excellent review of the literature before 1950 the interested reader is referred to the work of Zander and Weddell (1951a). In the latter paper, the authors published the results of a comprehensive study of the corneal innervation in rabbits, humans, and other vertebrates which has become the standard reference on the subject and on which much of the present day conception of the corneal innervation is based. These observations have since been confirmed and extended in a series of elegant studies in the rabbit (Robertson and Winkelmann, 1970; Tervo and Palkama, 1978b; Rózsa and Beuerman, 1982), cat (Chan-Ling, 1989) and human (Schimmelpfennig, 1982). The following description of corneal microscopic anatomy draws heavily on these and other accounts.

After piercing the sclera, ocular autonomic and sensory nerve fibres course towards the anterior eye segment in the so-called "suprachoroidal space", located between the sclera and the choroid. Smaller numbers of fibres run forward within the sclera. As the nerve bundles run in the suprachoroidal space, they branch and exchange axons with one another such that by the time they reach the corneoscleral limbus each bundle contains a mixture of sensory, sympathetic and parasympathetic fibres. Near the corneoscleral limbus, the nerves destined to reach the cornea move anteriorly and separate from those supplying the anterior uvea.

THE LIMBAL AND STROMAL PLEXUSES (Figure 2.2)

Before entering the cornea, the nerves contribute fibres to a series of complex pericorneal (limbal) plexuses whose exact number and arrangement vary according to species. These plexuses, containing mixtures of sensory and autonomic nerves, form dense ring-like fibre networks that completely surround the peripheral cornea. The majority of the fibres supply a rich vasomotor innervation to the limbal blood vessels; however, others course within the limbal stroma apparently unrelated to vascular elements.

Sensory and autonomic nerves that supply the cornea pass through the limbus and enter the peripheral cornea in one of several planes. Most fibres penetrate at about midstromal level in a series of prominent, radially-directed nerve bundles. Approximately 70–80 fascicles, containing 900–1500 axons, enter the human cornea, while 20-40 major fascicles are typically seen in other mammals (Zander and Weddell, 1951a; Millodot, Lim and Ruskell, 1978; Chan-Ling, 1989). Other, smaller nerve fascicles enter the cornea in the episcleral and conjunctival planes to supply the superficial stroma and epithelium, respectively, of the peripheral cornea (Zander and Weddell, 1951a; Lim and Ruskell, 1978; Chan-Ling, 1989).

All corneal sensory nerves derive from finely myelinated or unmyelinated axons (Figure 2.3). In rabbits, more than 70% of the nerves are unmyelinated (Beuerman et al., 1983); the rest are finely myelinated (A-δ) axons that shed their myelin sheath within 1–2 mm after entering the cornea (Zander and Weddell, 1951a; Lim and Ruskell, 1978; Rózsa and Beuerman, 1982). In the human cornea, most unmyelinated stromal axons are about 0.5 μm in diameter; however, a few may be as large as 2.5 μm (Müller, Pels and Vrensen, 1996; Müller et al., 1997). Occasional myelinated axons are present in the central cornea in some species (Rodger, 1950; Whitear, 1960; Wakui and Sugiura, 1965).

Soon after entering the corneal stroma, the main bundles shed their perineurium and continue as flattened, ribbon-like structures sandwiched between the connective tissue lamellae. Adjacent nerve trunks branch, subdivide and rejoin with one another continuously in a series of irregular bifurcations or trifurcations to form a plexiform, multilayered network that distributes relatively uniformly throughout all four corneal quadrants (Ishida et al., 1984). Individual stromal axons may travel as much as three-quarters of the way

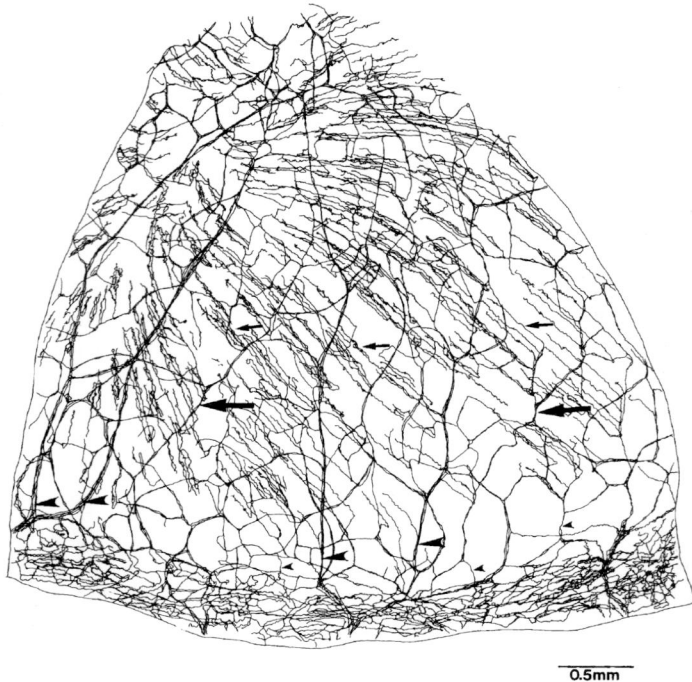

0.5mm

Figure 2.2   Pattern of corneal sensory innervation in one quadrant of an adult rat eye. At the bottom of the figure, a dense, ring-like network of nerve fibres supplies the corneoscleral limbus. Nerves enter the corneal stroma either in prominent, radially-directly nerve bundles (large arrowheads) or in superficially placed branches off the limbal plexus (small arrowheads). The nerves then branch and anastomose repeatedly to give rise to a dense stromal plexus (large arrows) and large numbers of radially-oriented basal epithelial leashes (arrows). The nerves illustrated in this whole-mount preparation have been stained immunohistochemically with antibodies against calcitonin gene-related peptide. See text for details. (Reprinted from Jones and Marfurt, 1991, with permission of John Wiley and Sons, Inc.).

across the cornea before ending (Zander and Weddell, 1951a,b). Not surprisingly, therefore, receptive fields of single corneal sensory axons may cover as much as 20–50% of the corneal surface (see below). The majority of the stromal nerves concentrate in the anterior one-third of the stroma, where they give rise to a dense subepithelial nerve plexus. The posterior stroma, on the other hand, is largely devoid of nerve fibres. A few workers have reported a sparse innervation of the corneal endothelium (Wolter, 1957; Léon-Felíu, Gómez-Ramos and Rodríguez-Echandía, 1978; ten Tusscher *et al.*, 1989); however, the majority of investigators have been unable to substantiate these findings.

The question of whether some sensory fibres "terminate" in the stroma remains unsettled. Some authors have described the presence of thin nerve fibres that are confined entirely in their distribution to the stroma; the latter fibres end as a series of bead-like varicosities or are tipped by a single terminal expansion (Zander and Weddell, 1951a;

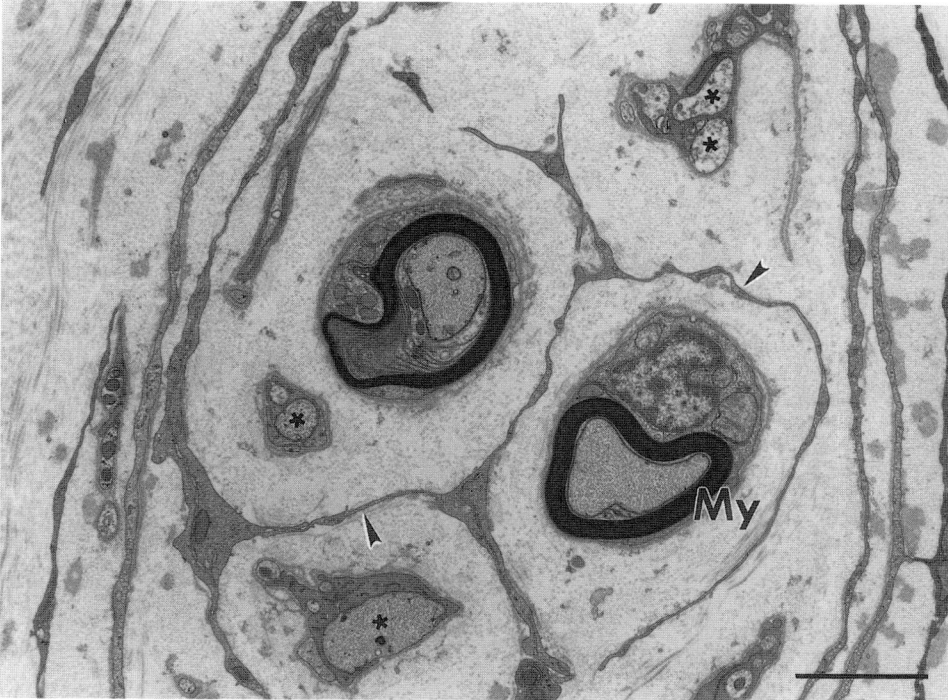

Figure 2.3   Cross section of a small nerve bundle in the peripheral stroma of a human cornea. Several unmyelinated axons (asterisks) and two myelinated axons (My) are present. The nerve fibres are surrounded by cytoplasmic protrusions of perineural cells (arrowheads). Scale bar = 2 μm. (Reprinted from Müller, Pels and Vrensen, 1996, with permission of Lippincott-Raven Publishers).

Chan-Ling, 1989). Ultrastructurally, some stromal axons lack complete Schwann cell investments and lie in direct contact with the stromal extracellular environment (Tervo and Palkama, 1978a). Other fibres run in close vicinity to stromal keratocytes, occasionally invaginating the keratocyte cytoplasm (Müller, Pels and Vrensen, 1996). Many stromal axons display linear arrays of closely spaced varicosities containing small accumulations of clear and dense-cored vesicles, mitochondria, and other organelles (Matsuda, 1968; Hoyes and Barber, 1976, 1977; Lim and Ruskell, 1978; Tervo and Palkama, 1978a,b; Tervo et al., 1979). Morphological similarities between these varicosities and sensory free nerve endings in other tissues has prompted speculation that the stromal varicosities subserve sensory transduction functions, including, detection of lamellar shearing forces (Lim and Ruskell, 1978) and generation of pain associated with acute elevations of intraocular pressure (Zuazo, Ibañez and Belmonte, 1986).

Intermingled with the stromal sensory nerve fibres are modest numbers of stromal sympathetic axons. The anatomy of the corneal sympathetic innervation has been investigated at the light microscopic level by histochemical fluorescence methods,

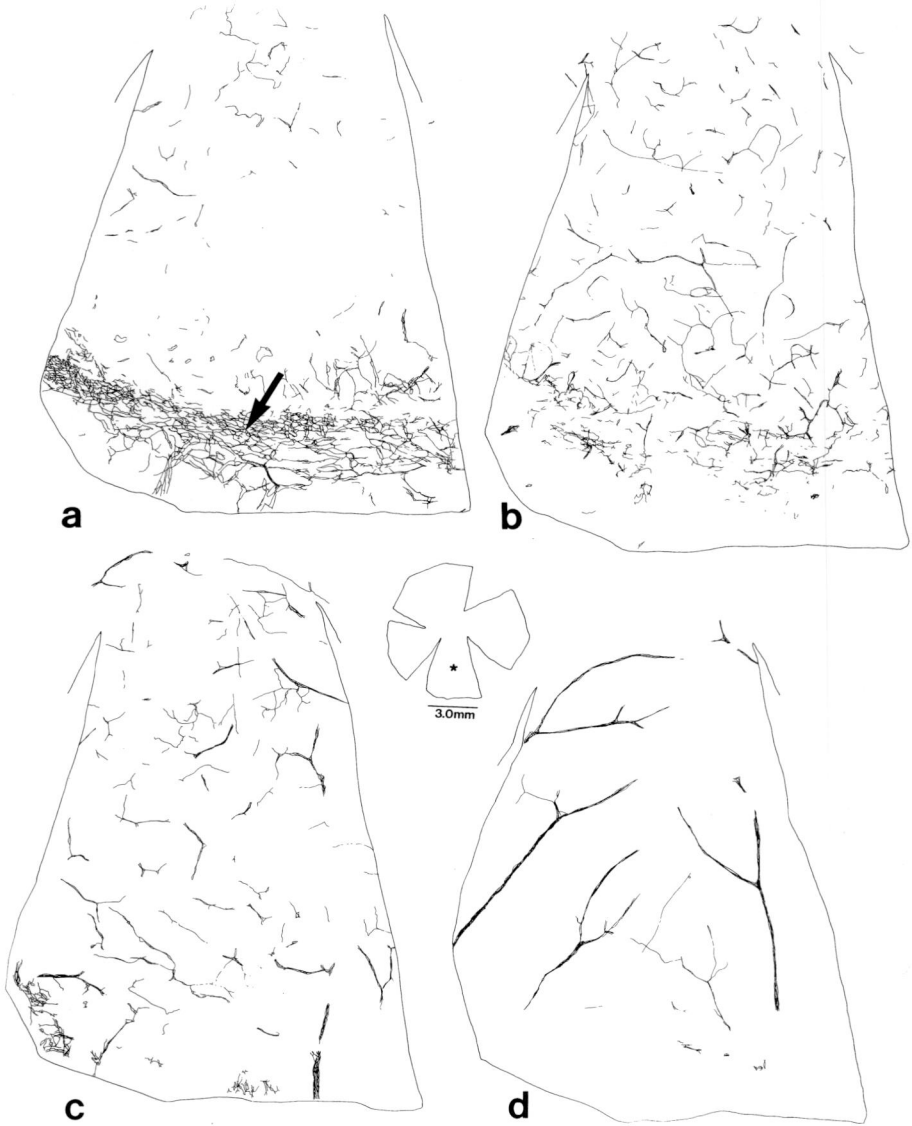

Figure 2.4   Sympathetic innervation of the guinea pig cornea. The illustration shows the distribution of tyrosine hydroxylase-immunoreactive nerves in serial anteroposterior sections (**a–d**, respectively) through the corneal quadrant indicated by the asterisk in the inset. A dense limbal plexus (arrow) is visible in section (**a**). (Reprinted from Marfurt and Ellis, 1993, with permission of John Wiley and Sons, Inc.).

immunohistochemistry, and nerve tracing techniques (Laties and Jacobowitz, 1964, 1966; Ehinger, 1966a,b,c; Ehinger and Sjöberg, 1971; Tervo and Palkama, 1976a,b; Klyce *et al.*, 1986; Marfurt, 1988; Marfurt and Ellis, 1993). Sympathetic nerves, like corneal sensory nerves, are concentrated in the anterior one-third of the stroma (Figure 2.4); however, overall sympathetic innervation density varies considerably among species (Marfurt, Kingsley and Echtenkamp, 1989). In general, sympathetic nerves supply a modest innervation to the rabbit, cat and guinea pig cornea, and a relatively sparse innervation to the mouse, rat, hamster, and dog cornea (Ehinger, 1966c; Laties and Jacobowitz, 1966; Marfurt and Ellis, 1993). Sympathetic innervation of the adult primate cornea, including human cornea, had been previously reported to be absent or insignificant (Ehinger, 1966b, 1971; Laties and Jacobowitz, 1966; Sugiura and Yamaga, 1968; Ehinger and Sjöberg, 1971; Toivanen *et al.*, 1987); however, more recent studies of human corneas using antibodies against tyrosine hydroxylase suggest a more substantial sympathetic innervation than has been previously recognized (Ueda *et al.*, 1989; Marfurt and Ellis, 1993).

Ultrastructurally, stromal sympathetic axons lack significant Schwann cell investments and may on occasion travel naked through the stromal matrix in direct contact with the extracellular environment (Tervo and Palkama, 1978a). The axons possess varicosities containing numerous small (300–500 Å), dense-cored vesicles thought to contain catecholamines; all of the axonal profiles with this morphology disappear after removal of the SCG (Tervo and Palkama, 1978a,b; Tervo *et al.*, 1979).

EPITHELIAL NERVE FIBRES

Sensory nerves enter the overlying corneal epithelium either from the subepithelial plexus or, in the peripheral cornea, directly from the conjunctiva. As the nerves penetrate Bowman's membrane, they shed their Schwann cell investments and continue into the epithelium as naked axon cylinders (Matsuda, 1968; Müller, Pels and Vrensen, 1996).

Intraepithelial sensory axons display a variety of morphologies; however, the two most common types of endings are complex basal epithelial "leashes", and simple, vertically oriented branch-like structures (Rózsa and Beuerman, 1982; MacIver and Tanelian, 1993a,b). Basal epithelial leash formations are a unique form of nerve specialization found only in the cornea. Each leash consists of a family of 2–15, tightly packed thin axons that run approximately parallel to one another in the deep part of the basal epithelial layer, or between the basal epithelium and Bowman's membrane (Figure 2.5). In most species, the leashes run in a predominantly radial (i.e., central) direction; however, in rabbits the leashes are oriented preferentially towards the nasal limbus (de Leeuw and Chan, 1989). Individual axons travel as far as several hundred microns in cats and rats (Chan-Ling, 1989; Jones and Marfurt, 1991) and up to 2 mm in humans (Schimmelpfennig, 1982). Occasional cross bridges interconnect adjacent axons. As the axons course horizontally through the basal epithelium, they give rise to an abundance of short, occasionally beaded, terminal axons which ascend vertically or obliquely into the more superficial epithelial layers before ending (Figure 2.6) (Rózsa and Beuerman, 1982; Ueda *et al.*, 1989).

Numerous other epithelial nerves originate as single axons directly from the subepithelial plexus, penetrate Bowman's membrane, and ascend vertically into the overlying

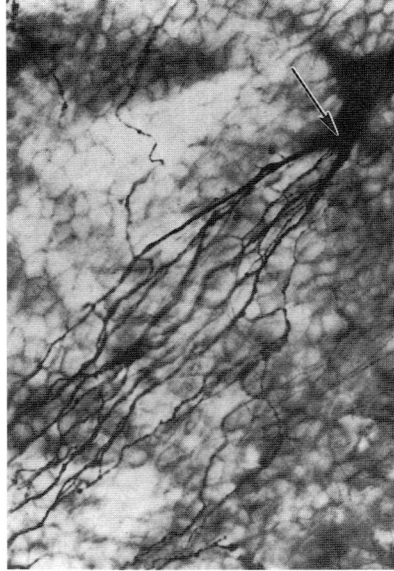

Figure 2.5 Leash formation in the basal epithelium of a rabbit cornea. The point at which the nerve penetrates the epithelium from the subepithelial plexus is indicated by the arrow. (Reprinted from Rózsa and Beuerman, 1982, with permission of Elsevier Science).

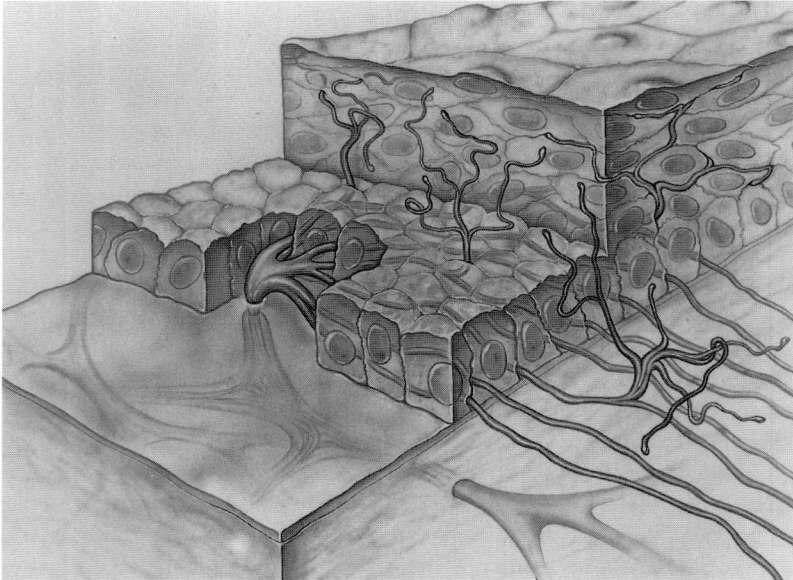

Figure 2.6 Innervation of the rabbit corneal epithelium. The innervation pattern shown here is based on camera lucida drawings and photomicrographs taken at successive focal levels. The size of the axon terminals relative to the dimensions of the epithelial cells is exaggerated for the sake of clarity. (Reprinted from Rózsa and Beuerman, 1982, with permission of Elsevier Science).

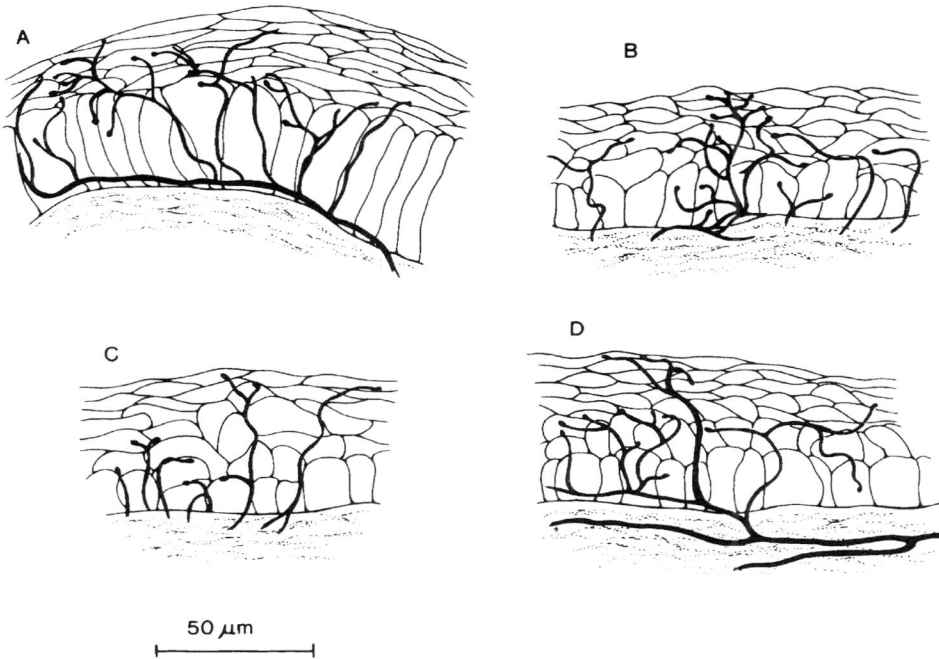

50 μm

Figure 2.7   Examples of terminal arrangements of intraepithelial axons in the rabbit cornea. (Reprinted from Rózsa and Beuerman, 1982, with permission of Elsevier Science).

epithelium (e.g., Rózsa and Beuerman, 1982; MacIver and Tanelian, 1993a,b). After a variable amount of additional branching, the fibres terminate in all layers of the epithelium (Figure 2.7).

Recently, MacIver and Tanelian (1993a,b) have described a third morphological type of intraepithelial axon in the rabbit cornea which outwardly resembles a basal epithelial leash formation, but which is located in the wing cell layer of the epithelium. This observation has not been confirmed by other workers.

The innervation density of the corneal epithelium is probably the highest of any surface epithelium. It has been estimated that there are approximately 5000–8000 nerve terminals per square millimeter of rabbit central corneal epithelium (Rózsa and Beuerman, 1982), and one terminal per 20 square micrometers of human corneal epithelium (Schimmelpfennig, 1982). Most of the nerve terminals are located in the wing and basal cell layers (Rózsa and Beuerman 1982); however, terminals may also extend to within a few microns of the corneal surface (Rózsa and Beuerman, 1982; Burton, 1992; MacIver and Tanelian, 1993a,b). Innervation density (Rózsa and Beuerman, 1982; Chan-Ling, 1989) is closely associated with corneal sensitivity to mechanical stimulation (Boberg-Ans, 1955; Cochet and Bonnet, 1960; Millodot and Larson, 1969; Draeger, 1984; Chan-Ling, 1989); both are maximal near the corneal apex and diminish progressively towards the corneoscleral limbus (Figure 2.8).

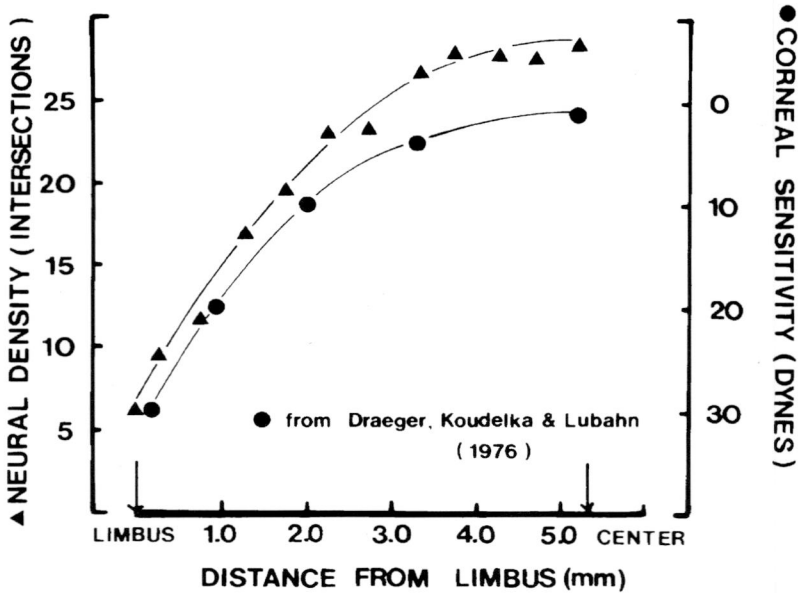

Figure 2.8 Comparison of psychophysical corneal sensitivity threshold values (●) and relative changes in neural density (▲) as a function of distance from the center of the cornea. The psychophysical data were replotted for illustration purposes. Each neural density value represents the mean of 30 corneas. Standard deviations were too small to plot. (Reprinted from Rózsa and Beuerman, 1982, with permission of Elsevier Science).

An undetermined percentage of intraepithelial nerve terminals are sympathetic. Intraepithelial sympathetic nerve fibres are relatively sparse in most species; however, a robust sympathetic innervation has been reported in the rabbit epithelium by horseradish peroxidase anterograde transport methods (Klyce et al., 1986).

Corneal epithelial nerves are not static structures but continuously elongate and undergo morphological rearrangements under normal physiological conditions. The dynamics of the human corneal epithelial plexus have recently been studied in vivo by scanning slit confocal microscopy (Masters and Thaer, 1994; Auran et al., 1995). These workers have shown that stromal nerve trunks, subepithelial nerve fibres, and epithelial perforation points remain stable in position and topography over extended periods of time and therefore provide reliable reference points for tracking nerve movement in the overlying basal epithelial plexus. By monitoring positional changes in distinctive epithelial nerve features (e.g., branch points, axon kinks, etc.), Auran and coworkers (1995) concluded that basal epithelial nerves slide centripetally, in concert with the basal epithelial cells, at a rate of 10–20 μm per day. They further concluded that neurite growth occurs by the addition of new nerve material at the site of nerve entry into the epithelium, rather than at distal growth cones. Physiological rearrangement of epithelial nerve terminal morphology has also been demonstrated in living mice corneas stained with nontoxic fluorescent

INITIAL OBSERVATION   ONE MONTH LATER

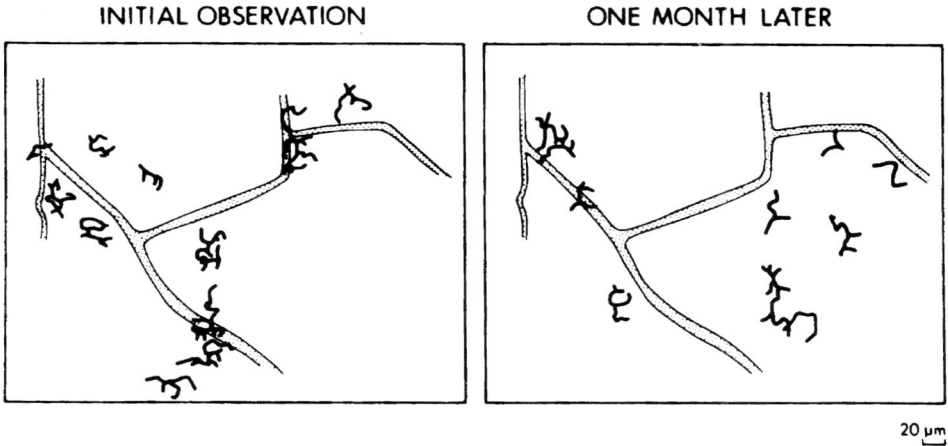

20 μm

Figure 2.9   Remodeling of nerve terminals in the living mouse cornea as shown by staining the nerve endings with a nontoxic fluorescent dye. The anatomical constancy of the stromal fibre bundles (stipple) allows the same region of the corneal surface to be examined at different times. After one month, nerve terminal rearrangement is so extensive that neither the location of terminal clusters nor their detailed structure bear any resemblance to the initial pattern of endings observed within the same corneal region. (Reprinted from Harris and Purves, 1989, with permission of the Society for Neuroscience).

dyes (Harris and Purves, 1989). According to the latter authors, corneal epithelial nerves undergo both short-term, passive rearrangements in terminal configuration caused by the outward migration of differentiating epithelial cells, and long-term nerve-directed reorganization (Figure 2.9).

ULTRASTRUCTURAL OBSERVATIONS

The fine structure of corneal free nerve endings and their morphological relationships to corneal epithelial cells have been described by several workers (Whitear, 1960; Matsuda, 1968; Hoyes and Barber, 1976, 1977; Tervo and Palkama, 1978b; Tervo et al., 1979; Ueda et al., 1989; Müller, Pels and Vrensen, 1996; Müller et al., 1997). Intraepithelial varicosities, ranging from 0.2 to 1.8 μm in diameter, are found in several locations within the epithelium, including between the basal epithelial cells and Bowman's membrane, between opposing cell membranes of adjacent epithelial cells, and invaginated within cytoplasmic infoldings of epithelial cells. The nerve endings are intimately apposed to the epithelial cell membrane, from which they are separated by a space of only 150–200 Å; however, membrane specializations and polarized accumulations of vesicles suggestive of "synaptic contacts" have not been described.

Two morphologically distinct types of sensory endings have been distinguished in some studies on the basis of organelle content (Matsuda, 1968; Hoyes and Barber, 1976, 1977; Ueda et al., 1989). One type contains numerous mitochondria, varying numbers of neurofilaments and microtubules, and occasional small, clear round vesicles. The second type is morphologically similar to a cholinergic nerve ending and contains abundant small, clear vesicles, occasional large or small dense-cored vesicles, and a few mitochondria. The large, dense-cored vesicles contain neuropeptides such as substance P (SP) and calcitonin gene-related peptide (CGRP) (Beckers *et al.*, 1992, 1993). The contents of the abundant small, clear vesicles remains unknown; however, studies in other systems suggest that they may contain excitatory amino acids (De Biasi and Rustioni, 1988). Both morphological types of terminals disappear following ophthalmic neurotomy and are, therefore, predominantly sensory in origin (Tervo *et al.*, 1979).

In contrast with these reports, work based on examination of serially sectioned human corneas suggests that corneal nerve terminals are morphologically homogeneous, and that perceived differences in organelle content reported by others reflect differences in the segment of the terminal that has been cut (Müller, Pels and Vrensen, 1996). In freshly fixed human eyes, all corneal nerve terminals contain varying numbers of clear and dense-cored vesicles, mitochondria and glycogen particles (Figure 2.10).

A third, less common type of epithelial axon terminal observed in some studies contains numerous small, dense-cored vesicles. This type of profile disappears following extirpation of the superior cervical ganglion and is therefore derived from corneal sympathetic nerves (Tervo and Palkama, 1978a,b; Tervo *et al.*, 1979).

NEUROCHEMISTRY OF CORNEAL NERVES

Corneal nerves contain a variety of cytoskeletal and cytoplasmic proteins, neuroenzymes, neurotransmitters, and neuropeptides. Many of these substances are non-specific nerve markers with widespread distributions in ocular sensory and autonomic nerves, whereas others are exclusively or predominantly associated with corneal sensory, sympathetic, or parasympathetic nerves.

GENERAL MARKERS

Substantial numbers of corneal sensory and autonomic nerves contain the cytoplasmic markers, protein gene product 9.5 (PGP 9.5) (Marfurt, Ellis and Jones, 1993; Marfurt, Jones and Thrasher, 1998) and peripherin (Marfurt, Jones and Thrasher, 1998) as well as the neuroenzymes, neuron specific enolase (NSE) (Ueda *et al.*, 1989) and acetylcholinesterase (AChE) (Laties and Jacobowitz, 1964, 1966; Petersen, Keat-Jin and Donn, 1965; Robertson and Winkelmann, 1970; Howard, Zadunaisky and Dunn, 1975; Tervo, 1976, 1977; Tervo and Palkama, 1978b). The presence of the latter enzyme suggested to early investigators a parasympathetic origin for these nerves; however, it is now known that AChE is a nonspecific marker that is expressed in large percentages of sensory and autonomic neurons. In the cornea, most AChE-positive nerves disappear after transection of the ophthalmic nerve and are therefore derived from the trigeminal ganglion (Tervo, 1976).

Many corneal nerves also stain immunohistochemically for growth associated protein-43 (GAP-43) (Martin and Bazan, 1992). GAP-43 is a neuron-specific protein found in

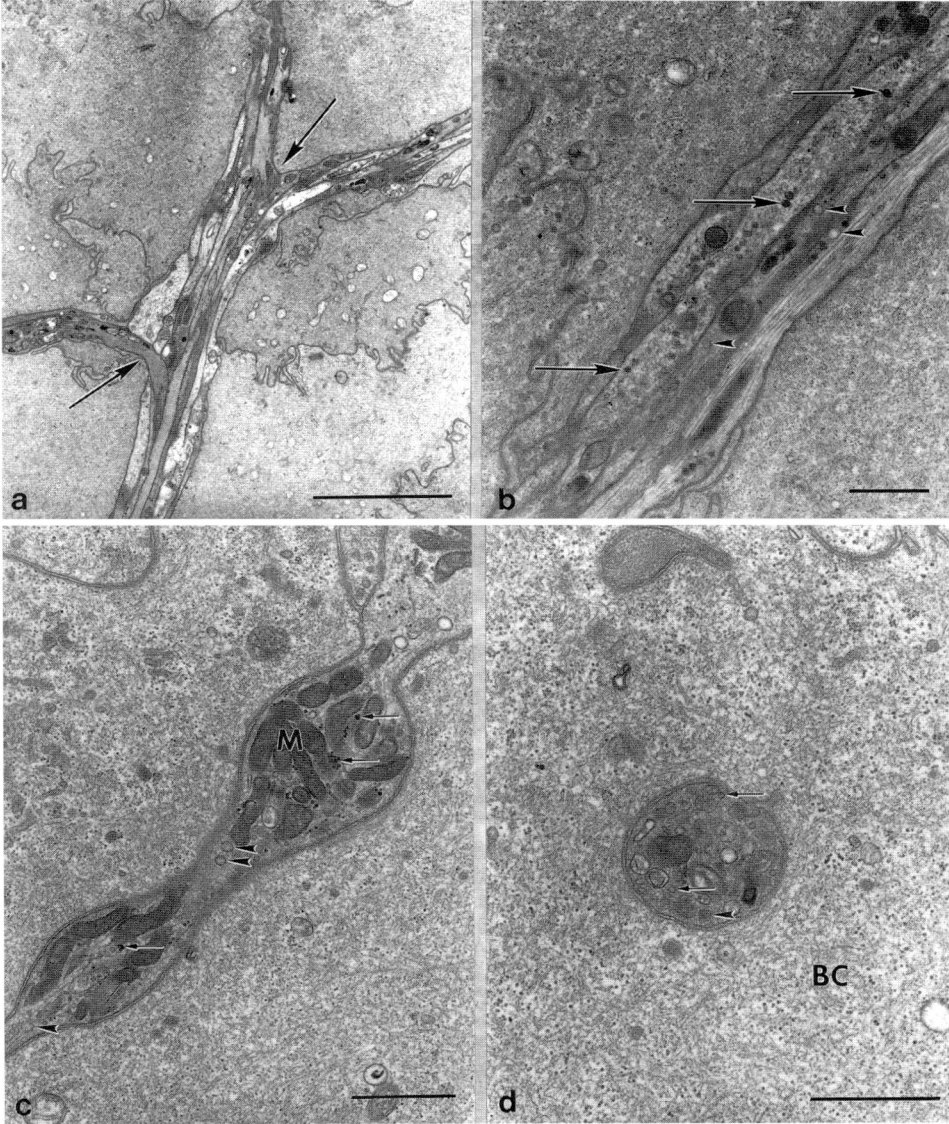

Figure 2.10 Ultrastructure of nerve fibres and axon terminals in the human corneal epithelium. **a**. Frontal section of a nerve bundle coursing horizontally through the basal epithelial cell layer. Bifurcations of the nerve bundle are indicated by arrows. **b**. Frontal section through a small bundle of intraepithelial axons. The axons contain numerous dense (arrows) and clear (arrowheads) vesicles. **c**. Two interconnected axon varicosities containing abundant dark mitochondria (M), vesicles (arrowheads) and a few glycogen particles (small arrows). **d**. Cross section of an axon containing many clear (arrows) and dense (arrowhead) vesicles. The axon is completely surrounded by basal cell epithelial cytoplasm (BC). Scale bar, **a** = 5 μm, **b–d** = 1 μm.

(Reprinted from Müller, Pels and Vrensen, 1996, with permission of Lippincott-Raven Publishers).

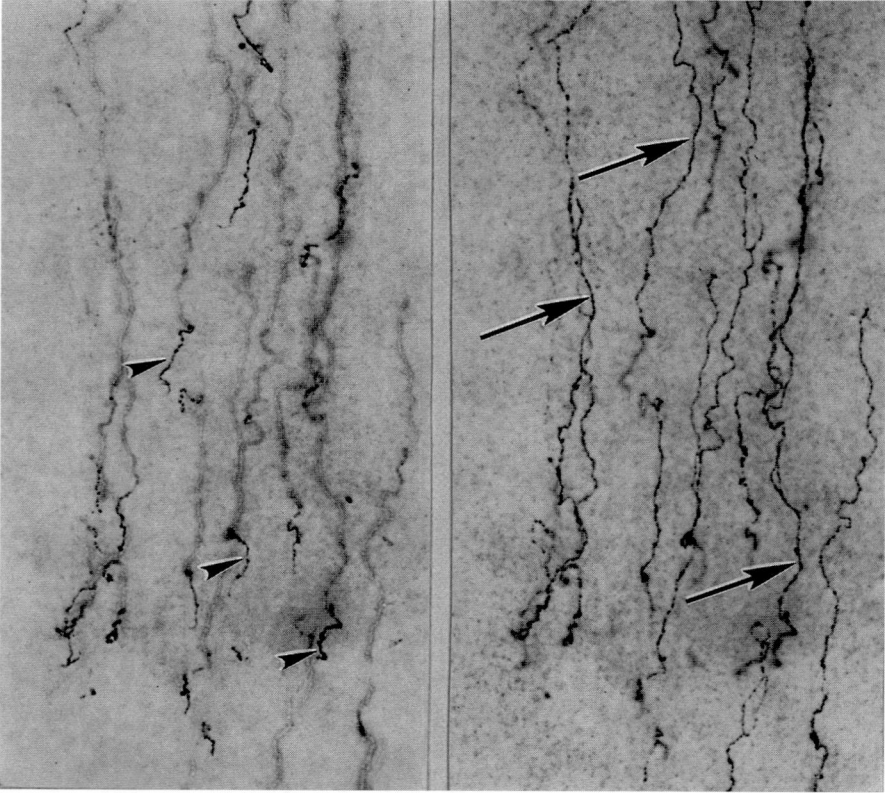

Figure 2.11 CGRP-immunoreactive nerve fibres in the rat corneal epithelium. The paired photomicrographs represent different focal planes through the same epithelial leash formation. As the leash axons (arrows) travel horizontally through the basal cell layer, they send numerous short, wavy terminal branches (arrowheads) into more superficial cell layers. (Reprinted from Jones and Marfurt, 1991, with permission of John Wiley and Sons, Inc.).

growth cones during development or regeneration of axons. The presence of this protein in mature corneal nerves is thought to be associated with the continuous process of intraepithelial axon elongation and terminal remodeling that occurs under normal physiological conditions (Harris and Purves, 1989; Auran et al., 1995) and during corneal wound healing (Lin and Bazan, 1995).

SENSORY NERVES

Corneal sensory nerves contain a variety of neuropeptides. Two peptides in particular, SP and CGRP (Figure 2.11), are found in large percentages of corneal sensory nerves (Miller et al., 1981; Tervo et al., 1981a,b; 1982a,b; 1983; Stone, Laties and Brecha, 1982; Ehinger et al., 1983; Sasaoka et al., 1984; Stone and Kuwayama, 1985; Stone et al., 1986; Kuwayama and Stone, 1987; Stone and McGlinn, 1988; Harti, Sharkey and Pierau, 1989;

Silverman and Kruger, 1989; Uusitalo, Krootila and Palkama, 1989; Jones and Marfurt, 1991, 1998). The majority of nerves that contain CGRP also contain SP (Kuwayama and Stone, 1987) and electron microscopic investigations of rat corneal nerves (Beckers *et al.*, 1992, 1993) suggest that the peptides may colocalize in the same vesicles (Gulbenkian *et al.*, 1986; Merighi *et al.*, 1988; Kummer, Fischer and Heym, 1989). Comparative radioimmunoassay studies reveal that corneal SP levels in small mammals are approximately 2–4 times higher than those in larger mammals (Gamse *et al.*, 1981; Unger *et al.*, 1981; Bucsics, Holzer and Lembeck, 1983; Elbadri *et al.*, 1991).

A robust population of corneal galanin-immunoreactive (-IR) nerves, rivaling in density those of the SP and CGRP fibre populations, innervates the rat eye (Jones and Marfurt, 1998). Smaller numbers of galanin-IR nerves have also been demonstrated in porcine corneas (Stone, McGlinn and Kuwayama, 1988). Approximately 95% of rat galanin-IR corneal fibres disappear after ophthalmomaxillary nerve transection, thus demonstrating their sensory origin (Jones and Marfurt, 1998). In the rodent eye, galanin-IR fibres apparently constitute a separate population of peptidergic fibres distinct from those that contain SP and CGRP because the former fibres (unlike ocular SP-IR and CGRP-IR nerves) are unaffected by neonatal capsaicin administration (Strömberg *et al.*, 1987).

Pituitary adenylate cyclase-activating peptide (PACAP), a neuropeptide with strong structural homology to vasoactive intestinal polypeptide (VIP), has been demonstrated in corneal sensory nerves in rats and rabbits (Moller *et al.*, 1993; Wang, Alm and Håkanson, 1995), but not in cats (Elsås, Uddman and Sundler, 1996). PACAP colocalizes extensively with CGRP and SP in ocular sensory nerves and in trigeminal ganglion neurons.

Many corneal sensory nerves contain the enzyme, fluoride-resistant acid phosphatase (FRAP) (Szönyi, Knyihar and Csillik, 1979; Silverman and Kruger, 1988a,b). FRAP-IR nerves also express a surface oligosaccharide which binds the isolectin I-B4 of *Griffonia simplicifolia* (GSA I-B4), but they do not contain CGRP (Silverman and Kruger, 1988b). The function of FRAP in corneal nerves is unclear; however, it has been hypothesized that the enzyme may play a role in transmitter metabolism (Szönyi, Knyihar and Csillik, 1979). Finally, some corneal sensory axons in rats and mice, and possibly in humans, contain tyrosine hydroxylase, the rate limiting enzyme in catecholamine synthesis (Ueda *et al.*, 1989; Marfurt and Ellis, 1993).

AUTONOMIC NERVES

Most, if not all, corneal sympathetic nerves contain the classical neurotransmitter, noradrenaline (e.g., Laties and Jacobowitz, 1964, 1966; Ehinger, 1966a,b,c) and the catecholamine synthesizing enzyme, tyrosine hydroxylase (Ueda *et al.*, 1989; Marfurt and Ellis, 1993). Many corneal sympathetic fibres also contain serotonin (Uusitalo *et al.*, 1982; Osborne, 1983; Palkama, Uusitalo and Lehtosalo, 1984; Osborne and Tobin, 1987) and neuropeptide Y (NPY) (Stone, 1986; Stone, Laties and Emson, 1986; Jones and Marfurt, 1998) (Figure 2.12).

As noted earlier, a modest parasympathetic innervation of the cornea has been demonstrated in cats and rats by horseradish peroxidase nerve tracing methods and/or selective ocular denervations coupled with immunohistochemistry. In the rat, corneal

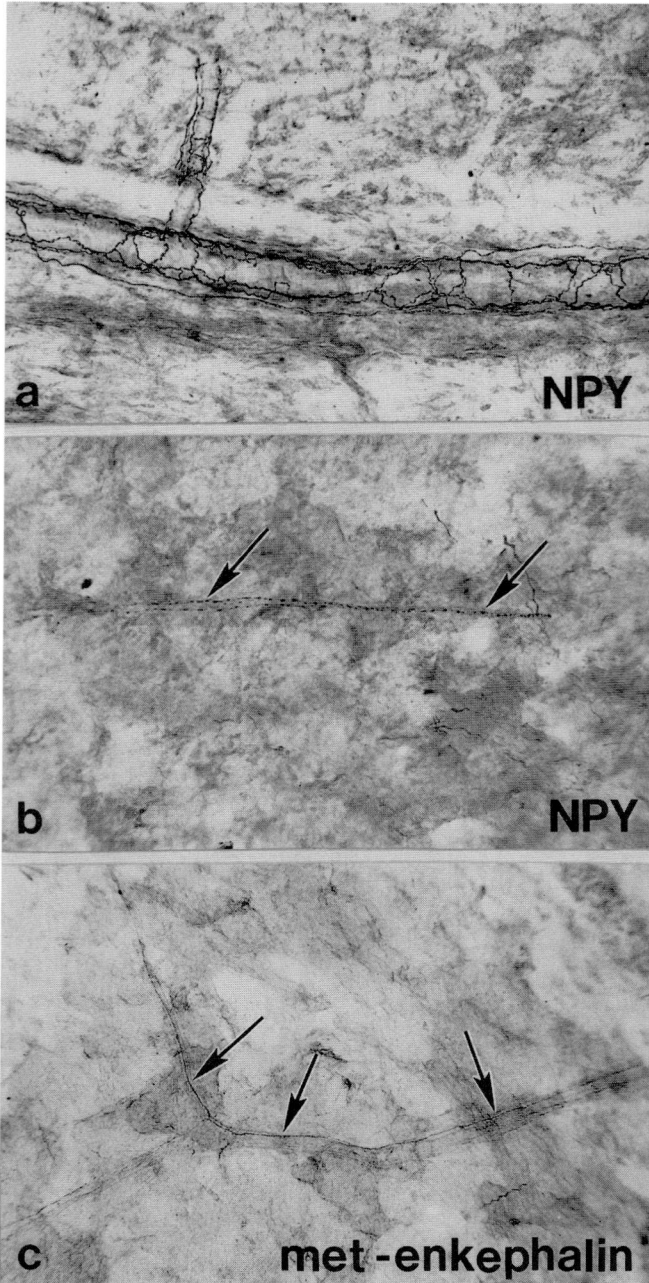

Figure 2.12 (a) Perivascular neuropeptide Y-immunoreactive fibres in the rat corneoscleral limbus. (b) NPY-immunoreactive fibres (arrows), and (c) met-enkephalin-immunoreactive fibres (arrows) in the rat corneal stroma. Courtesy of Dr. Mark Jones.

parasympathetic nerves express VIP, met-enkephalin, NPY, and galanin (Figure 2.12) (Jones and Marfurt, 1998). A few (presumably parasympathetic) rat corneal fibres also contain the enzyme, nicotinamide adenine dinucleotide phosphate (NADPH) diaphorase (Yamamoto et al., 1993). Measurable quantities of VIP (Unger et al., 1981; Elbadri et al., 1991) and the opioid-like peptides met-enkephalin, β-endorphin, and α-melanocyte-stimulating hormone (α-MSH) (Tinsley et al., 1989) have also been reported in corneas from larger mammals.

OTHER MARKERS

Several other peptides have been demonstrated by immunohistochemistry or radioimmunoassay in corneal nerve fibres, including cholecystokinin (Palkama, Uusitalo and Lehtosalo, 1986), brain natriuretic peptide (Yamamoto, McGlinn and Stone, 1991), vasopressin (Too et al., 1989) and neurotensin (Tinsley et al., 1989). At present, the origins, densities, and functions of these peptidergic nerve populations remain unknown.

The preceding account only hints at the complex neurochemistry of the corneal innervation. The percentage of corneal nerves that contain particular neurochemicals, and the extent to which individual neurochemicals colocalize and are co-released by subpopulations of corneal nerves, remain largely unknown. The impressive number of neuropeptides identified within corneal nerves, and recent demonstrations of peptide colocalization, suggest complex physiological interactions (Stone, Kuwayama and Laties, 1987). Until it is determined how particular neurochemicals correlate with specific sensory and "effector" functions of corneal nerves, the importance of each of these markers in corneal physiology is difficult to evaluate. It would be of great interest to learn, for example, if neurochemically distinct subpopulations of corneal nerve fibres (e.g., FRAP-IR and CGRP-IR) represent functionally distinct fibre populations for parallel processing of corneal nociceptive information, or if they subserve separate and distinct sensory, trophic, and inflammatory functions.

CORNEAL SENSORY INNERVATION: FUNCTIONAL CONSIDERATIONS

SENSORY MODALITIES

It is generally stated that only nociceptive sensations are evoked by naturally occurring corneal stimulation. However, the question of whether "pure" mechanical or "pure" thermal sensations (i.e., tactile or thermal sensations uncontaminated with any nociceptive component) can be perceived in human psychophysical experience remains unresolved. The issue has fostered a century-long, spirited debate in which "Workers seem to have arrayed themselves as contestants, those who subscribe to von Frey's theory [that nociception is the sole modality transduced by corneal receptors] on the one hand, those who do not on the other" (Lele and Weddell, 1956). The question has been difficult to answer due largely to methodological difficulties, including, high levels of test subject apprehension (Maurice, 1984).

According to some workers, mechanical (von Frey, 1894) and thermal (Nafe and Wagoner, 1937; Kenshalo, 1960; Beuerman and Tanelian, 1979) stimulation of the human cornea (carefully controlled to prevent spread to adjacent tissues) evokes only perceptions of pain. Identical stimuli perceived as painful when applied to the cornea are sensed as tactile, warm or cold when applied to adjacent regions of the conjunctiva, eyelid, or facial skin. Electrophysiological investigations in cats and rabbits suggest that some corneal afferent fibres are relatively modality-specific and that distinct fibre populations are activated preferentially by mechanical, thermal, or chemical stimulation (see below). It has been claimed that this modality-specificity is lost during subsequent central nervous system processing (Tanelian and Beuerman, 1984; MacIver and Tanelian, 1993a,b) and that all mechanical and thermal corneal stimuli are perceived as irritating or painful.

In contrast to the above point of view, other workers contend that graded punctate stimuli applied to the corneal surface with a smooth nylon monofilament under highly controlled testing conditions evoke pure (often liminal) sensations of touch (Lele and Weddell, 1956; Millodot, 1968; Draeger, 1984). As the pressure exerted by the nylon thread increases to many times the corneal touch threshold (Millodot, 1968), or if the stimulus remains in position for more than one second (Draeger, 1984), the sensation changes to pain. Other evidence in support of the existence of corneal tactile sensibility comes from observations in patients suffering from trigeminal neuralgia. Surgical interruption of the corneal central nociceptive pathways in these patients (i.e., trigeminal tractotomy) changes their perception of corneal stimulation from pain to touch (Rowbotham, 1939; Grant, Groff and Lewy, 1940; Falconer, 1949).

DETERMINANTS OF CORNEAL SENSITIVITY

Corneal sensitivity to mechanical stimulation is highest in the central cornea and decreases progressively through the peripheral cornea, limbus, and conjunctiva (Boberg-Ans, 1955; Cochet and Bonnet, 1960; Millodot and Larson, 1969; Draeger, 1984; Chan-Ling, 1989). Sensitivity is adversely affected by several factors, including age, extended contact lens wear, eyelid closure during sleep, iris color, menstruation and pregnancy, environmental and atmospheric factors, and most ocular and many non-ocular diseases. Several in depth reviews are available on this subject (Draeger, 1984; Millodot, 1984; Martin and Safran, 1988).

ELECTROPHYSIOLOGY

Receptive fields

Receptive fields of individual corneal afferent fibres are generally round, oval, or wedge-shaped in outline, variable in size, and exhibit considerable overlap. A typical mechanoreceptive field in the rabbit or cat cornea extends over 5–20% (10–20 mm^2) of the corneal surface, although individual receptive fields may be as small as 1 square millimeter or as much as 50–100 mm^2 (Tower, 1940; Lele and Weddell, 1959; Mark and Maurice, 1977; Giraldez, Geijo and Belmonte, 1979; Belmonte and Giraldez, 1981; Tanelian and Beuerman, 1984; Belmonte et al., 1991). The largest receptive fields cover

over 40–50% of the corneal surface. Receptive fields may also extend several millimeters beyond the cornea onto the adjacent limbus and bulbar conjunctiva. The large size and extensive overlap among adjacent receptive fields, coupled with convergence of sensory inputs onto common second order neurons in the trigeminal brainstem nuclear complex, explains why sensations arising from corneal stimulation are poorly localized (Lele and Weddell, 1956; Beuerman and Tanelian, 1979; Draeger, 1984).

Conduction velocities

All corneal sensory nerves originate from unmyelinated or finely myelinated fibres conducting in the C-fibre (0.25–2.0 m/s) or A-δ (greater than 2.0 m/s) range (Giraldez, Geijo and Belmonte, 1979; Belmonte and Giraldez, 1981; Belmonte *et al.*, 1991; Gallar *et al.*, 1993; MacIver and Tanelian, 1993a,b). The conduction velocities of individual nerves decrease significantly in the proximodistal direction (Belmonte *et al.*, 1991; MacIver and Tanelian, 1993a,b) as the result of progressive intracorneal tapering of the axon and, in the case of A-δ fibres, shedding of the myelin sheath at the corneoscleral limbus.

Modality specificity

The electrophysiological properties of corneal afferent nerve fibres in cats and rabbits have been well characterized (Figure 2.13) (see Belmonte and Gallar, 1996; Belmonte, Garcia-Hirschfeld and Gallar, 1997, for recent reviews). In general, corneal afferent neurons are either unimodal (responding to only a single modality of stimulation) or bimodal/polymodal (responding to more than one modality of stimulation).

Three types of corneal unimodal units, mechanosensitive, cold sensitive, and chemosensitive, have been described by various authors. Corneal mechanosensitive fibres are fast conducting, rapidly adapting, A-δ afferents that respond exclusively to mechanical stimulation (indentation) of the corneal surface. These fibres exhibit velocity and force sensitivity and are especially sensitive to moving stimuli on the corneal surface (Lele and Weddell, 1959; Belmonte and Giraldez, 1981; Tanelian and Beuerman, 1984; Belmonte *et al.*, 1991; MacIver and Tanelian, 1993a,b). Cold receptors are spontaneously active, slowly adapting C-fibres that respond with increased frequency of discharge to cooling of the corneal surface (Lele and Weddell, 1959; Dawson, 1962; Mark and Maurice, 1977; Tanelian and Beuerman, 1984; MacIver and Tanelian, 1993a,b) and which may be of special importance in signalling evaporative cooling of the cornea and initiation of the blink reflex (MacIver and Tanelian, 1993a,b). Whether the cold receptors reported in these studies are truly unimodal has been questioned since in the majority of cases their responses to other modalities of stimulation were not specifically tested; thus, some of them may actually be bimodal "cold nociceptors" (Gallar *et al.*, 1993, see below). Pure chemosensory fibres (i.e., sensory units that are insensitive to mechanical or thermal stimuli) have been described in the rabbit cornea (Tanelian, 1991; MacIver and Tanelian, 1993a,b). The latter fibres are excited by a variety of chemical substances, including, acetylcholine (ACh), nicotine, carbachol, glutamate and its agonist *N*-methyl-D-aspartate (NMDA), prostaglandin E_1, and bradykinin (MacIver and Tanelian, 1993a,b). Similar "pure" chemosensory units have not been identified in the cat, and their existence in the rabbit eye has been questioned on methodological grounds (Belmonte and Gallar, 1996).

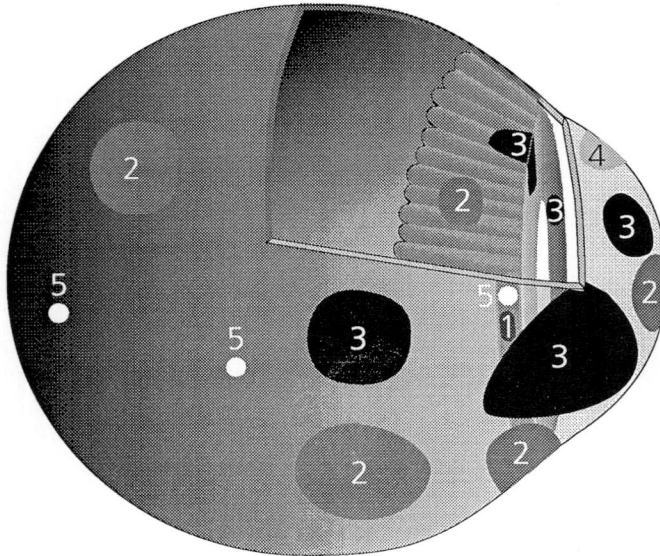

Figure 2.13 Functional classes of corneal sensory units in the cat eye. In the upper part of the figure, the characteristics of the impulse discharge, either spontaneous (On-going activity) or evoked by different types of stimulating energy (Stimuli) have been represented together with the functional classification of the peripheral terminals (Receptor type). The lower part of the figure shows the relative size and location in the eye of the receptive fields of different functional types of ocular sensory neurons. (Reprinted from Belmonte, Garcia-Hirschfeld and Gallar, 1997, with permission of Elsevier Science Ltd.)

Polymodal units are A-δ and C-fibres that respond to all three modalities of stimulation, i.e., to mechanical stimulation, heating of the cornea over 39°C, and select forms of chemical stimulation (e.g., acetic acid, hypertonic saline, and capsaicin) (Belmonte et al., 1991). In contrast, cold is ineffective in exciting these fibres. Sensitivity to protons and to mechanical stimuli in the polymodal fibres may be subserved by separate transduction mechanisms (Belmonte et al., 1991). Bimodal "mechanoheat nociceptor" A-δ fibres have been described by several investigators. Strictly speaking, the latter fibres respond to high threshold mechanosensitive stimuli and to heat, but not to chemical stimuli (Lele and Weddell, 1959; Belmonte and Giraldez, 1981; Belmonte et al., 1991; MacIver and Tanelian, 1993a,b). It has been hypothesized that the latter fibres may actually be polymodal nociceptors, but with atypically high thresholds of activation to chemical stimuli (Belmonte and Gallar, 1996). Finally, cold nociceptors are bimodal units that respond to thermal and chemical stimulation, but not to mechanical stimulation (Gallar et al., 1993).

Attempts to correlate the electrophysiological properties of corneal nerve fibres with specific epithelial terminal morphologies have been reported by MacIver and Tanelian (1993a,b). These authors stained morphologically distinct subpopulations of living free nerve endings in rabbit corneal epithelium with vital fluorescent dyes and then characterized them electrophysiologically, in vitro. According to their findings, A-δ directionally-sensitive mechanoreceptive endings comprise small clusters of elongated, wavy, thin axons that run horizontally and roughly parallel to one another for long distances. Morphologically, they resemble basal epithelial leashes, but are located more superficially (about 10–40 μm from the corneal surface). In contrast, C-fibre endings are identified by electrical stimulation and form clusters of short (less than 50 μm) poorly branched, vertical, tree-like endings.

TROPHIC FUNCTIONS OF CORNEAL SENSORY NERVES

In addition to their well known sensory functions, corneal afferent nerves exert important nutritive, or trophic, influences on their target organ that contribute to the survival and normal function of the tissue. Damage to the ocular sensory nerves by surgery, trauma, herpes simplex infection or disease produces a degenerative condition in the cornea known as neuroparalytic, or neurotrophic, keratitis (Paton, 1926; Pannabecker, 1944; Davies, 1970). This trophic influence of the trigeminal nerve, first described in rabbit corneas over 170 years ago (Magendie, 1824) has since been demonstrated in numerous mammals following ophthalmic nerve transection or trigeminal ganglion electrocoagulation (Zander and Weddell, 1951b; Rodger, 1953; Sigelman and Friedenwald, 1954; Moses and Feldman, 1969; Alper, 1975; Huhtala, Johansson and Saari, 1975; Lim, 1976; Schimmelpfennig and Beuerman, 1982; Knyazev, Knyazeva and Nikiforov, 1990; Knyazev, Knyazeva and Tolochko, 1991). Keratitis also develops in rats and mice following partial destruction of their corneal innervation by neonatal administration of capsaicin (Figure 2.14) (Keen et al., 1982; Buck et al., 1983; Fujita et al., 1984; Shimizu et al., 1984; Herbort, Weissman and Payan, 1989; Abelli, Geppetti and Maggi, 1993). In capsaicin-treated animals, the severity of the keratitis diminishes in intensity when the nerves reinnervate the cornea (Ogilvy and Borges, 1990; Ogilvy, Silverberg and Borges, 1991; Marfurt, Ellis and Jones, 1993).

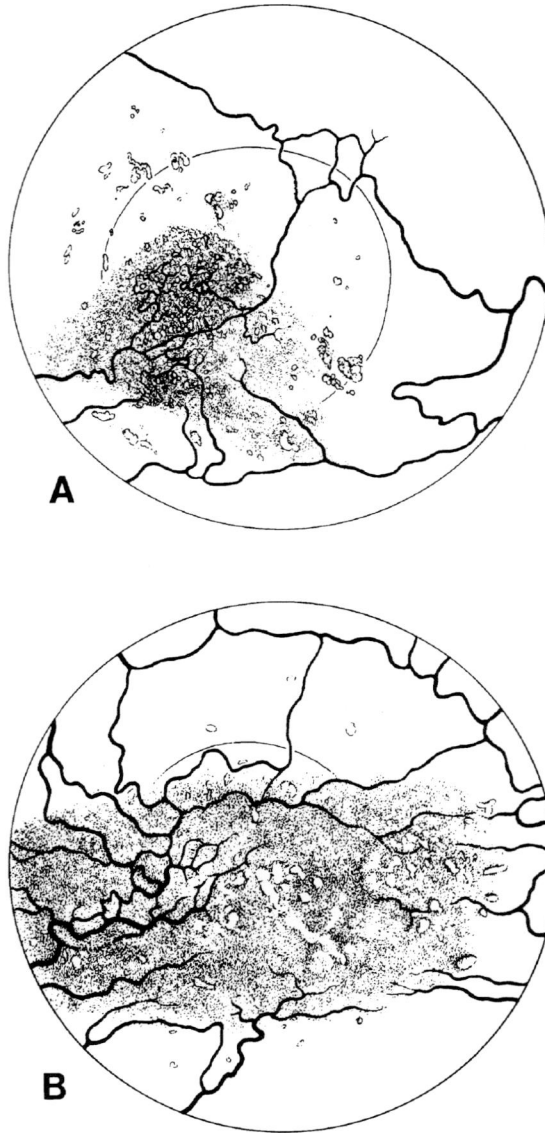

Figure 2.14 Schematic drawings of young adult mouse corneas illustrating moderate (A) and severe (B) neurotrophic keratitis 11–14 weeks after neonatal injection of the sensory neurotoxin, capsaicin. (Reprinted from Herbort, Weissman and Payan, 1989, with permission of FASEB).

Clinically, the development of corneal disturbances in neuroparalytic keratitis follows a characteristic pattern (Paton, 1926; Duke-Elder and Leigh, 1965). Within 24–36 h after nerve injury, the corneal surface becomes stippled and hazy due to progressive epithelial cell death and surface sloughing. Multiple punctate epithelial defects that stain with fluorescein appear first in the centre of the cornea, coalesce into larger areas of cell loss, and rapidly spread into more peripheral areas. If left untreated, massive exfoliation of the epithelium occurs, leaving only a thin, perilimbal ring of intact epithelium. Eventually, the denuded stromal surface becomes dry, milky, and hazy, and the entire cornea may become opaque (edematous) with significantly impaired vision. If secondary infection sets in, corneal ulceration and perforation may follow.

Studies in experimental animals have contributed significantly to our understanding of the morphological and pathophysiological events that take place in neuroparalytic keratitis. Deprived of their normal sensory innervation, corneal epithelial cells become abnormally rounded, exhibit decreased numbers of surface membrane interdigitations, and a loss of tonofilaments (Sugiura, Kawanabe and Kotashima, 1964; Sugiura and Matsuda, 1967; Lim, 1976; Beuerman, Schimmelpfennig and Burstein, 1979). Superficial epithelial cells become swollen (Gilbard and Rossi, 1990; Knyazev, Knyazeva and Nikiforov, 1990; Knyazev, Knyazeva and Tolochko, 1991) and their surface microvilli are lost or become truncated in appearance (Beuerman, Schimmelpfennig and Burstein, 1979; Gilbard and Rossi, 1990). Intercellular spaces become dilated and epithelial permeability to topical fluorescein increases (Beuerman and Schimmelpfennig, 1980). Total epithelial thickness, cell number, mitotic rate, and epithelial glycogen content are all decreased (Sigelman and Friedenwald, 1954; Mishima, 1957; Alper, 1975; Mackie, 1978; Holden et al., 1982; Beuerman, Tanelian, and Schimmelpfennig, 1988; Gilbard and Rossi, 1990). Massive numbers of leukocytes infiltrate the epithelium and stroma (Knyazev, Knyazeva and Tolochko, 1991). Finally, corneal epithelial oxygen uptake rates are decreased and hypoxic swelling responses are altered (Holden et al., 1982; Vannas et al., 1985).

Corneas deprived of their normal sensory innervation also demonstrate an impaired ability to heal after corneal injuries. Surgical destruction of the rabbit trigeminal ganglion inhibits the rate at which corneal epithelial cells resurface standardized epithelial abrasions (Figures 2.15 and 2.16) (Beuerman and Schimmelpfennig, 1980; Schimmelpfennig and Beuerman, 1982; Araki et al., 1994). The slowed rate of re-epithelialization may be related, in part, to the loss of essential cytoskeletal structures necessary for optimal cell migration and adhesion (Beuerman, Schimmelpfennig and Burstein, 1979). Wounded corneas in sensory-denervated eyes also exhibit high numbers of exfoliating surface cells, abnormal migratory cell orientations, and recurrent, spontaneous epithelial erosions (Araki et al., 1994).

Many theories have been proposed to explain the pathogenesis of neuroparalytic keratitis, including, desiccation of the corneal surface due to diminished lacrimal secretions, loss of corneal sensation leading to an absence of normal protective (blink) reflexes, abnormal epithelial cell metabolism with subsequent failure to resist the effects of trauma, drying, and infection, and the loss of trophic impulses supplied by corneal sensory nerves (Paton, 1926; De Haas,1962; Duke-Elder and Leigh, 1965). Many authors feel that the actual cause is probably represented by a combination of these factors.

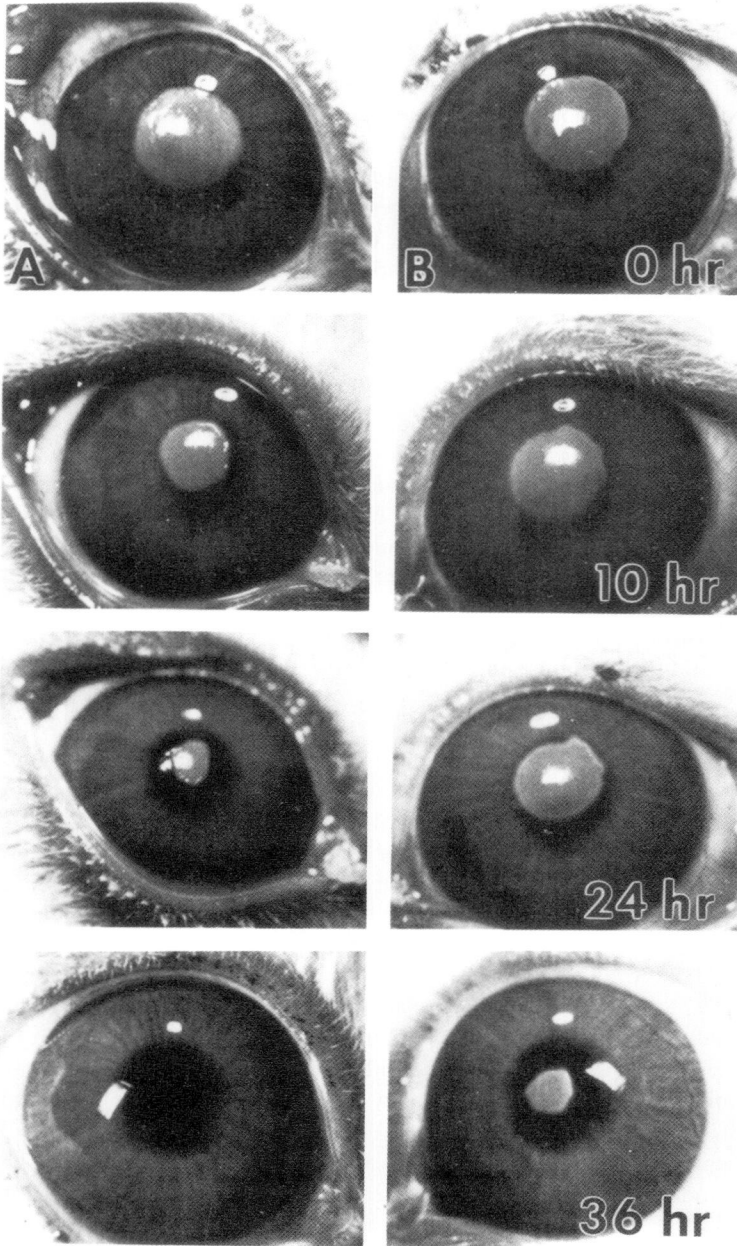

Figure 2.15 Delayed epithelial wound healing in the rabbit cornea resulting from ocular sensory denervation. The photomicrographs illustrate the healing rates of standardised 4 mm diameter abrasions in control (A, left column) and denervated (B, right column) corneas from the same animal. The fluorescein-stained wounds have been photographed at time of wounding (0 hr), and at 10, 24 and 36 hours after wounding. The rate of wound closure is significantly delayed in the denervated eye. (Reprinted from Schimmelpfennig and Beuerman, 1982, with permission of Springer-Verlag).

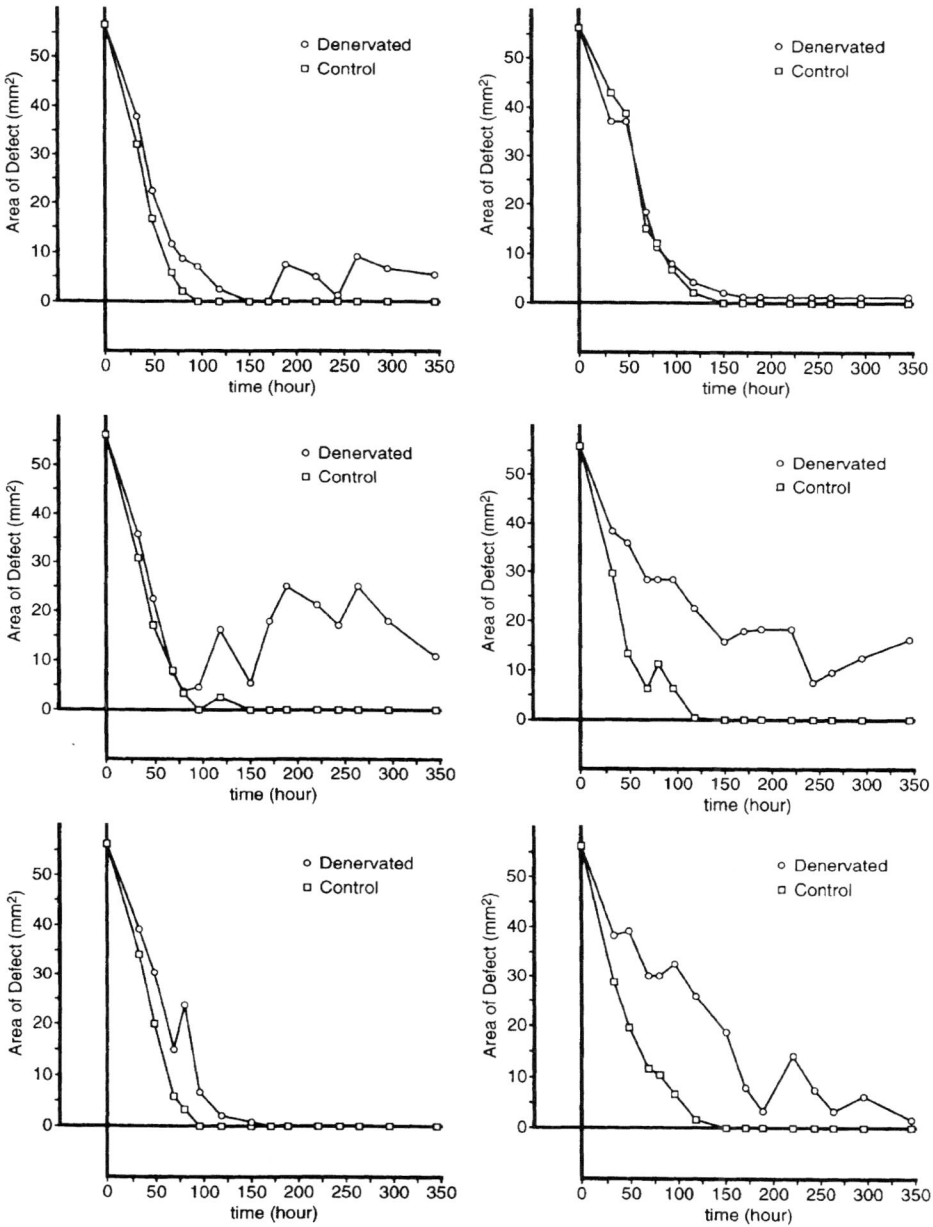

Figure 2.16 Time course of rabbit corneal epithelial wound healing after bilateral keratectomy in control (□) and sensory-denervated (○) corneas from six different animals. The epithelial defects in all of the control corneas are completely re-surfaced by 4–6 days; however, 14 days later five out of six (83%) denervated corneas show persistent epithelial defects. (Reprinted from Araki *et al.*, 1994, with permission of Oxford University Press).

A trophic role for corneal nerves is supported by several observations. For example, studies *in vivo* have shown that corneal abnormalities develop in denervated eyes even if preventative measures are taken to prevent desiccation and trauma by suturing the eyelids (Alper, 1975; Huhtala, Johansson and Saari, 1975; Shimizu *et al.*, 1984; Ogilvy and Borges, 1990). Co-culture of trigeminal ganglion neurons with rabbit corneal epithelial cells stimulates epithelial cell growth, proliferation and differentiation (Garcia-Hirschfeld, Lopez-Briones and Belmonte, 1994), and epithelial production of type VII collagen (Baker *et al.*, 1993), an important determinant of cell-basement membrane adhesion.

The mechanisms by which corneal sensory nerves exert their trophic functions are unknown; however, the release of axonally transported substances, particularly neuropeptides, has been postulated. In support of this hypothesis, retrobulbar injection of capsaicin results in depletion of neuropeptides in the rat eye and produces a neuroparalytic keratitis indistinguishable from that seen after total ophthalmic nerve transection (Knyazev, Knyazeva and Nikiforov, 1990). It is tempting to speculate that SP and CGRP subserve ongoing trophic and regulatory processes in the corneal epithelium, and that when released from corneal sensory nerves (either tonically or in response to corneal wounding) stimulate corneal epithelial cells as part of the normal process of tissue maintenance and physiological renewal.

SP and CGRP receptors have been demonstrated on corneal and limbal epithelial cells by autoradiographic analyses (Kieselbach *et al.*, 1990; Denis *et al.*, 1991; Heino *et al.*, 1995). Cultured rabbit corneal epithelial cells possess approximately 2.43×10^4 SP binding sites per cell, nearly all of which are of the NK_1 subtype (Nakamura *et al.*, 1997b).

Substance P stimulates cell growth and proliferation in primary cultures of rabbit corneal epithelial cells grown in media supplemented with growth factors and hormones (Garcia-Hirschfeld, Lopez-Briones and Belmonte, 1994) and in established rabbit corneal epithelial cells (SIRC cells) grown in a serum-free medium (Reid *et al.*, 1993). In contrast, addition of CGRP alone to either of the above cell lines has no significant effect on DNA synthesis. Of interest, when SP and CGRP are added concurrently to the culture medium, they exert synergistic effects on cell proliferation in the SIRC cells (Reid *et al.*, 1993), but antagonistic effects on proliferation in the primary cell culture (Garcia-Hirschfeld, Lopez-Briones and Belmonte, 1994) (Figure 2.17).

Recent evidence suggests that SP released from corneal sensory nerve fibres may also promote epithelial cell migration during corneal wound healing. Systemic and topical administration of capsaicin in rabbits depletes ocular neuropeptide levels and impairs the healing rate of standardized corneal epithelial abrasions (Gallar *et al.*, 1990; Murphy *et al.*, 1990). Topical application of SP alone fails to stimulate wound healing in injured rabbit corneas (Nishida *et al.*, 1996; Kingsley and Marfurt, 1997; Nakamura *et al.*, 1997a,b); however, when SP and insulin-like growth factor-1 (IGF-1; Nishida *et al.*, 1996; Nakamura *et al.*, 1997a,b) or SP and epidermal growth factor (Nakamura *et al.*, 1997c) are added together, the neuropeptide and humoral factor act synergistically to stimulate cell migration in a dose-dependent fashion. This effect is mediated by NK_1, but not NK_2 or NK_3, tachykinin receptors (Nakamura *et al.*, 1997b). SP and IGF-1 also act synergistically to promote epithelial cell attachment to extracellular matrix proteins (Nishida *et al.*, 1996). Of clinical interest, topical applications of SP and IGF-1 may

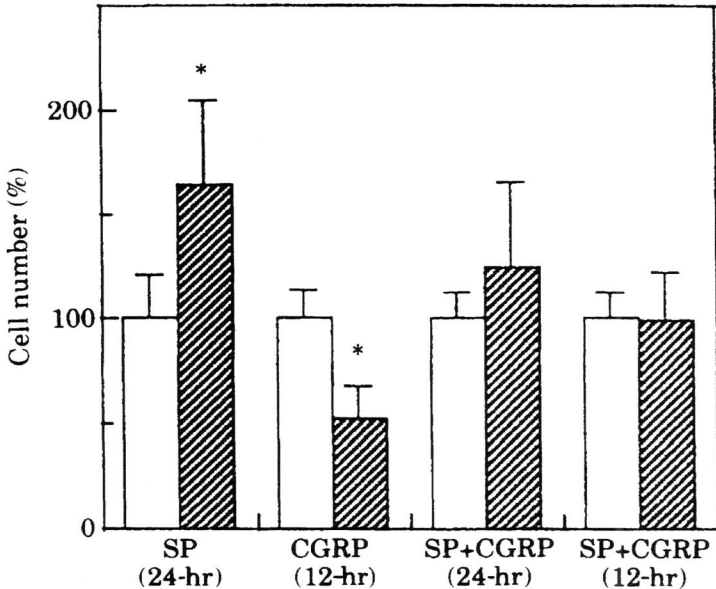

Figure 2.17 Effects of application of substance P (SP), calcitonin gene-related peptide (CGRP) and a combination of both, on rabbit corneal epithelial cell proliferation, *in vitro*. The neuropeptides (0.1 μM–1.0 μm) were applied to cultures during 12-hour or 24-hour experiments. Cell number is expressed as a percentage of control (untreated) values. Results are mean ± S.E.M. of data. *indicates $P < 0.02$. (Reprinted from Garcia-Hirschfeld, Lopez-Briones and Belmonte, 1994, with permission of Academic Press Ltd.).

promote re-epithelialization in patients with severe neurotrophic and anhidrotic keratopathy (Brown *et al.*, 1997).

The ability of CGRP to stimulate corneal epithelial cell migration is also currently under investigation in several laboratories; however, at the time of this writing the results are inconsistent and vary according to the model of cell migration employed. For example, CGRP has been reported to stimulate migration of rabbit corneal epithelial cells in a whole-cornea organ culture system (Mikulec and Tanelian, 1996), and of SV-40 transformed human corneal epithelial cells in blindwell (chemotaxis) chambers (Lee *et al.*, 1996). In contrast, CGRP has no effect on epithelial cell migration when added to cultured rabbit corneal blocks (Nakamura *et al.*, 1997c) or when applied topically to *n*-heptanol wounded rabbit corneas *in vivo* (Kingsley and Marfurt, unpublished observations).

"OTHER" FUNCTIONS OF CORNEAL SENSORY NEUROPEPTIDES

The extraordinary density of corneal SP-immunoreactive nerves, coupled with the high nociceptive sensibility of the cornea, suggested to early investigators a possible role for SP in peripheral nociceptive transduction mechanisms; however, this hypothesis is now considered unlikely. Topical application of SP antagonists, or depletion of ocular SP levels by capsaicin administration or by herpes simplex virus inoculation, do not eliminate corneal sensitivity or the blink reflex response (Holmdahl *et al.*, 1981; Lembeck and Donnerer, 1981; Keen *et al.*, 1982; Tullo *et al.*, 1983; Bynke, Håkanson and Sundler, 1984). These findings demonstrate that impulse transmission in the peripheral parts of corneal SP fibres proceeds without involvement of SP, or, alternatively, that the sensory fibres responsible for corneal nociception are distinct from those containing SP.

In addition to their important roles in corneal trophism and wound healing, SP and CGRP contribute to protective local tissue responses to noxious stimulation of the eye. When the cornea is injured, corneal nerves are stimulated and SP and CGRP are released from anterior uveal nerve fibres by axon-reflex mechanisms. The resulting neurogenic inflammatory response, or "ocular response to injury", consists of miosis, vasodilation, breakdown of the blood-aqueous barrier, protein extravasation, and a transient elevation in intraocular pressure. The physiological actions of SP and CGRP in this response are subject to considerable interspecies differences and may involve additional mediators such as prostaglandins; however, in most species SP mediates the miotic response and CGRP the vascular reactions leading to raised intraocular pressure and breakdown of the blood aqueous barrier (for reviews see Waldrep, 1989; Unger, 1990; Bill, 1991).

CORNEAL SYMPATHETIC INNERVATION: FUNCTIONAL CONSIDERATIONS

Corneal sympathetic nerves, like corneal sensory nerves, help promote the anatomical and physiological barrier functions of the corneal epithelium. Studies of corneal sympathetic nerves in animal models have revealed at least four important functions: (i) modulation of epithelial ion transport processes; (ii) regulation of epithelial cell proliferation and mitosis; (iii) modulation of cell migration during epithelial wound healing; (iv) interactions with sensory fibres to exert trophic influences on the cornea.

MODULATION OF EPITHELIAL ION TRANSPORT

Corneal epithelial cells express large numbers of α- and β-adrenoceptors on their cell membranes (Candia and Neufeld, 1978; Neufeld *et al.*, 1978; Cavanagh and Colley, 1982; Walkenbach *et al.*, 1985; Elena *et al.*, 1987, 1990; Walkenbach *et al.*, 1991) and activation of these receptors has been linked to a variety of intracellular processes. In isolated frog and rabbit corneas, catecholamines stimulate epithelial Cl^- transport. According to the model of Klyce and coworkers (Klyce, Beuerman and Crosson, 1985; Klyce and Crosson, 1985) (Figure 2.18), noradrenaline released from rabbit corneal sympathetic nerve fibres activates β-adrenoceptors and results in the stimulation of adenylate cyclase, enhanced production of intracellular adenosine-$3'5'$-cyclic monophosphate (cAMP), and activation of cAMP-dependent protein kinases (Klyce, Neufeld and Zadunaisky, 1973; Walkenbach and LeGrand, 1981; Walkenbach, LeGrand and Barr, 1981; Reinach and Kirchberger, 1983). These cAMP-dependent protein kinases, in turn, catalyze protein phosphorylation and induce conformational changes in ion channels, leading to increased Cl^- permeability of the surface epithelial membranes and a net Cl^- transport from stroma to tears (Chalfie,

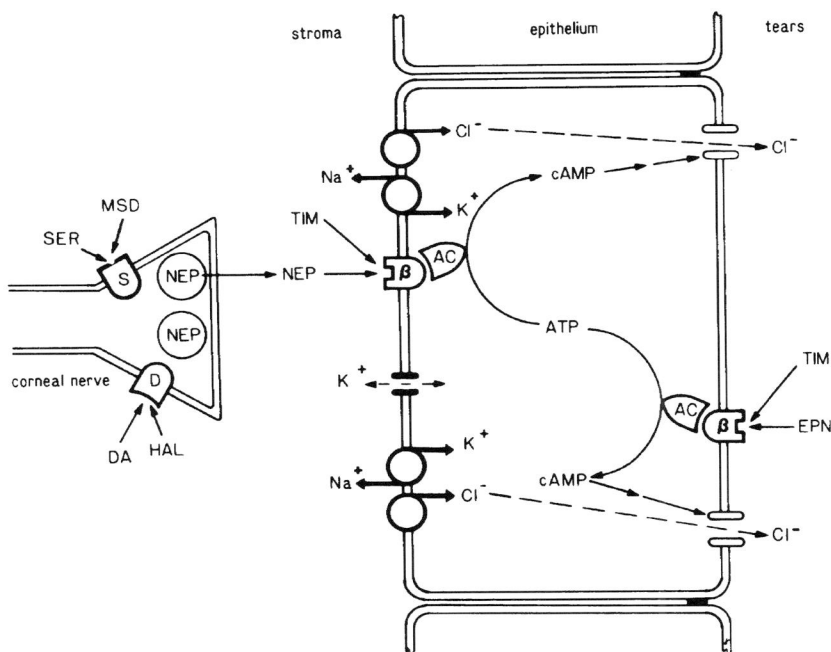

Figure 2.18 Proposed model for the mechanism by which various biogenic amines stimulate chloride ion transport across the rabbit corneal epithelium. The scheme is consistent with experimental observations that both sympathectomy and timolol treatment block epithelial responsiveness to serotonin and dopamine. The probable source of serotonin or dopamine is the sympathetic fibres that innervate the cornea. Abbreviations: EPN = adrenaline (epinephrine), NEP = noradrenaline (norepinephrine), TIM = timolol, SER= serotonin, MSD = methysergide, DA = dopamine, HAL = haloperidol, β = β-adrenoceptor, S = serotonin receptor, D = dopamine receptor, AC = adenylate cyclase. (Reprinted from Klyce and Crosson, 1985, with permission of Oxford University Press).

Neufeld and Zadunaisky, 1972; Klyce, Neufeld and Zadunaisky, 1973; Montoreano, Candia and Cook, 1976; Klyce and Wong, 1977; Fischer *et al.*, 1978; Wiederholt *et al.*, 1983). Additional effects on ion transport may be mediated by α_1-adrenoceptor stimulation and the hydrolysis of phosphatidylinositol into the second messenger molecules 1,2-diacylglycerol and inositol 1,4,5-triphosphate (Akhtar, 1987).

Additional work in the isolated rabbit corneal model has demonstrated that serotonin (Klyce *et al.*, 1982; Jumblatt and Neufeld, 1983; Neufeld, Ledgard and Yoza, 1983; Marshall and Klyce, 1984; Neufeld *et al.*, 1984) and dopamine (Crosson, Beuerman and Klyce, 1984) also regulate epithelial Cl⁻ transport, most likely via indirect, presynaptic mechanisms. According to Klyce and coworkers (Klyce and Crosson, 1985), serotonin and dopamine released from corneal sympathetic nerve fibres bind to corresponding receptors located on the preterminal portions of the same, or neighbouring, sympathetic nerve fibres. Receptor-mediated events then produce stimulation of chloride ion transport by facilitating the release of noradrenaline.

In theory, it is tempting to speculate that sympathetic stimulation of corneal epithelial chloride transport is physiologically relevant and assists the corneal endothelium in the moment-to-moment control of stromal hydration and corneal transparency. Indeed, work in amphibian corneas suggests a positive correlation between epithelial chloride ion transport and corneal transparency (Zadunaisky and Lande, 1971; Beitsch, Beitsch and Zadunaisky, 1974). In mammals, however, the ion transport capability of the corneal epithelium under normal conditions is only 3–4% that of the corneal endothelium (Klyce, 1982). Clinically, drugs with α- or β-adrenergic activity do not produce corneal stromal hydration problems in patients being treated for primary open-angle glaucoma (Bartels, 1994). Nevertheless, the fact that the corneal epithelium is sympathetically innervated suggests that it may assume some heightened importance during conditions of ocular stress (Klyce and Crosson, 1985). Indeed, patients with unilateral Horner's Syndrome often present with increased corneal stromal or epithelial thickness in the sympathetically denervated eye (Neilson, 1983; Sweeney et al., 1985) and are less able to control fluid accumulation during hypoxia (Sweeney et al., 1985).

REGULATION OF EPITHELIAL CELL PROLIFERATION

Sympathetic nerves exert pronounced regulatory effects on corneal epithelial cell proliferation (DNA synthesis) and mitosis; however, whether the effects are stimulatory or inhibitory has been the subject of much contention. On the one hand, co-culture of SCG neurons with rabbit corneal epithelial cells increases epithelial cell proliferation by 250% (Garcia-Hirschfeld, Lopez-Briones and Belmonte, 1994). Similarly, ocular sympathetic denervation decreases corneal epithelial proliferation by 30–50% (Figure 2.19) (Jones and Marfurt, 1996) and decreases epithelial mitotic index and mitotic rate by 25–70% (Friedenwald and Buschke, 1944a; Mishima, 1957; Voaden, 1971). On the basis of these findings, it is tempting to speculate that ocular sympathetic nerves release trophic substances (including, noradrenaline) that stimulate corneal epithelial proliferation, and that the decrease in epithelial mitotic activity seen following chronic sympathetic denervation is due to the loss of these substances. In support of this hypothesis, noradrenaline stimulates epithelial cell proliferation in sympathetically-denervated rat eyes *in vivo* (Jones and Marfurt, 1996) and in human SV-40 transformed epithelial cells *in vitro* (Murphy et al., 1998).

In contrast, the results of other studies support an inhibitory role for sympathetic innervation on corneal epithelial proliferation and mitosis. Addition of noradrenaline to rabbit corneal epithelial cells *in vitro* inhibits tritiated thymidine incorporation (Cavanagh and Colley, 1982). Topical adrenaline administration (Friedenwald and Buschke, 1944a) and electrical stimulation of the cervical sympathetic trunk (Mishima, 1957) inhibit epithelial mitotic rate in rat and rabbit corneal epithelium, respectively. Extirpation of the rabbit superior cervical ganglion is followed by a sharp decrease in mitotic activity 19–20 hours postoperatively (Mishima, 1957; Butterfield and Neufeld, 1977). This decrease has been hypothesized to be caused by the massive release of noradrenaline from degenerating ocular nerve terminals (Fogle and Neufeld, 1979). The reason for the inconsistent nature of past results in this area remains unknown; however, methodological differences, species differences, and the tendency to equate proliferation with mitosis, may all contribute.

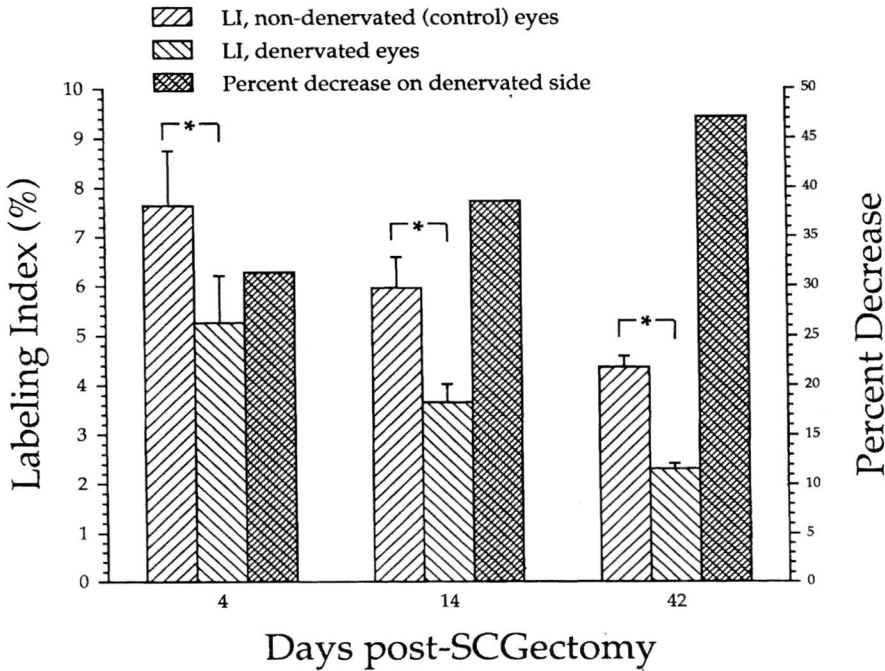

Figure 2.19 The effect of ocular sympathectomy on the labelling index (LI) of the rat corneal epithelium. Four, 14 and 42 days after unilateral removal of the superior cervical ganglion, there is a significant decrease in the ipsilateral epithelial LI. Labelling indices were determined by bromodeoxyuridine (BrdU) incorporation and immunohistochemistry. Values are means ± S.E.M. *Indicates $P < 0.01$. (Reprinted from Jones and Marfurt, 1996, with permission of Lippincott-Raven Publishers.)

EPITHELIAL WOUND HEALING

When the corneal epithelium is injured, intact cells at the wound margin flatten out and migrate in sheet-like fashion over the denuded area to resurface the defect. Factors which influence the rate of cell migration are multiple and complex and cannot be adequately reviewed here; however, there is increasing evidence that the sympathetic nervous system may play a role in this process. Corneal epithelial wound healing in rabbits is inhibited by sympathetic stimulation (Perez *et al.*, 1987), and accelerated by sympathetic denervation (Beuerman *et al.*, 1985). These data suggest that sympathetic nerves release some substance (possibly noradrenaline) which inhibits the rate of cell migration at the wound margin. In support of this theory, systemic ephedrine or adrenaline inhibit postinjury cell movements in injured rat corneas (Friedenwald and Buschke, 1944b) and topical adrenaline and β-adrenergic blocking agents decrease and increase, respectively, the rate of re-epithelialization in wounded rabbit corneas (Krejci and Harrison, 1970; Reidy *et al.*, 1994). In marked contrast to these findings, other workers have concluded that catecholamines either have no effect (Jumblatt and Neufeld, 1981) or facilitate (Liu, Basu and Trope, 1987; Trope, Liu and Basu, 1988; Liu, Trope and Basu, 1990) the rate of re-epithelialization. NE over a wide range of concentrations has also been reported to stimulate migration of transformed human corneal epithelial cells through membrane filters in blindwell chambers (Murphy *et al.*, 1998).

TROPHIC FUNCTIONS

Whether sympathetic nerves exert trophic effects on the cornea is uncertain. Removal of the SCG in experimental animals is apparently without acute, macroscopic effects on corneal transparency or epithelial integrity (e.g., Zander and Weddell, 1951b; Rodger, 1953; Knyazev, Knyazeva and Tolochko, 1991); however, the matter has not been extensively investigated. Clinically, patients with unilateral Horner's syndrome exhibit only minor abnormalities of corneal physiology. When subjected to a hypoxic "Stress Test", sympathetically-denervated human eyes show significantly more epithelial graying and microcytic edema, and diminished rates of deswelling (Sweeney *et al.*, 1985).

A small, but intriguing body of literature suggests that maintenance of normal corneal integrity depends on a critical balance of neuronal activity between ocular sympathetic and sensory nerves. For example, the development of neuroparalytic keratitis following trigeminal nerve injury may be prevented or alleviated by removing the superior cervical ganglion (Spalitta, 1894; cited in Paton, 1926) or by cutting the preganglionic inputs to the SCG via stellate ganglionectomy (Harris, 1940; Baker and Gottlieb, 1959). In animal studies, the development of capsaicin-induced, neurotrophic corneal lesions in neonatal rats is prevented or inhibited if the ocular sympathetic nerves are also destroyed (Shimizu *et al.*, 1987; Abelli, Geppetti and Maggi, 1993).

MODULATION OF CORNEAL SENSITIVITY

There has been some speculation that corneal sympathetic fibres modulate the sensitivity of the cornea (Laties and Jacobowitz, 1966), possibly by influencing the release of SP from sensory nerve terminals (Morgan, DeGroat and Janetta, 1987). At present, there is little experimental evidence to support this interesting hypothesis. Bernard (1851) observed that extirpation of the cat SCG appears to leave the animals with hypersensitivity of the skin and cornea; however, patients with Horner's syndrome show no difference in corneal touch threshold between the normal and the sympathetically denervated eye (Sweeney *et al.*, 1985). Studies on peripheral pain mechanisms in other systems suggest that sympathetic nerve stimulation has no effect on the excitability of healthy, primary afferent nociceptors (Shea and Perl, 1985).

THE CORNEAL CHOLINERGIC SYSTEM

The mammalian corneal epithelium contains all of the necessary components of a functional cholinergic system (van Alphen, 1957; Petersen, Keat-Jin and Donn, 1965; Williams and Cooper, 1965; Howard, Wilson and Dunn, 1973; Mindel and Mittag, 1976), including, high levels of ACh, its synthetic enzyme, choline acetyltransferase (ChAT), its degradative enzyme, AChE, and cholinergic receptors. There are apparently two sources of ACh and its associated enzymes in the epithelium: one dependent on the corneal innervation and

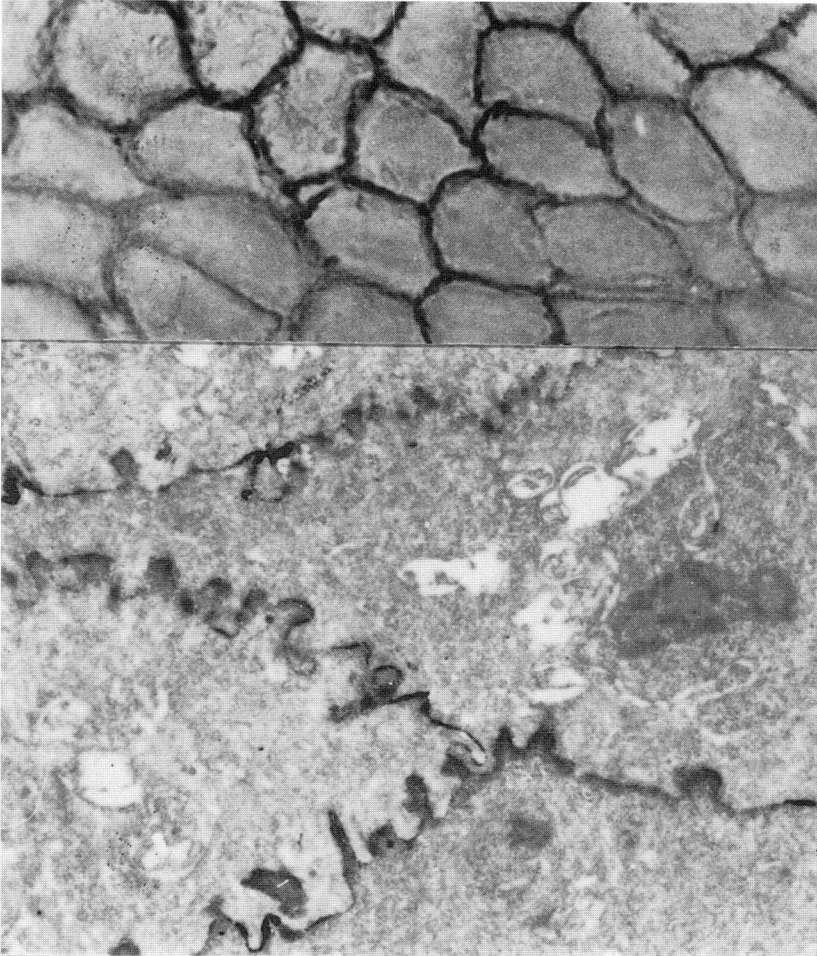

Figure 2.20 Light (top panel) and electron (lower panel) microscopic demonstrations of AChE activity associated with the cell surface membranes and extracellular spaces of the rat corneal epithelium. Acetylthiocholine iodide and iso-OMPA technique. (Reprinted from Tervo, 1977, with permission of Springer-Verlag).

one independent of the innervation. Levels of corneal ACh and its enzymes have been reported to either decrease (von Brücke, Hellauer and Umrath, 1949; Petersen, Keat-Jin and Donn, 1965; Fitzgerald and Cooper, 1971) or stay the same (Mindel and Mittag, 1977) following ocular sensory denervation. A non-neuronal source for part of the corneal cholinergic system is supported by the observation that cultured epithelial cells grown in the absence of nerves continue to express the enzymes ChAT and AChE (Gnädinger, Heiman and Markstein, 1973) and by histological demonstrations of AChE in corneal epithelial cell membranes and intercellular spaces (Figure 2.20) (Howard, Zadunaisky and Dunn, 1975; Tervo, 1976; Tervo and Palkama, 1978b).

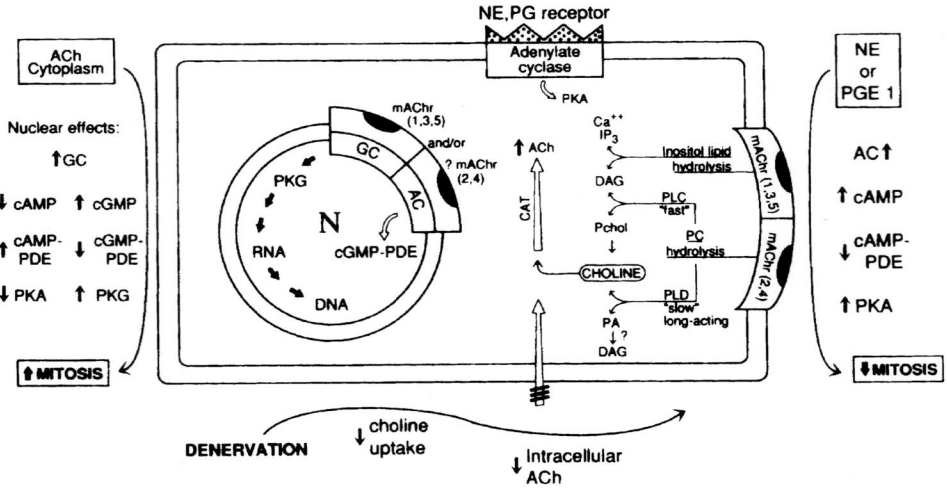

Figure 2.21 A model for cholinergic regulation of epithelial cellular growth. (Reprinted from Lind and Cavanagh, 1993, with permission of Lippincott-Raven Publishers).

Pharmacological and radioligand binding studies have demonstrated the presence of muscarinic (but not nicotinic) cholinergic receptors on intact corneal epithelial cells of rabbits and humans (Cavanagh and Colley, 1982; Colley and Cavanagh, 1982; Walkenbach and Ye, 1991). Work by Cavanagh and coworkers (Colley *et al.*, 1985; Lind and Cavanagh, 1993) suggests that the receptors are expressed not only on the surface of the epithelial cells but also on the epithelial nuclear membrane.

FUNCTIONAL CONSIDERATIONS

A definitive role for the cholinergic system in corneal physiology remains elusive; however, several possible functions have been proposed. Corneal ACh may play a role in corneal epithelial cell growth, proliferation, and the healing of epithelial defects (Figure 2.21). The addition of cholinergic agonists to corneal epithelial cell cultures increases intracellular guanosine-3'5'-cyclic monophosphate (cGMP) production (Cavanagh and Colley, 1982; Colley and Cavanagh, 1982; Walkenbach and Ye, 1991) stimulates [^{3}H]-thymidine and [^{14}C]-leucine incorporation (Cavanagh and Colley, 1982), enhances nuclear DNA and RNA polymerase activities (Colley *et al.*, 1985), activates the phosphoinositide cycle (Proia *et al.*, 1986; Baratz *et al.*, 1987) and promotes healing of persistent epithelial defects (Cavanagh, 1975).

ACh has also been implicated in the regulation of corneal epithelial ion transport processes. Depletion of endogenous corneal ACh levels by administration of ChAT inhibitors decreases Na^+ and Cl^- transport in frog corneal epithelium (Pesin and Candia, 1982) and increases Na^+ transport into the corneal stroma of preswollen rabbit corneas (Stevenson and Wilson, 1975).

ACh has also been postulated to play a role in corneal sensory transduction mechanisms. In most mammals, there is a positive correlation between corneal sensitivity and epithelial ACh content. Species that possess high corneal ACh content require less stimulation, using von Frey hairs, to elicit a blink reflex than do species with low ACh content (Hellauer, 1950). Corneal ACh content, ChAT activity and corneal sensitivity are highest in the central corneal epithelium and lowest in the peripheral epithelium (Hellauer, 1950; Mindel and Mittag, 1976, 1977; Wilson and McKean, 1986). Corneal ACh content and corneal sensitivity decrease proportionately in aged individuals, after eyelid suturing, and following administration of anticholinergic drugs (Hellauer, 1950; Fitzgerald and Cooper, 1971; Mindel and Mittag, 1976, 1977; Millodot and O'Leary, 1979). Application of ACh to the corneal surface evokes action potentials in rabbit C-fibre afferents (Tanelian, 1991) and causes eye pain in humans (Jancso, Jancso-Gabor and Takats, 1961). Taken collectively, these data suggest that a certain critical level of ACh is requisite to corneal sensitivity, and that ACh released in the vicinity of corneal nerve terminals following corneal injury may produce pain. It should be cautioned, however, that much of the evidence in support of this hypothesis is indirect, and that a role for ACh in corneal sensory transduction processes is not universally accepted.

DEVELOPMENT OF THE CORNEAL INNERVATION

Development of the corneal innervation has been described for several mammalian and avian species with similar findings (Kitano, 1957; Lúkas and Dolezel, 1975; Ozanics, Rayborn and Sagun, 1977; Tervo, Tervo and Palkama, 1978; Tervo and Tervo, 1981; Bee, 1982; Bee et al., 1986). The innervation of the embryonic chick cornea has been especially well described by Bee and coworkers (Bee, 1982; Bee et al., 1982, 1986) and can be subdivided into two distinct phases. In Phase I (embryonic days 6–10), the ocular nerves approach (but do not penetrate) the ventrotemporal region of the corneal limbus. The nerves then separate into two groups which extend ventrally and dorsally around the cornea before meeting each other to form a prominent ring-like latticework of encircling nerve fibres (Figure 2.22). The pathway followed by the elongating axons is determined in part by the nerves' ability to recognize tissue-derived guidance cues (e.g., glycosaminoglycans) in the extracellular matrix (Bee et al., 1982). Soon thereafter, the nerves penetrate the peripheral corneal stroma and begin growing towards the central zone. Over the next several months, the fibres increase in number and branching complexity. Axonal bifurcations occur within distinct concentric zones which are, for unknown reasons, conducive to nerve branching (Bee et al., 1986).

In the human embryo, ciliary nerves reach the optic cup about midway through the first trimester, and first enter the cornea at about 3 months (Kitano, 1957). By 4 months, stromal nerves are present in large numbers beneath Bowman's membrane, and from 4 to 9 months the nerves undergo a gradual maturation process to reach the adult-like arrangement at time of birth.

Axonal penetration into the corneal epithelium first occurs at about 5 months in humans (Kitano, 1957), a little earlier in monkeys (Ozanics, Rayborn and Sagun, 1977), and apparently not until birth in the rat (Singh and Beuerman, 1979; Tervo and Tervo, 1981). The mechanisms that direct the developing axons into the epithelium are largely unknown;

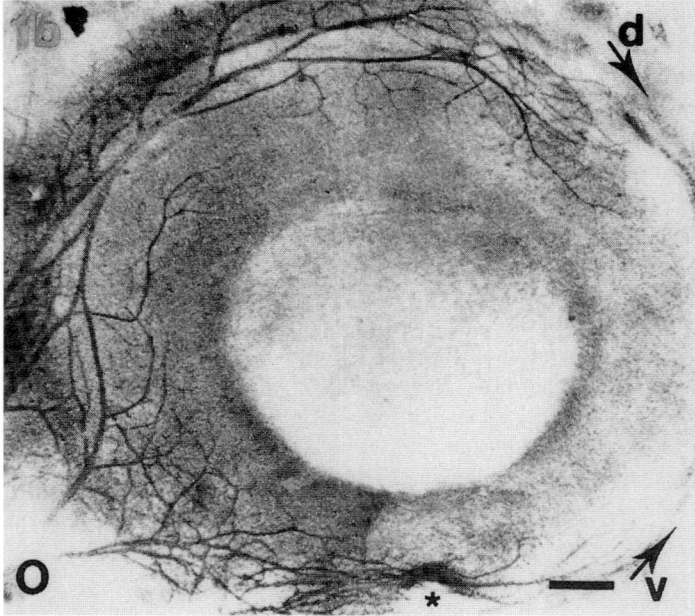

Figure 2.22 Initial stage of migration and positioning of corneal nerves in the avian cornea. The micrograph illustrates a whole-mounted cornea from a 7 day-old chick embryo. Numerous large nerve fascicles approach (but do not enter) the cornea and migrate both dorsally (d) and ventrally (v) from their origin (O). Eventually, the nerve groups will meet one another and form a complete ring around the periphery of the cornea. Scale bar = 200 μm. (Reprinted from Bee, 1982, with permission of Academic Press, Inc.).

however, it is known that embryonic and mature epithelial cells produce diffusible neuronotropic "factors" which promote neurite extension (Chan and Haschke, 1982, 1985; Emoto and Beuerman, 1987; Pavlidis, Steuhl and Thanos, 1994). At the electron microscopic level, the first recognizable epithelial nerve terminals contain small numbers of large, dense-cored vesicles, a few mitochondria and, at a slightly later stage, smaller vesicles (Ozanics, Rayborn and Sagun, 1977). The number of vesicles in the nerve terminals then increases progressively as a function of time (Tervo and Tervo, 1981). The mitochondria in the primitive terminals are smaller, more dense, and differ in internal configuration from the mitochondria in surrounding corneal epithelial cells, suggesting high levels of neural metabolic activity (Ozanics, Rayborn and Sagun, 1977).

In the rat, corneal nerve development continues for several weeks after birth (Jones and Marfurt, 1991). At the time of birth, the intraepithelial segments of the nerves are morphologically simple and support few intraepithelial terminals. Over the next three weeks, the epithelial segments increase significantly in length, branching complexity, degree of radial orientation, and richness of terminal elements, finally attaining their adult-like morphology on about postnatal day 21 (Figure 2.23).

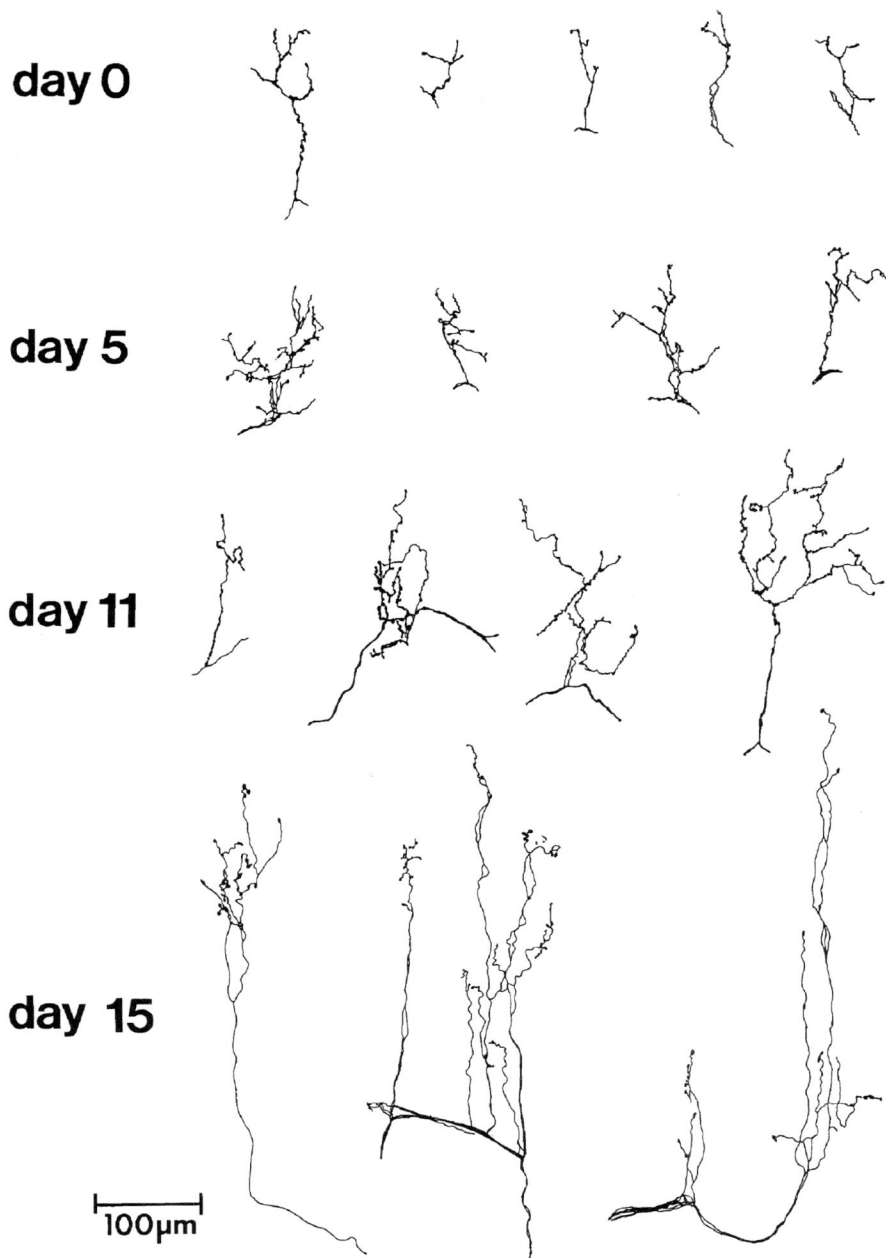

Figure 2.23 Postnatal maturation of intraepithelial axons in the developing rat cornea. On the day of birth (day 0), intraepithelial axons are short and morphologically uncomplicated. Over the next two weeks, the axons elongate and increase in overall complexity; immature versions of adult leash formations are recognizable at about day 15. (Reprinted from Jones and Marfurt, 1991, with permission of John Wiley and Sons, Inc.)

Neuropeptides are first observed immunohistochemically in the developing rat cornea approximately 3–4 days prior to birth (Sakiyama *et al.*, 1984). SP is seen in developing chick corneal nerves only after the nerves penetrate the epithelium, prompting speculation that SP expression is initiated either by physical contact with the epithelial cells or in response to the elaboration of one or more epithelium-derived neuronotrophic factors (Bee *et al.*, 1988; Corvetti, Pignocchino and Sisto Daneo, 1988).

Functionally, it has been suggested that corneal nerves and their neurochemicals may trigger certain aspects of embryonic and postnatal corneal development. For example, dehydration of the chick cornea and the change from corneal opacity to transparency correlates with the arrival of corneal nerves in the stroma (Bee, 1982; Clarke and Bee, 1996). SP-IR nerves in the chick corneal epithelium increase in number and branching complexity between embryonic days 12 and 17 (Bee *et al.*, 1988; Corvetti, Pignocchino and Sisto Daneo, 1988); at the same time, the corneal epithelium begins to differentiate (Hay, 1979). Whether the latter events are causal or coincidental remains to be shown. It has been hypothesized that the acceleration of corneal epithelial differentiation seen postnatally may be due to enhanced neurotransmitter release from sensory nerve terminals as the result of repeated blinking of the newly opened lids (Watanabe, Tisdale and Gipson, 1993). Neonatal administration of capsaicin in mice and rats significantly depletes the normal SP and CGRP content of the developing corneas; several weeks later these corneas develop epithelial abnormalities and chronic neuroparalytic keratitis (Fujita *et al.*, 1984; Marfurt, Ellis and Jones, 1993). In contrast to the situation in chick and rat eyes, differentiation in the rabbit embryonic cornea clearly begins prior to the invasion of nerve fibres (Lúkas and Dolezel, 1975).

The development of the corneal sympathetic innervation has been described by several investigators on the basis of histochemical fluorescence and electron microscopic observations. Fluorescent adrenergic nerves are first observed in the developing human cornea in the second trimester and in the rat, guinea pig and rabbit cornea near the beginning of the third trimester (Ehinger and Sjöberg, 1971; Lúkas and Dolezel, 1975; Tervo, Tervo and Palkama, 1978). Morphologically, embryonic adrenergic nerves are smoother than their adult counterparts and lack varicosities (Ehinger and Sjöberg, 1971; Tervo, 1977), suggesting a certain level of functional immaturity. The number of stromal and epithelial fibres, and the intensity with which the nerves fluoresce, increase progressively throughout development, reaching a maximum shortly before, or immediately after, birth (Ehinger, 1966c; Ehinger and Sjöberg, 1971; Tervo, 1977; Tervo, Tervo and Palkama, 1978; Toivanen *et al.*, 1987). The numbers of adrenergic fibres then decline progressively, even precipitously, in the weeks immediately following birth to quickly reach adult levels. Most of this fibre loss is probably due to extensive postnatal cell death in the neonatal SCG (Vidovic and Hill, 1988), due to an inability of the sympathetic nerves to successfully compete with the corneal sensory nerve fibres for available growth factors. Alternatively, postnatal increases in corneal size (Tervo, Tervo and Palkama, 1978) or accelerated rates of neurotransmitter release by mature nerve fibres may also explain why adult corneas appear to contain fewer nerves than fetal corneas.

ACKNOWLEDGEMENTS

The author expresses his deepest appreciation to Drs. Carlos Belmonte, Mark Jones, Robert Kingsley, and Gordon Ruskell for their invaluable comments on earlier versions of this manuscript. I would also like to thank the following individuals for generously providing original photomicrographs of their work for use in this paper: Kaoru Araki-Sasaki, Carlos Belmonte, Roger Beuerman, Juana Gallar, Carl Herbort, Mark Jones, Linda Müller, and Timo Tervo. This work was supported by NIH grant EY05717.

REFERENCES

Abelli, L., Geppetti, P. and Maggi, C.A. (1993). Relative contribution of sympathetic and sensory nerves to thermal nociception and tissue trophism in rats. *Neuroscience*, **57**, 739–745.

Akhtar, R.A. (1987). Effects of norepinephrine and 5-hydroxytryptamine on phosphoinositide-PO_4 turnover in rabbit cornea. *Experimental Eye Research*, **44**, 849–862.

Alper, M.G. (1975). The anesthetic eye: an investigation of changes in the anterior ocular segment of the monkey caused by interrupting the trigeminal nerve at various levels along its course. *Transactions of the American Ophthalmological Society*, **873**, 323–365.

Araki, K., Ohashi, Y., Kinoshita, S., Kozaburo, Y., Kuwayama, Y. and Tano, Y. (1994). Epithelial wound healing in the denervated cornea. *Current Eye Research*, **13**, 203–211.

Arvidson, B. (1977). Retrograde axonal transport of horseradish peroxidase from cornea to trigeminal ganglion. *Acta Neuropathologica*, **38**, 49–52.

Auran, J.D., Koester, C.J., Kleiman, N.J., Rapaport, R., Bomann, J.S., Wirotsko, M., Florakis, G.J. and Koniarek, J.P. (1995). Scanning slit confocal microscopic observation of cell morphology and movement within the normal human anterior cornea. *Ophthalmology*, **102**, 33–41.

Baker, G.S. and Gottlieb, C.M. (1959). The prevention of corneal ulceration in the denervated eye by cervical sympathectomy: An experimental study in cats. *Proceedings of the Staff Meetings of the Mayo Clinic*, **34**, 474–478.

Baker K.S., Anderson, S.C., Romanowski, E.G., Thoft, R.A. and SundarRaj, N. (1993). Trigeminal ganglion neurons affect corneal epithelial phenotype. Influence on type VII collagen expression *in vitro*. *Investigative Ophthalmology and Visual Science*, **34**, 137–144.

Baratz, K.H., Proia, A.D., Klintworth, G.K., and Lapetina, E.G. (1987). Cholinergic stimulation of phosphatidylinositol hydrolysis by rat corneal epithelium in vitro. *Current Eye Research*, **6**, 691–701.

Bartels, S.P. (1994). Adrenergic agents. In *Principles and Practice of Ophthalmology: Basic Sciences*, edited by D.M. Albert and F.A. Jakobiec, pp. 993–1012. Philadelphia: W.B. Saunders Co.

Beckers, H.J.M., Klooster, J., Vrensen, G. and Lamers, W. (1992). Ultrastructural identification of trigeminal nerve endings in the rat cornea and iris. *Investigative Ophthalmology and Visual Science*, **33**, 1979–1986.

Beckers, H.J.M., Klooster, J., Vrensen, G. and Lamers, W. (1993). Substance P in rat corneal and iridial nerves: an ultrastructural immunohistochemical study. *Ophthalmic Research*, **25**, 192–200.

Bee, J.A. (1982). The development and pattern of innervation of the avian cornea. *Developmental Biology*, **92**, 5–15.

Bee, J.A., Unruh, N., Sommerfeld, D.L. and Conrad, G.W. (1982). Avian corneal innervation: inhibition of nerve ring formation by 6-diazo-5-oxo-l-norleucine. *Developmental Biology*, **92**, 123–132.

Bee, J.A., Hay, R.A., Lamb, E.M., Devore, J.J. and Conrad, G.W. (1986). Positional specificity of corneal nerves during development. *Investigative Ophthalmology and Visual Science*, **27**, 38–43.

Bee, J.A., Kuhl, U., Edgar, D. and von der Mark, K. (1988). Avian corneal nerves: co-distribution with collagen type IV and acquisition of substance P immunoreactivity. *Investigative Ophthalmology and Visual Science*, **29**, 101–107.

Beitsch, B.R., Beitsch, I. and Zadunaisky, J.A. (1974). The stimulation of chloride transport by prostaglandins and their interaction with epinephrine, theophylline, and cyclic AMP in the corneal epithelium. *Journal of Membrane Biology*, **19**, 381–396.

Belmonte, C. and Gallar, J. (1996). Corneal nociceptors. In *Neurobiology of Nociceptors*, edited by C. Belmonte and F. Cervero, pp. 146–183. Oxford: Oxford University Press.

Belmonte, C. and Giraldez, F. (1981). Responses of cat corneal sensory receptors to mechanical and thermal stimulation. *Journal of Physiology*, **321**, 355–368.

Belmonte, C., Gallar, J., Pozo, M.A. and Rebollo, I. (1991). Excitation by irritant chemical substances of sensory afferent units in the cat's cornea. *Journal of Physiology*, **437**, 709–725.

Belmonte, C., Garcia-Hirschfeld, J., and Gallar, J. (1997). Neurobiology of ocular pain. *Progress in Retinal and Eye Research,* **16**, 117–156.

Bernard, C. (1851). Influence du grand sympathique sur la sensibilite et sur la calorification. *Compte Rendu de la Sociéte de Biologie (Paris),* **3**, 163–164.

Beuerman R.W. and Schimmelpfennig, B. (1980). Sensory denervation of the rabbit cornea affects epithelial properties. *Experimental Neurology,* **69**, 196–201.

Beuerman, R.W. and Tanelian, D.L. (1979). Corneal pain evoked by thermal stimulation. *Pain,* **7**, 1–14.

Beuerman, R.W., Schimmelpfennig, and Burstein, N. (1979). Anatomy of the denervated corneal epithelium. *Investigative Ophthalmology and Visual Science (Suppl),* **18**, 126.

Beuerman, R.W., Klyce, S., Kooner, S., Tanelian, D. and Rózsa, A. (1983). Dimensional analysis of rabbit ciliary nerve. *Investigative Ophthalmology and Visual Science (Suppl.),* **24**, 261.

Beuerman, R.W., Klyce, S., Vigo, M., Dupuy, B. and Crosson, C. (1985). Modulation of latency and velocity in corneal epithelial wound healing. *Investigative Ophthalmology and Visual Science (Suppl),* **26**, 91.

Beuerman, R.W., Tanelian, D.L. and Schimmelpfennig, B. (1988). Nerve tissue interactions in the cornea. In *The Cornea: Transactions of the World Congress on the Cornea III,* edited by H.D. Cavanagh, pp. 59–62. New York: Raven Press.

Bill, A. (1991). Effects of some neuropeptides on the uvea. *Experimental Eye Research,* **53**, 3–11.

Boberg-Ans, J. (1955). Experience in clinical examination of corneal sensitivity. *British Journal of Ophthalmology,* **39**, 709–726.

Brown, S.M., Lamberts, D.W., Reid, T.W., Nishida, T. and Murphy, C.J. (1997). Neurotrophic and anhidrotic keratopathy treated with substance P and insulinlike growth factor 1. *Archives of Ophthalmology,* **115**, 926–927.

Buck, S.H., Walsh, J.H., Davis, T.P., Brown, M.R., Yamamura, H.I. and Burks, T.F. (1983). Characterization of the peptide and sensory neurotoxic effects of capsaicin in the guinea pig. *Journal of Neuroscience,* **3**, 2064–2074.

Bucsics, A., Holzer, P. and Lembeck, F. (1983). The substance P content of peripheral tissues in several mammals. *Peptides,* **4**, 451–455.

Burton, H. (1992). Somatic sensations from the eye. In *Adler's Physiology of the Eye,* edited by W.M. Hart, Jr., pp. 71–100. St. Louis: Mosby-Year Book, Inc.

Butterfield, L.C. and Neufeld, A.H. (1977). Cyclic nucleotides and mitosis in the rabbit cornea following superior cervical ganglionectomy. *Experimental Eye Research,* **25**, 427–433.

Bynke G., Håkanson, R., and Sundler, F. (1984). Is substance P necessary for corneal nociception? *European Journal of Pharmacology,* **101**, 253–258.

Candia, O.A. and Neufeld, A.H. (1978). Topical epinephrine causes a decrease in density of β-adrenergic receptors and catcholamine-stimulated chloride transport in the rabbit cornea. *Biochimica et Biophysica Acta,* **543**, 403–408.

Cavanagh, H.D. (1975). Herpetic ocular disease: therapy of persistent epithelial defects. *International Ophthalmology Clinics,* **15**, 67–88.

Cavanagh, H.D. and Colley, A.M. (1982). Cholinergic, adrenergic, and PGE_1 effects on cyclic nucleotides and growth in cultured corneal epithelium. *Metabolic, Pediatric and Systemic Ophthalmology,* **6**, 63–74.

Chalfie, M., Neufeld, A.H. and Zadunaisky, J.A. (1972). Action of epinephrine and other cyclic AMP-mediated agents on the chloride transport of the frog cornea. *Investigative Ophthalmology,* **11**, 644–650.

Chan, K.W. and Haschke, R.H. (1982). Isolation and culture of corneal cells and their interactions with dissociated trigeminal neurons. *Experimental Eye Research,* **35**, 137–156.

Chan, K.W. and Haschke, R.H. (1985). Specificity of a neurotrophic factor from rabbit corneal epithelial cultures. *Experimental Eye Research,* **41**, 687–699.

Chan-Ling, T. (1989). Sensitivity and neural organization of the cat cornea. *Investigative Ophthalmology and Visual Science,* **30**, 1075–1082.

Clarke, N.D. and Bee, J.A. (1996). Innervation of the chick cornea analyzed *in vitro. Investigative Ophthalmology and Visual Science,* **37**, 1761–1771.

Cochet, P. and Bonnet, R. (1960). L'esthésie cornéenne. *Clinique Ophthalmologique,* **4**, 3–27.

Colley, A.M. and Cavanagh, H.D. (1982). Binding of [^{3}H]dihydroalprenolol and [^{3}H]quinuclidinyl benzilate to intact cells of cultured corneal epithelium. *Metabolic, Pediatric and Systemic Ophthalmology,* **6**, 75–86.

Colley, A.M., Cavanagh, H.D., Drake, L.A. and Law, M.L. (1985). Cyclic nucleotides in muscarinic regulation of DNA and RNA polymerase activity in cultured corneal epithelial cells in the rabbit. *Current Eye Research*, **4**, 941–950.

Corvetti, G., Pignocchino, P. and Sisto Daneo, L. (1988). Distribution and development of substance P immunoreactive axons in the chick cornea and uvea. *Basic and Applied Histochemistry*, **32**, 187–192.

Crosson, C.E., Beuerman, R.W. and Klyce, S.D. (1984). Dopamine modulation of active ion transport in rabbit corneal epithelium. *Investigative Ophthalmology and Visual Science*, **25**, 1240–1245.

Davies, M.S. (1970). Corneal anaesthesia after alcohol injection of the trigeminal sensory root. Examination of 100 anaesthetic corneae. *British Journal of Ophthalmology*, **54**, 577–586.

Dawson, W.M. (1962). Chemical stimulation of the peripheral trigeminal nerve. *Nature*, **196**, 341–345.

De Biasi, S. and Rustioni, A. (1988). Glutamate and substance P coexist in primary afferent terminals in the superficial laminae of spinal cord. *Proceedings of the National Academy of Sciences USA*, **85**, 7820–7824.

De Hass, E.B.H. (1962). Desiccation of cornea and conjunctiva after sensory denervation. Significance of dessiccation for pathogenesis of neuroparalytic keratitis. *Archives of Ophthalmology*, **67**, 439–452.

de Leeuw, A.M. and Chan, K.Y. (1989). Corneal nerve regeneration. Correlation between morphology and restoration of sensitivity. *Investigative Ophthalmology and Visual Science*, **30**, 1980–1990.

Denis, P., Fardin, V., Nordmann, J.P., Elena, P.P., Laroche, L., Saraux, H. and Rostene, W. (1991). Localization and characterization of substance P binding sites in rat and rabbit eyes. *Investigative Ophthalmology and Visual Science*, **32**, 1894–1901.

Draeger, J. (1984). *Corneal Sensitivity. Measurement and Clinical Importance.*, pp. 40–55. New York-Wien: Springer-Verlag.

Duke-Elder, S.S. and Leigh, A.G. (1965). Diseases of the outer eye. In *System of Ophthalmology, Vol. VIII*. edited by S. Duke-Elder, pp. 803–811. St. Louis: C. V. Mosby.

Ehinger, B. (1966a). Distribution of adrenergic nerves in the eye and some related structures in the cat. *Acta Physiologica Scandinavica*, **66**, 123–128.

Ehinger, B. (1966b). Adrenergic nerves to the eye and to related structures in man and in the cynomolgus monkey (*Macaca irus*). *Investigative Ophthalmology*, **5**, 42–52.

Ehinger, B. (1966c). Connections between adrenergic nerves and other tissue components in the eye. *Acta Physiologica Scandinavica*, **67**, 57–64.

Ehinger, B. (1971). A comparative study of the adrenergic nerves to the anterior eye segment of some primates. *Zeitschrift für Zellforschung*, **116**, 157–177.

Ehinger B. and Sjöberg, N.O. (1971). Development of the ocular nerve supply in man and guinea-pig. *Zeitschrift für Zellforschung*, **118**, 579–592.

Ehinger, B., Sundler, F., Tervo, K., Tervo, T. and Tornqvist, K. (1983). Substance P fibres in the anterior segment of the rabbit eye. *Acta Physiologica Scandinavica*, **118**, 215–218.

Elbadri, A.A., Shaw, C., Johnston, C., Archer, D. and Buchanan, K. (1991). The distribution of neuropeptides in the ocular tissues of several mammals: a comparative study. *Comparative Biochemistry and Physiology*, **100C**, 625–627.

Elena, P.-P., Kosina-Boix, M., Moulin, G. and Lapalus, P. (1987). Autoradiographic localization of β-adrenergic receptors in rabbit eye. *Investigative Ophthalmology and Visual Science*, **28**, 1436–1441.

Elena, P.-P., Denis, P., Kosina-Boix, M., Saraux, H. and Lapalus, P. (1990). β-Adrenergic binding sites in the human eye: an autoradiographic study. *Journal of Ocular Pharmacology*, **6**, 143–149.

Elsås, T., Uddman, R. and Sundler, F. (1996). Pituitary adenylate cyclase-activating peptide-immunoreactive nerve fibres in the cat eye. *Graefes Archive for Clinical and Expeimental Ophthalmology*, **234**, 573–580.

Emoto, I. and R.W. Beuerman. (1987). Stimulation of neurite growth by epithelial implants into corneal stroma. *Neuroscience Letters*, **82**, 140–144.

Falconer, M.A. (1949). Intramedullary trigeminal tractotomy and its place in the treatment of facial pain. *Journal of Neurology, Neurosurgery and Psychiatry*, **12**, 297–311.

Fischer, F.H., Schmitz, L., Hoff, W., Schartl, S., Liegl, O. and Wiederholt, M. (1978). Sodium and chloride transport in the isolated human cornea. *Pflügers Archiv*, **373**, 179–188.

Fitzgerald, G.G. and Cooper, J.R. (1971). Acetylcholine as a possible sensory mediator in rabbit corneal epithelium. *Biochemical Pharmacology*, **20**, 2741–2748.

Fogle, J.A. and Neufeld, A.H. (1979). The adrenergic and cholinergic corneal epithelium. *Investigative Ophthalmology and Visual Science*, **18**, 1212–1215.

Friedenwald, J.S. and Buschke, W. (1944a). The effects of excitement, of epinephrine and of sympathectomy on the mitotic activity of the corneal epithelium in rats. *American Journal of Physiology*, **141**, 689–694.

Friedenwald, J.S. and Buschke, W. (1944b). The influence of some experimental variables on the epithelial movements in the healing of corneal wounds. *Journal of Celular and Comparative Physiology*, **23**, 95–107.

Fujita, S., Shimizu, T., Izumi, K., Fukuda, T., Sameshima, M. and Ohba, N. (1984). Capsaicin-induced neuroparalytic keratitis-like corneal changes in the mouse. *Experimental Eye Research,* **38**, 165–175.

Gallar J., Pozo, M.P., Rebollo, I. and Belmonte, C. (1990). Effects of capsaicin on corneal wound healing. *Investigative Ophthalmology and Visual Science,* **31**, 1968–1973.

Gallar, J., Pozo, M.A., Tuckett, R.P. and Belmonte, C. (1993). Response of sensory units with unmyelinated fibres to mechanical, thermal and chemical stimulation of the cat's cornea. *Journal of Physiology,* **468**, 609–622.

Gamse, R., Leeman, S., Holzer, P. and Lembeck, F. (1981). Differential effects of capsaicin on the content of somatostatin, substance P, and neurotensin in the nervous system of the rat. *Naunyn-Schmiedberg's Archives of Pharmacology,* **317**, 140–148.

Garcia-Hirschfeld, J., Lopez-Briones, L.G. and Belmonte, C. (1994). Neurotrophic influences on corneal epithelial cells. *Experimental Eye Research,* **59**, 597–605.

Gilbard, J.P. and Rossi, S.R. (1990). Tear film and ocular surface changes in a rabbit model of neurotrophic keratitis. *Ophthalmology,* **97**, 308–312.

Giraldez, F., Geijo, E. and Belmonte, C. (1979). Response characteristics of corneal sensory fibers to mechanical and thermal stimulation. *Brain Research,* **177**, 571–576.

Gnädiger, M.C., Heiman, R. and Markstein, R. (1973). Choline acetyltransferase in corneal epithelium. *Experimental Eye Research,* **15**, 395–399.

Grant, F.C., Groff, R.A. and Lewy, F.H. (1940). Section of the descending spinal root of the fifth cranial nerve. *Archives of Neurology and Psychiatry,* **43**, 498–509.

Gulbenkian, S., Merighi, A., Wharton, J., Varndell, I. and Polak, J. (1986). Ultrastuctural evidence for the coexistence of calcitonin gene-related peptide and substance P in secretory vesicles of peripheral nerves in the guinea pig. *Journal of Neurocytology,* **15**, 535–542.

Harris, L.W. and Purves, D. (1989). Rapid remodeling of sensory endings in the corneas of living mice. *Journal of Neuroscience,* **9**, 2210–2214.

Harris, W. (1940). An analysis of 1,433 cases of paroxysmal trigeminal neuralgia (trigeminal-tic). and the end-results of gasserian alcohol injection. *Brain,* **63**, 209–224.

Harti, G., Sharkey, K.A. and Pierau, Fr.-K. (1989). Effects of capsaicin in rat and pigeon on peripheral nerves containing substance P and calcitonin gene-related peptide. *Cell and Tissue Research,* **256**, 465–474.

Hay, E.D. (1979). Development of the vertebrate cornea. *International Review of Cytology,* **63**, 263–322.

Heino, P., Oksala, O., Luhtala, J. and Uusitalo, H. (1995). Localization of calcitonin gene-related peptide binding sites in the eye of different species. *Current Eye Research,* **14**, 783–790.

Hellauer, H.F. (1950). Sensibilitat und acetylcholingehalt der hornhaut verschiedener tiere und des menschen. *Zeitschrift für Vergleichende Physiologie,* **32**, 303–310.

Herbort, C.P., Weissman, S.S. and Payan, D.G. (1989). Role of peptidergic neurons in ocular herpes simplex infection in mice. *FASEB Journal,* **3**, 2537–2541.

Holden, B.A., Polse, K., Fonn, D. and Mertz, G. (1982). Effects of cataract surgery on corneal function. *Investigative Ophthalmology and Visual Science,* **22**, 343–350.

Holmdahl, G., Håkanson, R., Leander, S., Rosell, S., Folkers, K. and Sundler, F. (1981). A substance P antagonist, (D-Pro[2], D-Trp[7,9])-SP, inhibits inflammatory responses in the rabbit eye. *Science,* **214**, 1029–1031.

Howard, R.O., Wilson, W.S. and Dunn, B. (1973). Quantitative determination of choline acetylase, acetylcholine, and acetylcholinesterase in the developing rabbit cornea. *Investigative Ophthalmology,* **12**, 418–425.

Howard, R.O., Zadunaisky, J.A. and Dunn, B.J. (1975). Localization of acetylcholinesterase in the rabbit cornea by light and electron microscopy *Investigative Ophthalmology,* **14**, 592–603.

Hoyes, A.D. and Barber, P. (1976). Ultrastructure of the corneal nerves in the rat. *Cell and Tissue Research,* **172**, 133–144.

Hoyes, A.D. and Barber, P. (1977). Ultrastructure of corneal receptors. In *Pain in the Trigeminal Region,* edited by D.J. Anderson and B. Matthews, pp. 1–12. Amsterdam, New York: Elsevier Biomedical Press.

Huhtala, A., Johansson, G., and Saari, M. (1975). Myelinated nerves of guinea pig iris after denervation of the ophthalmic division of the trigeminal nerve. *Ophthalmic Research,* **7**, 354–362.

Ishida, N., del Cerro, M., Rao, G.M., Mathe, M. and Aquavello, J.V. (1984). Corneal stromal innervation. A quantitative analysis of distribution. *Ophthalmic Research,* **16**, 139–144.

Jancso, N., Jancso-Gabor, A. and Takats, I. (1961). Pain and inflammation induced by nicotine, acetylcholine and structurally related compounds and their prevention by desensitizing agents. *Acta Physiologica,* **19**, 113–132.

Jones, M. and Marfurt, C.F. (1991). Calcitonin gene-related peptide (CGRP). and corneal innervation: A developmental study in the rat. *Journal of Comparative Neurology,* **313**, 132–150.

Jones, M. and Marfurt, C.F. (1996). Sympathetic stimulation of corneal epithelial proliferation in wounded and non-wounded rat eyes. *Investigative Ophthalmology and Visual Science*, **37**, 2535–2547.

Jones, M. and Marfurt, C.F. (1998). Peptidergic innervation of the rat cornea. *Experimental Eye Research*, **66**, 437–448.

Jumblatt, M.M. and Neufeld, A.H. (1981). Characterization of cyclic AMP-mediated wound closure of the rabbit corneal epithelium. *Current Eye Research*, **1**, 189–195.

Jumblatt, N.M. and Neufeld, A.H. (1983). β-Adrenergic and serotonergic responsiveness of rabbit corneal epithelial cells in culture. *Investigative Ophthalmology and Visual Science*, **24**, 1139–1143.

Karvounis, P.C. and Frangos, E. (1972). Some new aspects of the nerve supply of the cornea and conjunctiva. *Monogaphs of Human Genetics*, **6**, 206.

Keen, P., Tullo, A.B., Blyth, W.A. and Hill, T.J. (1982). Substance P in the mouse cornea: effects of chemical and surgical denervation. *Neuroscience Letters*, **29**, 231–235.

Keller, J. T., Bubel, H.C., Wander, A.H. and Tierney, B.E. (1991). Somatotopic distribution of corneal afferent neurons in the guinea pig trigeminal ganglion. *Neuroscience Letters*, **121**, 247–250.

Kenshalo, D.R. (1960). Comparison of thermal sensitivity of the forehead, lip, conjunctiva and cornea. *Journal of Applied Physiology*, **15**, 987–991.

Kieselbach, G.F., Ragaut, R., Knaus, H.G., Konig, P. and Wiedermann, C.J. (1990). Autoradiographic analysis of binding sites for [^{25}I]Bolton-Hunter-substance P in the human eye. *Peptides*, **11**, 655–659.

Kingsley, R. and Marfurt, C.F. (1997). Topical substance P and corneal epithelial wound closure in the rabbit. *Investigative Ophthalmology and Visual Science*, **38**, 388–395.

Kitano, S. (1957). An embryological study of the human corneal nerves. *Japanese Journal of Ophthalmology*, **1**, 48–55.

Klyce, S.D. (1982). Cl transport in rabbit cornea. In *Chloride Transport in Biological Membranes*, edited by J.A. Zadunaisky, pp. 199–221. New York: Academic Press, Inc.

Klyce, S.D. and Crosson, C. (1985). Transport processes across the rabbit corneal epithelium: a review. *Current Eye Research*, **4**, 323–331.

Klyce, S.D. and Wong, R.K.S. (1977). Site and mode of adrenaline action on chloride transport across the rabbit corneal epithelium. *Journal of Physiology*, **266**, 777–799.

Klyce, S.D., Neufeld, A.H. and Zadunaisky, J.A. (1973). The activation of chloride transport by epinephrine and Db cyclic-AMP in the cornea of the rabbit. *Investigative Ophthalmology*, **12**, 127–139.

Klyce, S.D., Palkama, K.A., Harkonen, M., Marshall, W.S., Huhtaniitty, S., Mann, K.P. and Neufeld, A.H. (1982). Neural serotonin stimulates chloride transport in rabbit corneal epithelium. *Investigative Ophthalmology and Visual Science*, **23**, 181–192.

Klyce, S.D., Beuerman, R.W. and Crosson, C.E. (1985). Alteration of corneal epithelial ion transport by sympathectomy. *Investigative Ophthalmology and Visual Science*, **26**, 434–442.

Klyce, S., Jenison, G.L., Crosson, C.E. and Beuerman, R.W. (1986). Distribution of sympathetic nerves in the rabbit cornea. *Investigative Ophthalmology and Visual Science (Suppl.)*, **27**, 354.

Knyazev, G.G., Knyazeva, G.B. and Nikiforov, A.F. (1990). Neuroparalytic keratitis and capsaicin. *Acta Physiologica Hungarica*, **75**, 29–34.

Knyazev, G.G., Knyazeva, G.B. and Tolochko, Z.S. (1991). Trophic functions of primary sensory neurons: are they really local? *Neuroscience*, **42**, 555–560.

Krejci, L. and Harrison, R. (1970). Epinephrine effects on corneal cells in tissue culture. *Archives of Ophthalmology*, **83**, 451–454.

Kummer, W., Fischer, A. and Heym, C. (1989). Ultrastucture of CGRP- and SP-LI nerve fibers in the carotid body and carotid sinus of the guinea-pig. *Histochemistry*, **92**, 433–439.

Kuwayama, Y. and Stone, R.A. (1987). Distinct substance P and calcitonin gene-related peptide immunoreactive nerves in the guinea pig eye. *Investigative Ophthalmology and Visual Science*, **28**, 1947–1954.

Laties A. and Jacobowitz, D. (1964). A histochemical study of the adrenergic and cholinergic innervation of the anterior segment of the rabbit eye. *Investigative Ophthalmology*, **3**, 592–600.

Laties, A.M. and Jacobowitz, D. (1966). A comparative study of the autonomic innervation of the eye in monkey, cat, and rabbit. *Anatomical Record*, **156**, 383–396.

Lavail, J.H., Welkin, J. and Spencer, L. (1993). Immunohistochemical identification of trigeminal ganglion neurons that innervate the mouse cornea: relevance to intracellular spread of herpes simplex virus. *Journal of Comparative Neurology*, **327**, 133–140.

Lee, C.H., Nelson, M., Barney, N.P., Brightbill, F.S., Araki-Sasaki, K., Reid, T. *et al.* (1996). Migration and proliferation characteristics of SV-40 transformed human corneal epithelial cells in response to neuropeptides. *Investigative Ophthalmology and Visual Science*, **37**, S861.

Lele, P.P. and Weddell, G. (1956). The relationship between neurohistology and corneal sensibility. *Brain*, **79**, 119–154.

Lele, P.P. and Weddell, G. (1959). Sensory nerves of the cornea and cutaneous sensibility. *Experimental Neurology*, **1**, 334–359.

Lembeck, F. and Donnerer, J. (1981). Time course of capsaicin-induced functional impairments in comparison with changes in neuronal substance P content. *Naunyn-Schmiedeberg's Archive of Pharmacology*, **316**, 240–243.

Léon-Felíu, E., Gómez-Ramos, P. and Rodríguez-Echandía, E.L. (1978). Endothelial nerve fibres in the cornea of the frog *Rana ridibunda*. *Experientia*, **34**, 1352–1353.

Lim, C.H. (1976). Innervation of the cornea of monkeys and the effects of denervation. *British Journal of Physiological Optics*, **31**, 38–42.

Lim, C.H. and Ruskell, G.L. (1978). Corneal nerve access in monkeys. *Albrecht von Graefes Archiv für Klinische und Exxperimentalle Ophthalmolaogie*. **208**, 15–23.

Lin, N. and Bazan, H. (1995). Protein kinase C substrates in corneal epithelium during wound healing: the phosphorylation of growth associated protein-43 (GAP-43). *Experimental Eye Research,* **61**, 451–460.

Lind, G.J. and Cavanagh, H.D. (1993). Nuclear muscarinic acetylcholine receptors in corneal cells from rabbit. *Investigative Ophthalmology and Visual Science,* **34**, 2943–2952.

Liu, G.S., Basu, P.K. and Trope, G.E. (1987). Ultrastructural changes of the rabbit corneal epithelium and endothelium after timoptic treatment. *Graefes Archives of Clinical and Experimental Ophthalmology*, **225**, 325–330.

Liu, G.S., Trope, G.E. and Basu, P.K. (1990). α-Adrenoceptors and regenerating corneal epithelium. *Journal of Ocular Pharmacology*, **6**, 101–112.

Lukás Z. and Dolezel, S. (1975). A histochemical study on the development of the innervation of rabbit cornea. *Folia Morphologica*, **3**, 272–274.

MacIver, M.B. and Tanelian, D.L. (1993a). Structural and functional specialization of Aδ and C fiber free nerve endings innervating rabbit corneal epithelium. *Journal of Neuroscience*, **13**, 4511–4524.

MacIver, M.B. and Tanelian, D.L. (1993b). Free nerve endings terminal morphology is fiber type specific for Ad and C fibers innervating rabbit corneal epithelium. *Journal of Neurophysiology*, **69**, 1779–1783.

Mackie I.A. (1978). Role of the corneal nerves in destructive disease of the cornea. *Transactions of the Ophthalmological Society*, **98**, 343–347.

Magendie, J. (1824). De l'influence de la cinquième paire de nerfs sur la nutrition et les fonctins de l'oeil. *Journal of Physiology (Paris)*, **4**, 176–182.

Marfurt, C.F. (1981). The somatotopic organization of the cat trigeminal ganglion as determined by the horseradish peroxidase technique. *Anatomical Record*, **201**, 105–118.

Marfurt, C.F. (1988). Sympathetic innervation of the rat cornea as demonstrated by the retrograde and anterograde transport of horseradish peroxidase-wheat germ agglutinin. *Journal of Comparative Neurology*, **268**, 147–160.

Marfurt, C.F. and Del Toro, D. (1987). The corneal sensory pathway in the rat: a horseradish peroxidase tracing study. *Journal of Comparative Neurology*, **261**, 450–459.

Marfurt, C.F. and Echtenkamp, S. (1988). Central projections and trigeminal ganglion somatotopy of corneal afferent neurons in the monkey, *Macaca fascicularis*. *Journal of Comparative Neurology*, **272**, 370–382.

Marfurt, C.F. and Ellis, L.C. (1993). Immunohistochemical localizaton of tyrosine hydroxylase in corneal nerves. *Journal of Comparative Neurology*, **336**, 517–531.

Marfurt, C.F., Kingsley, R.E. and Echtenkamp, S.F. (1989). Sensory and sympathetic innervation of the mammalian cornea. A retrograde tracing study. *Investigative Ophthalmology and Visual Science,* **30**, 461–471.

Marfurt, C.F., Ellis, L.C. and Jones, M.A. (1993). Sensory and sympathetic nerve sprouting in the rat cornea following neonatal administration of capsaicin. *Somatosensory and Motor Research*, **10**, 377–398.

Marfurt, C.F., Jones, M.A. and Thrasher, K. (1998). Parasympathetic innervation of the rat cornea. *Experimental Eye Research,* **66**, 421–435.

Mark, D. and Maurice, D. (1977). Sensory recording from the isolated cornea. *Investigative Ophthalmology and Visual Science,* **16**, 541–545.

Marshall, W.S. and Klyce, S.D. (1984). Cellular mode of serotonin action on Cl⁻ transport in the rabbit corneal epithelium. *Biochemica et Biophysica Acta*, **778**, 139–143.

Martin, R.E. and Bazan, H.E.P. (1992). Growth-associated protein (GAP-43). and nerve cell adhesion molecule in sensory nerves of the cornea. *Experimental Eye Research,* **55**, 307–314.

Martin, X.Y. and Safran, A.B. (1988). Corneal hypoesthesia. *Survey of Ophthalmology*, **33**, 28–40.

Masters, B.R. and Thaer, A.A. (1994). *In vivo* human corneal confocal microscopy of identical fields of subepithelial nerve plexus, basal epithelial, and wing cells at different times. *Microscopy Research and Technique*, **29**, 350–356.

Matsuda, H. (1968). Electron microscopic study on the corneal nerve with special reference to its endings. *Japanese Journal of Ophthalmology*, **12**, 163–173.

Maurice, D.M. (1984). The cornea and sclera. In *The Eye. Volume IB. Vegetative Physiology and Biochemistry*, edited by H. Davson, pp. 1–158. Orlando: Academic Press.

Merighi, A., Polak, J.M., Gibson, S.J., Gulbenkian, S., Valentino, K.L. and Peirone, S.M. (1988). Ultrastructural studies on CGRP-TKs- and somatostatin-immunoreactive neurons in rat dorsal root ganglia: evidence for the colocalization of different peptides in single secretory granules. *Cell and Tissue Research*, **254**, 101–109.

Mikulec, A.A. and Tanelian, D.L. (1996). CGRP increases the rate of corneal re-epithelialization in an *in vitro* whole mount preparation. *Journal of Ocular Pharmacology and Therapeutics*, **12**, 417–423.

Miller, A., Costa, M., Furness, J.B. and Chubb, I.W. (1981). Substance P immunoreactive sensory nerves supply the rat iris and cornea. *Neuroscience Letters*, **23**, 243–249.

Millodot, M. (1968). Psychophysical scaling of corneal sensitivity. *Psychonomic Science*, **12**, 401–402.

Millodot, M. (1984). A review of research on the sensitivity of the cornea. *Ophthalmic and Physiological Optics*, **4**, 305–318.

Millodot, M. and Larson, W. (1969). New measurements of corneal sensitivity: a preliminary report. *American Journal of Optometry*, **46**, 261–265.

Millodot, M. and O'Leary, D.J. (1979). Loss of corneal sensitivity with lid closure in humans. *Experimental Eye Research*, **29**, 417–421.

Millodot, M., Lim, C.H. and Ruskell, G.L. (1978). A comparison of corneal sensitivity and nerve density in albino and pigmented rabbits. *Ophthalmic Research*, **10**, 307–311.

Mindel, J.S. and Mittag, T.W. (1976). Choline acetyltransferase in ocular tissues of rabbits, cats, cattle, and man. *Investigative Ophthalmology*, **15**, 808–814.

Mindel, J.S. and Mittag, T.W. (1977). Variability of choline acetyltransferase in ocular tissues of rabbits, cats, cattle and humans. *Experimental Eye Research*, **24**, 25–33.

Mishima, S. (1957). The effects of the denervation and the stimulation of the sympathetic and the trigeminal nerve on the mitotic rate of the corneal epithelium in the rabbit. *Japanese Journal of Ophthalmology*, **1**, 65–73.

Moller, K., Zhang, Y.-Z., Håkanson, R., Luts, A., Sjölund, B., Uddman, R. and Sundler, F. (1993). Pituitary adenylate cyclase activating peptide is a sensory neuropeptide: immunocytochemical and immunochemical evidence. *Neuroscience*, **57**, 725–732.

Montoreano, R., Candia, O.A. and Cook, P. (1976). Alpha- and beta-adrenergic receptors in regulation of ionic transport in frog cornea. *American Journal of Physiology*, **230**, 1487–1493.

Morgan, C.W., Nadelhaft, I. and DeGroat, W.C. (1978). Anatomical localization of corneal afferent cells in the trigeminal ganglion. *Neurosurgery*, **2**, 252–258.

Morgan, C., DeGroat, W.C. and Janetta, P.J. (1987). Sympathetic innervation of the cornea from the superior cervical ganglion. An HRP study in the cat. *Journal of the Autonomic Nervous System (Suppl)*, **20**, 179–183.

Morgan, C.W., Janetta, P.J. and DeGroat, W.C. (1987). Organization of corneal afferent axons in the trigeminal nerve root entry zone in the cat. *Experimental Brain Research*, **768**, 411–416.

Moses, R.A. and Feldman, M.F. (1969). Ocular lesion following fifth-nerve injury in rats. *American Journal of Ophthalmoogy*, **68**, 1082–1088.

Müller, L.J., Pels, L. and Vrensen, G.F.J.M. (1996). Ultrastructural organization of human corneal nerves. *Investigative Ophthalmology and Visual Science*, **37**, 476–488.

Müller, L.J., Vrensen, G.F.J.M., Pels, L., Nunes Cardozo, B. and Willekens, B. (1997). Architecture of human corneal nerves. *Investigative Ophthalmology and Visual Science*, **38**, 985–994.

Murphy, C.J., Mannis, M.J., Malfroy, B. and Reid, T.K. (1990). Neuropeptide depletion impairs corneal epithelial wound healing. *Investigative Ophthalmology and Visual Science (Suppl)*, **31**, 266.

Murphy, C.J., Campbell, S., Marfurt, C.F. and Araki-Sasaki, K. (1998). Effect of norepinephrine on proliferation, migration, and adhesion of SV-40 transformed human corneal epithelial cells. *Cornea*, **17**, 529–536.

Nafe, J.P. and Wagoner, K.S. (1937). The insensitivity of the cornea to heat and pain derived from high temperatures. *American Journal of Psychology*, **49**, 631–635.

Nakamura, M., Ofuji, K., Chikama, T. and Nishida, T. (1997a). Combined effects of substance P and insulin-like growth factor-1 on corneal epithelial wound closure of rabbit *in vivo*. *Current Eye Research*, **16**, 275–278.

Nakamura, M., Ofuji, K., Chikama, T. and Nishida, T. (1997b). The NK_1 receptor and its participation in the synergistic enhancement of corneal epithelial migration by substance P and insulin-like growth factor-1. *British Journal of Pharmacology*, **120**, 547–552.

Nakamura, M., Nishida, T., Ofuji, K., Reid, T.W., Mannis, M.J. and Murphy, C.J. (1997c). Synergistic effect of substance P with epidermal growth factor on epithelial migration in rabbit cornea. *Experimental Eye Research*, **65**, 321–329.

Neilson, P.J. (1983). The central corneal thickness in patients with Horner's Syndrome. *Acta Ophthalmologica,* **61**, 467–473

Neufeld, A.H., Zawistowski, K., Page, E. and Bromberg, B. (1978). Influences on the density of α-adrenergic receptors in the cornea and iris-ciliary body of the rabbit. *Investigative Ophthalmology and Visual Science,* **17**, 1069–1075.

Neufeld, A.H., Ledgard, S.E. and Yoza, B.K. (1983). Changes in responsiveness of the α-adrenergic and serotonergic pathways of the rabbit corneal epithelium. *Investigative Ophthalmology and Visual Science,* **24**, 527–534.

Neufeld, A.H., Jumblatt, M.M., Esser, K.A., Cintron, C. and Beuerman, R.W. (1984). Beta-adrenergic and serotonergic stimulation of rabbit corneal tissues and cultured cells. *Investigative Ophthalmology and Visual Science,* **25**, 1235–1239.

Nishida, T., Nakamura, M., Ofuji, K., Reid, T.W., Mannis, M.J. and Murphy, C.J. (1996). Synergistic effects of substance P with insulin-like growth factor-1 on epithelial migration of the cornea. *Journal of Cellular Physiology,* **169**, 159–166.

Nishimori, T., Sera, M., Suemune, S., Yoshida, A., Tsuru, K., Tsuiki, Y., Akisaka, T., Okamoto,T., Dateoka-Y. and Shigenaga, Y. (1986). The distribution of muscle primary afferents from the masseter nerve to the trigeminal sensory nuclei. *Brain Research,* **372**, 375–381.

Norn, M.S. (1975). Conjunctival sensitivity in pathological cases. *Acta Ophthalmologica,* **53**, 450–457.

Ogilvy, C.S. and Borges, L.F. (1990). Changes in corneal innervation during postnatal development in normal rats and in rats treated at birth with capsaicin. *Investigative Ophthalmology and Visual Science,* **31**, 1810–1815.

Ogilvy, C.S., Silverberg, K.R. and Borges, L.F. (1991). Sprouting of corneal sensory fibers in rats treated at birth with capsaicin. *Investigative Ophthalmology and Visual Science,* **32**, 112–121.

Osborne, N.N. (1983). The occurrence of serotonergic nerves in the bovine cornea. *Neuroscience Letters,* **35**, 15–18.

Osborne, N.N. and Tobin, A.B. (1987). Serotonin-accumulating cells in the iris-ciliary body and cornea of various species. *Experimental Eye Research,* **44**, 731–746.

Ozanics, V., Rayborn, M. and Sagun, D. (1977). Observations on the morphology of the developing primate cornea: epithelium, its innervation and anterior stroma. *Journal of Morphology,* **153**, 263–298.

Palkama, A., Uusitalo, H. and Lehtosalo, J. (1984). Immunohistochemical evidence for serotoninergic nerves in the cornea, iris, and ciliary processes. *Investigative Ophthalmology and Visual Science (Suppl),* **25**, 261.

Palkama, A., Uusitalo, U. and Lehtosalo, J. (1986). Innervation of the anterior segment of the eye: with special reference to functional aspects. In *Neurohistochemistry: Modern Methods and Applications,* edited by P. Panula, H. Aaivarienta and S. Soinila, pp. 587–615. New York: Alan R. Liss.

Pannabecker, C.L. (1944). Keratitis neuroparalytica. Corneal lesions following operations for trigeminal neuralgia. *Archives of Ophthalmology,* **32**, 456–463.

Paton, L. (1926). The trigeminal and its ocular lesions. *British Journal of Ophthalmology,* **10**, 305–342.

Pavlidis, C., Steuhl, K.-P. and Thanos, S. (1994). Growth of trigeminal neurites and interactions with corneal cells in embryonic chick organ cultures. *International Journal of Developmental Neuroscience,* **12**, 587–602.

Perez, E., Lopez-Briones, L., Gallar, J. and Belmonte, C. (1987). Effects of chronic sympathetic stimulation on corneal wound healing. *Investigative Ophthalmology and Visual Science,* **28**, 221–224.

Pesin, S.R. and Candia, O.A. (1982). Acetylcholine concentration and its role in ionic transport by the corneal epithelium. *Investigative Ophthalmology and Visual Science,* **22**, 651–659.

Petersen, R.A., Keat-Jin, L. and Donn, A. (1965). Acetylcholinesterase in the rabbit cornea. *Archives of Ophthalmology,* **73**, 370–377.

Proia, A.D., Chung, S.M., Klintworth, G. and Lapetina, E.G. (1986). Cholinergic stimulation of phosphatidic acid formation by rat cornea *in vitro. Investigative Ophthalmology and Visual Science,* **27**, 905–908.

Reid, T.W., Murphy, C.J., Iwahashi, C.K., Foster, B.A. and Mannis, M.J. (1993). Stimulation of epithelial cell growth by the neuropeptide substance P. *Journal of Cellular Biochemistry,* **52**, 476–485.

Reidy, J.J., Zarzour, J., Thompson, H.W., and Beuerman, R.W. (1994). Effect of topical (-blockers on corneal epithelial wound healing in the rabbit. *British Journal of Ophthalmology,* **78**, 377–380.

Reinach, P.S. and Kirchberger, M.A. (1983). Evidence for catecholamine-stimulated adenylate cyclase activity in frog and rabbit corneal epithelium and cyclic-AMP dependent protein kinase and its protein substrates in frog cornea epithelium. *Experimental Eye Research,* **37**, 327–336.

Robertson, M.R. and Winkelmann, R.K. (1970). A whole-mount cholinesterase technique for demonstrating corneal nerves: observations in the albino rabbit. *Investigative Ophthalmology,* **9**, 710–715.

Rodger, F.C. (1950). The pattern of the corneal innervation in rabbits. *British Journal of Ophthalmology,* **33**,107–113.

Rodger, F.C. (1953). Source and nature of nerve fibers in cat cornea. *A.M.A. Archives of Neurology and Psychiatry*, **70**, 206–223.

Rowbotham, G.F. (1939). Observations on the effect of trigeminal denervation. *Brain*, **62**, 364–380.

Rózsa, A.J. and Beuerman, R.W. (1982). Density and organization of free nerve endings in the corneal epithelium of the rabbit. *Pain*, **14**, 105–120.

Ruskell, G.L. (1970). The orbital branches of the pterygopalatine ganglion and their relationship with internal carotid nerve branches in primates. *Journal of Anatomy*, **106**, 323–339.

Ruskell, G.L. (1974). Ocular fibres of the maxillary nerve in monkeys. *Journal of Anatomy*, **118**, 195–203.

Sakiyama, T., Kuwayama, Y., Ishimoto I., Sasaoka, A., Shiosaka, S., Tohyama, M., Manabe, R. and Shiotani, Y. (1984). Ontogeny of substance P-containing structures in the ocular tissue of the rat: an immuno-histochemical analysis. *Brain Research*, **315**, 275–281.

Sasaoka, A., Ishimoto, I., Kuwayama, Y., Sakiyama, T., Manabe, R., Shiosaka, S., Inagaki, S. and Tohyama, M. (1984). Overall distribution of substance P nerves in the rat cornea and their three-dimensional profiles. *Investigative Ophthalmology and Visual Science,* **25**, 351–356.

Schimmelpfennig, B. (1982). Nerve structures in human central corneal epithelium. *Graefes Archive for Clinical and Experimental Ophthalmology*, **218**, 14–20.

Schimmelpfennig, B. and Beuerman, R.W. (1982). A technique for controlled sensory denervation of the rabbit cornea. *Graefes Archive for Clinical and Experimental Ophthalmology*, **218**, 287–293.

Shea, V.K. and Perl, E.R. (1985). Failure of sympathetic stimulation to affect responsiveness of rabbit polymodal nociceptors. *Journal of Neurophysiology*, **54**, 513–519.

Shimizu, T., Fujita, S., Izumi, K., Koja, T., Ohba, N. and Fukuda, T. (1984). Corneal lesions induced by the systemic administration of capsaicin in neonatal mice and rats. *Naunyn-Schmiederberg's Archives of Pharmacology*, **326**, 347–351.

Shimizu, T., Izumi, K., Fujita, S., Koja, T., Sorimachi, M., Ohba, N. and Fukuda,T. (1987). Capsaicin-induced corneal lesions in mice and the effects of chemical sympathectomy. *Journal of Pharmacology and Experimental Therapeutics*, **243**, 690–695.

Sigelman, S. and Friedenwald, J.S. (1954). Mitotic and wound-healing activities of the corneal epithelium. Effect of sensory denervation. *Archives of Ophthalmology*, **52**, 46–57.

Silverman, J.D. and Kruger, L. (1988a). Acid phosphatase as a selective marker for a class of small sensory ganglion cells in several mammals: spinal cord distribution, histochemical properties, and relation to fluoride-resistant acid phosphatase (FRAP). of rodents. *Somatosensory Research*, **5**, 219–246.

Silverman, J.D. and Kruger, L. (1988b). Lectin and neuropeptide labelling of separate populations of dorsal root ganglion neurons and associated "nociceptor" thin axons in rat testis and cornea whole-mount preparations. *Somatosensory Research*, **5**, 259–267.

Silverman, J.D. and Kruger, L. (1989). Calcitonin gene-related peptide-immunoreactive innervation of the rat head with emphasis on specialized sensory structures. *Journal of Comparative Neurology*, **280**, 303–330.

Singh, T. and Beuerman, R.W. (1979). Development of innervation in the rat cornea. *Investigative Ophthalmology and Visual Science (Suppl)*, **18**, 125.

Stevenson, R.W. and Wilson, W.S. (1975). The effect of acetylcholine and eserine on the movement of Na$^+$ across the corneal epithelium. *Experimental Eye Research,* **21**, 235–244.

Stone, R.A. (1986). Neuropeptide Y and the ocular innervation of rat, guinea pig, cat and monkey. *Neuroscience*, **17**, 1207–1216.

Stone, R.A. and Kuwayama, Y. (1985). Substance P-like immunoreactive nerves in the human eye. *Acta Ophthalmologica*, **103**, 1207–1211.

Stone, R.A. and McGlinn, A. (1988). Calcitonin gene-related peptide immunoreactive nerves in human and rhesus monkey eyes. *Investigative Ophthalmology and Visual Science*, **29**, 305–310.

Stone, R.A., Laties, A.M. and Brecha, N.C. (1982). Substance P-like immunoreactive nerves in the anterior segment of the rabbit, cat and monkey eye. *Neuroscience*, **7**, 2459–2468.

Stone, R.A., Laties, A.M. and Emson, P.C. (1986). Neuropeptide Y and the ocular innervation of rat, guinea pig, cat and monkey. *Neuroscience*, **17**, 1207–1216.

Stone, R.A., Kuwayama, Y., Terenghi, G. and Polak, J.M. (1986). Calcitonin gene-related peptide: occurrence in corneal sensory nerves. *Experimental Eye Research,* **43**, 279–283.

Stone, R.A., Kuwayama, Y. and Laties, A. (1987). Regulatory peptides in the eye. *Experentia*, **43**, 791–800.

Stone, R., McGlinn, A. and Kuwayama, Y. (1988). Galanin-like immunoreactive nerves in the porcine eye. *Experimental Eye Research,* **46**, 457–461.

Strömberg, I., Björklund, H., Melander, T., Rökaeus, Å., Hökfelt, T. and Olson, L. (1987). Galanin-immunoreactive nerves in the rat iris: alterations induced by denervations. *Cell and Tissue Research*, **250**, 267–275.

Sugimoto, T., Takemura, M. and Wakisaka, S. (1988). Cell size analysis of primary neurons innervating the cornea and tooth pulp of the rat. *Pain*, **32**, 373–381.

Sugiura, S. and Matsuda, H. (1967). Electron microscopic observations on the corneal epithelium after sensory denervation. *Japanese Journal of Ophthalmology*, **11**, 207–212.

Sugiura, S. and Yamaga, C. (1968). Studies on the adrenergic nerve of the cornea. *Acta Societatis Ophthalmologicar Japonicae*, **72**, 872–879.

Sugiura, S., Kawanabe, K. and Kotashima, T. (1964). Morphologic changes of the polygonal cell of the corneal epithelium after sensory denervation. *Acta Societatis Ophthalmologicar Japonicae*, **68**, 912–920.

Sweeney, D.F., Vannas, A., Holden, B., Tervo, T. and Telaranta, T. (1985). Evidence for sympathetic neural influence on human corneal epithelial function. *Acta Ophthalmologica.*, **63**, 215–220.

Szönyi, G., Knyihar, E. and Csillik, B. (1979). Extra-lysosomal, fluoride-resistant acid phosphatase-active neuronal system subserving nociception in the rat cornea. *Zeitschrift für Mikroskopisch-Anatomische Forschung*, **93**, 974–981.

Tanelian, D.L. (1991). Cholinergic activation of a population of corneal afferent nerves. *Experimental Brain Research*, **86**, 414–420.

Tanelian, D.L. and Beuerman, R.W. (1984). Responses of rabbit corneal nociceptors to mechanical and thermal stimulation. *Experimental Neurology*, **84**, 165–178.

ten Tusscher, M.P.M., Klooster, J. and Vrensen, G.F.J.M. (1988). The innervation of the rabbit's anterior eye segment: a retrograde tracing study. *Experimental Eye Research*, **46**, 717–730.

ten Tusscher, M.P.M., Klooster, J., Van der Want, J., Lamers, W. and Vrensen, G. (1989). The allocation of nerve fibres to the anterior eye segment and peripheral ganglia of rats. I. The sensory innervation. *Brain Research*, **494**, 95–104.

Tervo, T. (1976). Histochemical demonstration of cholinesterase activity in the cornea of the rat and the effect of various denervations on the corneal nerves. *Histochemistry*, **47**, 133–143.

Tervo, T. (1977). Consecutive demonstration of nerves containing catecholamine and acetylcholinesterase in the rat cornea. *Histochemistry*, **50**, 291–299.

Tervo, T. and Palkama, A. (1976a). Sympathetic nerves to the rat cornea. *Acta Ophthalmologica*, **54**, 75–84.

Tervo, T. and Palkama, A. (1976b). Adrenergic innervation of the rat corneal epithelium. *Investigative Ophthalmology*, **15**, 147–150.

Tervo, T. and Palkama, A. (1978a). Ultrastructure of the corneal nerves after fixation with potassium permanganate. *Anatomical Record*, **190**, 851–860.

Tervo, T. and Palkama, A. (1978b). Innervation of the rabbit cornea. A histochemical and electron-microscopic study. *Acta Anatomica*, **102**, 164–175.

Tervo, K. and Tervo, T. (1981). The ultrastructure of rat corneal nerves during development. *Experimental Eye Research*, **33**, 393–402.

Tervo K., Tervo, T. and Palkama, A. (1978). Pre- and postnatal development of catecholamine-containing and cholinesterase-positive nerves of the rat cornea and iris. *Anatomy and Embryology*, **154**, 267–284.

Tervo, T., Joó, F., Huikuri, K., Toth, I. and Palkama, A. (1979). Fine structure of sensory nerves in the rat cornea: an experimental nerve degeneration study. *Pain*, **6**, 57–70.

Tervo, K., Tervo, T., Eränkö, L., Eränkö, O. and Culleo, C. (1981a). Immunoreactivity for substance P in the Gasserian ganglion, ophthalmic nerve and anterior segment of the rabbit eye. *Histochemical Journal*, **13**, 435–443.

Tervo, K., Tervo, T., Eränkö, L. and Eränkö, O. (1981b). Substance P immunoreactive nerves in the rodent cornea. *Neuroscience Letters*, **25**, 95–97.

Tervo, K., Tervo, T., Eränkö, L., Eränkö, O., Valtonen, S. and Cuello, A.C. (1982a). Effect of sensory and sympathetic denervation on substance P immunoreactivity in nerve fibers of the rabbit eye. *Experimental Eye Research*, **34**, 577–585.

Tervo, K., Tervo, T., Eränkö, L., Vannas, A., Cuello, A.C. and Eränkö, O. (1982b). Substance P-immunoreactive nerves in the human cornea and iris. *Investigative Ophthalmology and Visual Science*, **23**, 671–674.

Tervo, T., Tervo, K., Eranko, L., Vannus, A., Eranko, O. and Culleo, A.C. (1983). Substance P immunoreaction and acetylcholinesterase activity in the cornea and gasserian ganglion. *Ophthalmic Research*, **15**, 280–288.

Tinsley, P.W., Fridland, G.H., Killmar, J.T. and Desiderio, D.M. (1989). Purification, characterization and localization of neuropeptides in the cornea. *Peptides*, **9**, 1373–1379.

Toivanen M., Tervo, T., Partanen, M., Vannas, A. and Hervonen, A. (1987). Histochemical demonstration of adrenergic nerves in the stroma of human cornea. *Investigative Ophthalmology and Visual Science*, **28**, 398–400.

Too, H.P., Todd, K., Lightman, S.L., Horn, A., Unger, W.G. and Hanley, M.R. (1989). Presence and actions of vasopressin-like peptides in the rabbit anterior uvea. *Regulatory Peptides*, **25**, 259–266.

Tower, S.S. (1940). Unit for sensory reception in cornea, with notes on nerve impulses from sclera, iris and lens. *Journal of Neurophysiology*, **3**, 486–500.

Trope, G.E., Liu, G.S. and Basu, P.K. (1988). Toxic effects of topically administered betagan, betoptic, and timoptic on regenerating corneal epithelium. *Journal of Ocular Pharmacology*, **4**, 359–366.

Tullo, A., Keen, P., Blyth, W., Hill, T. and Easty, D. (1983). Corneal sensitivity and substance P in experimental herpes simplex keratitis in mice. *Investigative Ophthalmology and Visual Science*, **24**, 596–598.

Ueda, S., del Cerro, M., LoCascio, J.A. and Aquavella, J.V. (1989). Peptidergic and catecholaminergic fibers in the human corneal epithelium: an immunohistochemical and electron microscopic study. *Acta Ophthalmologica (Suppl)*, **67**, 80–90.

Unger, W.G. (1990). Review: mediation of the ocular response to injury. *Journal of Ocular Pharmacology*, **6**, 337–353.

Unger, W.G., Butler, J.M., Cole, D.F., Bloom, S.R. and McGregor, G.P. (1981). Substance P, vasoactive intestinal polypeptide (VIP). and somatostatin levels in ocular tissue of normal and sensorily denervated rabbit eyes. *Experimental Eye Research*, **32**, 797–801.

Uusitalo, H., Lehtosalo, J., Laakso, J., Harkonen, M. and Palkama, A. (1982). Immunohistochemical and biochemical evidence for 5-hydroxytryptamine containing nerves in the anterior part of the eye. *Experimental Eye Research*, **35**, 671–675.

Uusitalo, H., Krootila, K. and Palkama, A. (1989). Calcitonin gene-related peptide (CGRP). immunoreactive sensory nerves in the human and guinea pig uvea and cornea. *Experimental Eye Research*, **48**, 467–475.

van Alphen, G.W. (1957). Acetylcholine synthesis in corneal epithelium. *Archives of Ophthalmology*, **58**, 449–451.

Vannas, A., Holden B.A., Sweeney D.F. and Polse K.A. (1985). Surgical incision alters the swelling response of the human cornea. *Investigative Ophthalmology and Visual Science*, **26**, 864–868.

Vidovic, M. and Hill, C.E. (1988). Withdrawal of collaterals of sympathetic axons to the rat eye during postnatal development: the role of function. *Journal of the Autonomic Nervous System*, **22**, 57–65.

Voaden, M.J. (1971). The effects of superior cervical ganglionectomy and/or bilateral adrenalectomy on the mitotic activity of the adult rat cornea. *Experimental Eye Research*, **12**, 337–341.

von Brücke, H., Hellauer, H.F. and Umrath, K. (1949). Azetylcholin und aneuringehat der hornhaut und seine beziehungen zur nervenversorgung. *Ophthalmologica*, **117**, 19–35.

von Frey, M. (1894). Berichte uber die verhaindlungen der koniglich sachsichen. *Gestame Wissenschaftliche*, **46**, 185–196.

Vonderahe, A.R. (1928). Corneal and scleral anesthesia of the lower half of the eye in a case of trauma of the superior maxillary nerve. *Archives of Neurology and Psychiatry (Chicago)*, **20**, 836–837.

Wakui, K. and Sugiura, S. (1965). Corneal nerve canals and the less-myelinated nerve fiber bundles. *Japanese Journal of Ophthalmology*, **9**, 199–211.

Waldrep, J. (1989). Neurogenic inflammation of the cornea. In *Healing Processes in the Cornea*, edited by R.W. Beuerman, C.E. Crosson and H.E. Kaufman, pp. 27–43. The Woodlands, Texas: Portfolio Publishing Co.

Walkenbach, R.J. and LeGrand, R.D. (1981). Regulation of cyclic AMP-dependent protein kinase and glycogen synthase by cyclic AMP in the bovine cornea. *Experimental Eye Research*, **33**, 111–120.

Walkenbach, R.J. and Ye, G.-S. (1991). Muscarinic cholinoceptor regulation of cyclic guanosine monophosphate in human corneal epithelium. *Investigative Ophthalmology and Visual Science*, **32**, 610–615.

Walkenbach, R.J., LeGrand, R.D. and Barr, R.E. (1981). Distribution of cyclic AMP-dependent protein kinase in the bovine cornea. *Experimental Eye Research*, **32**, 451–459.

Walkenbach, R.J., Chao, W.T.H., Bylund, D.B. and Gibbs, S.R. (1985). Characterization of β-adrenergic receptors in fresh and primary cultured bovine corneal epithelium. *Experimental Eye Research*, **40**, 15–24.

Walkenbach, R.J., Ye, G.S., Reinach, P.S. and Boney, F. (1991). α_1-Adrenoceptors in human corneal epithelium. *Investigative Ophthalmology and Visual Science*, **32**, 3067–3072.

Wang, Z.-Y., Alm, P. and Håkanson, R. (1995). Distribution and effects of pituitary adenylate cyclase-activating peptide in the rabbit eye. *Neuroscience*, **69**, 297–308.

Watanabe, H., Tisdale, A.S. and Gipson, I.K. (1993). Eyelid opening induces expression of a glycocalyx glycoprotein of rat ocular surface epithelium. *Investigative Ophthalmology and Visual Science*, **34**, 327–338.

Watson, C. and Vijayan, N. (1995). The sympathetic innervation of the eyes and face: a clinicoanatomic review. *Clinical Anatomy*, **8**, 262–272.

Whitear, M. (1960). An electron microscope study of the cornea in mice, with special reference to the innervation. *Journal of Anatomy*, **94**, 387–409.

Wiederholt, M., Schmidt, D.K., Eggebrecht, R., Zimmermann, J. and Fischer, F.H. (1983). Adrenergic regulation of sodium and chloride transport in the isolated cornea of rabbit and man. *Graefes Archives for Clinical and Experimental Ophthalmology*, **220**, 240–244.

Williams, J.D. and Cooper, J.R. (1965). Acetylcholine in bovine corneal epithelium. *Biochemical Pharmacology*, **14**, 1286–1289.

Wilson, W.S. and McKean, C.E. (1986). Regional distribution of acetylcholine and associated enzymes and their regeneration in corneal epithelium. *Experimental Eye Research*, **43**, 235–242.

Wolter, J.R. (1957). Innervation of the corneal endothelium of the eye of the rabbit. *Archives of Ophthalmology*, **58**, 246–250.

Yamamoto, R., McGlinn, A. and Stone, R.A. (1991). Brain natriuretic peptide-immunoreactive nerves in the porcine eye. *Neuroscience Letters*, **122**, 151–153.

Yamamoto, R., Bredt, D.S., Snyder, S.H. and Stone, R.A. (1993). The localization of nitric oxide synthase in the rat eye and related cranial ganglia. *Neuroscience*, **54**, 189–200.

Zadunaisky, J.A. and Lande, M.A. (1971). Active chloride transport and the control of corneal transparency. *American Journal of Physiology*, **221**, 1837–1844.

Zander, E. and Weddell, G. (1951a). Observations of the innervation of the cornea. *Journal of Anatomy*, **85**, 68–99.

Zander, E. and Weddell, G. (1951b). Reaction of corneal nerve fibres to injury. *British Journal of Ophthalmology*, **35**, 61–88.

Zuazo, A., Ibañez, J. and Belmonte, C. (1986). Sensory nerve responses elicited by experimental ocular hypertension. *Experimental Eye Research*, **43**, 759–769.

3 Control of Accommodation

Stuart J. Judge[1,*] and D. Ian Flitcroft[2]

[1]*University Laboratory of Physiology, Parks Road, Oxford, OX1 3PT, UK*
[2]*Institute of Ophthalmology, 60 Eccles Street, Dublin 7, Ireland*

In animals with high acuity, accommodation is necessary to avoid a loss of visual resolution when viewing nearby objects. The characteristics of human accommodation behaviour are described. The sensory guidance of accommodation is discussed, with an emphasis on factors about which there is some idea of their neural substrate. The present state of knowledge of the anatomy and neurophysiology of accommodation control are surveyed.

KEY WORDS: accommodation; chromatic aberration; vergence; visual cortex

INTRODUCTION

Fixed-focus visual systems are adequate for some purposes. For example, there seems to be no good evidence for accommodation in rodents (Hughes, 1977). Presumably animals such as rats and rabbits have no need to vary accommodation because their visual acuity is low, and even a large defocus of the retinal image does not materially affect their spatial discrimination. Humans and other foveate animals, on the other hand, have high acuity and a behavioural repertoire that includes the need to make fine spatial discriminations at a range of distances, and hence accommodation is necessary.

The issue can be considered quantitatively by noting that the precision with which accommodation needs to be controlled depends on the depth of focus of the eye. Green, Powers and Banks (1980) have suggested that the depth of focus, ΔD (in dioptres), of the eye of any animal can be estimated by the formula:

$$\Delta D = \pm\, 7.03/pv,$$

*Correspondence: Dr S.J. Judge, University Laboratory of Physiology, Parks Road, Oxford, OX1 3PT, UK. Tel: +44-1865-272508; Fax: +44-1865-272469.

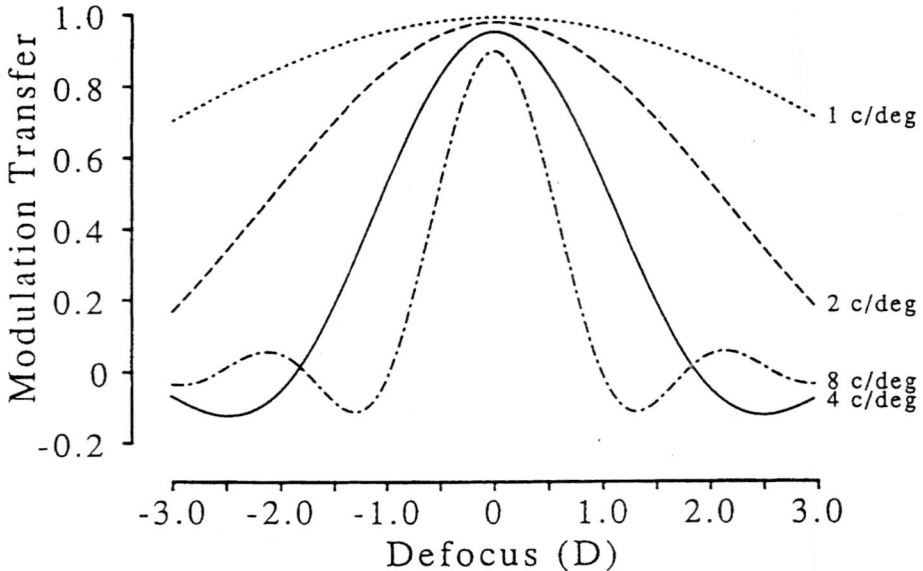

Figure 3.1 Calculated effects of defocus on the monochromatic modulation transfer of the diffraction limited eye. A pupil diameter of 3 mm and a wavelength of 550 nm were used for these calculations. (Reprinted from Flitcroft, 1990, with permission from Cambridge University Press).

where p is pupil diameter in millimetres and v is cut-off spatial frequency in c/degree. To take one example, this formula predicts a depth of focus of the rat eye of ±13.2 D, which matches the tolerance (±14 D) of rat retinal ganglion cells to defocus.

It is a commonplace that small errors in refraction are practically troublesome to us, and if we examine the way in which the modulation transfer of the human eye is affected by defocus (Figure 3.1) it is easy to see that one reason why this is so is that mid- and high spatial frequencies are sharply attenuated by defocus.

AMPLITUDE OF ACCOMMODATION

The amplitude of accommodation has been objectively determined in a wide range of species. It is particularly large in diving birds, such as the cormorant, that need to see well both in air and water, where much of the power of the corneal surface is lost (Sivak, 1980). The sea otter also has a very large amplitude of accommodation, for the same reason (Murphy et al., 1990). In other birds the amplitude of accommodation may be substantial e.g. 9 D in the pigeon, and 17 D in the chicken (Schaeffel and Howland, 1987). Primates also have large accommodation amplitudes: at least 18 D in young macaques (Smith and Harwerth, 1984); more than 20 D in marmosets (Troilo, Howland and Judge, 1993), and about 14 D in young humans. Of those animals that have been extensively used for studies of the pathways controlling accommodation, the cat has the smallest and most uncertain amplitude of accommodation. Surveying various studies of accommodation reporting amplitudes between less than 2 D and as much as 12 D, Thibos and Levick (1982) commented that "the scatter of values is uncomfortably large". Some have doubted that cats accommodate at all. Certainly Fisher (1971) considered that the mechanical properties of the cat lens made in impossible for the cat to accommodate by changing lens shape. It seems reasonably clear that some cats do not accommodate when actively fixating nearby objects: for example, Howland (personal communication) was unable to see any evidence of accommodation in two red tabby cats fixating and pouncing on ribbon toys at a range of distances. An infra-red photorefraction technique was used; half a dioptre would have been evident. Of course, to say that some cats do not accommodate is not to assert that none do: the whole subject should be re-examined, if cats continue to be used for studies of vision and accommodation. One obvious possibility is that there is genetic variation, with in-bred house cats having poorer accommodation than animals that have to catch mice to stay alive!

One can ask the question of how accommodation amplitude of cats and macaques compares when tested in the same way. The method that seems to have been most carefully applied to both species is the indirect one of examining how close a viewing distance can be used (or how strong a power of negative lens can be placed in front of the eye) before visual acuity falls from its best value. In cats this suggests a near point of 25 cm and an accommodation amplitude of no more than 4 D, assuming cats are emmetropic (Bloom and Berkley, 1977); in young macaques, at least 18 D (Smith and Harwerth, 1984) — or more than four times the cat amplitude.

One point that should be made is that in photopic illumination the cat eye has a small depth of focus, and so accommodation would seem to be necessary. Using the formula of Green, Powers and Banks (1980) with a cut-off spatial frequency of 9 c/degree (Jacobson, Franklin and McDonald, 1976; Hall and Mitchell, 1991) and a pupil width of between 2.5 and 6 mm (Jacobson, Franklin and McDonald, 1976) gives a depth of focus of between 0.13 and the 0.3 D, with the larger figure more probable because according to Jacobson, Franklin and McDonald (1976) it was notable that pupils constricted as discrimination was made.

Neither the various mechanisms of accommodation in vertebrates surveyed by Walls (1942), Hughes (1977) and Sivak (1980), nor the issue of the cause or causes of presbyopia which have been surveyed recently by Glasser and Campbell (1998) will be discussed here. The most recent general review of accommodation would appear to be Ciuffreda (1991).

DESCRIPTION OF BEHAVIOUR

STATIC STIMULUS-RESPONSE RELATIONSHIP

Accommodation is conventionally measured in dioptres. A target at distance x metres from the eye is said to present an accommodation demand of $1/x$ dioptres, and if the plane of focus is distance y metres from the eye (strictly speaking from the principal plane of the eye, which is about 2 mm behind the corneal apex) the accommodation response is

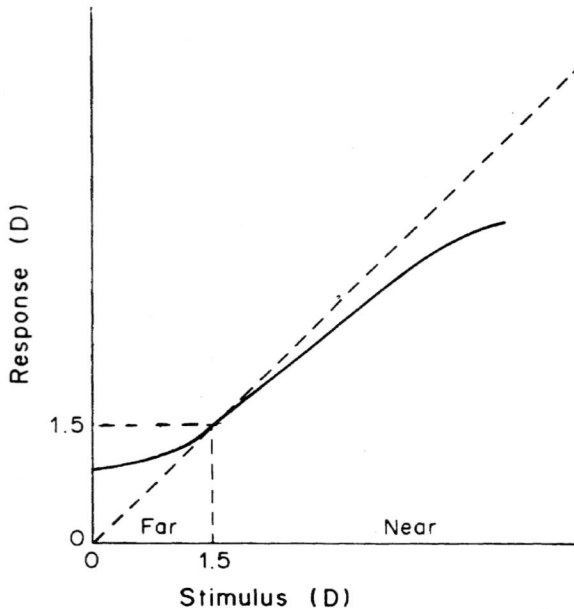

Figure 3.2   Schematic stimulus-response curve for human accommodation. (Reprinted from McBrien and Millodot, 1986 [after Charman], with permission).

$1/y$ dioptres. Figure 3.2 shows a schematic stimulus-response curve for the relationship between accommodation demand and accommodation response in monocular viewing. Three features are obvious: the slope of the linear portion of the relationship is less than unity (McBrien and Millodot (1986) found a mean value of 0.91 in forty subjects between the age of 18 and 23, with slightly higher values in hyperopes than in myopes); accommodation demand and response are equal not at 0 D (infinite target distance) but at some intermediate value, typically 1.5 D or so; and response soft saturates for high dioptric demand before reaching a limit called the near point. There is therefore a limited range over which accommodation is tolerably accurate. Note also that when viewing targets at large distances, the error of focus has the opposite sense to that at close distances — it is myopic rather than hyperopic. The extent of the linear range depends on age: there is a progressive loss of accommodation range with age (*presbyopia*) at a rate of about 2–3 D per decade of life (see, Hamasaki, Ong and Marg, 1956).

Accommodation requires cone stimulation and is not present in scotopic illumination (Campbell, 1954). The slope of the accommodation stimulus-response curve (and therefore the accuracy of static accommodation) depends on pupil diameter, because the depth of focus increases as pupil diameter decreases (Ward and Charman, 1985). The slope also depends on the spatial frequency content, with slope greatest for intermediate spatial frequencies between 1 and 8 c/degree (Owens, 1980). It was claimed that higher spatial frequencies (up to 30 c/degree) were useful for "fine-tuning" of accommodation (Charman and Tucker, 1977), but subsequent work in the same laboratory has failed to verify this with more realistic stimulus contrast (Ward, 1987).

TONIC ACCOMMODATION

In darkness or an empty field, accommodation is at an intermediate value (Whiteside, 1952) so there is an internal bias in the accommodation control system that does not depend on visual stimulation, and the null point of the stimulus-response curve, where stimulus matches response, is at this dark-focus level (Johnson, 1976). The dark-focus value is also called the level of tonic accommodation.

Adaptation of tonic accommodation

Schor, Kotulak and Tsuetaki (1986) reported, and Wolfe and O'Connell (1987) confirmed, that in some subjects tonic after-effects of near accommodation were present, and that such effects were masked by darkness. Many studies, reviewed by Rosenfield *et al.* (1994) have now been made of adaptation of tonic accommodation. In particular, McBrien and Millodot (1988) have shown that adaptation is greater in late onset myopes than in emmetropes or early onset myopes.

CONSENSUALITY

Accommodation has usually been thought always to be consensual in humans (Ball, 1952). Subsequent studies have confirmed this view (Thorn, Greenan and Heath, 1983) and dynamic binocular recordings of accommodation have demonstrated highly correlated accommodative responses in the two eyes (Winn, 1987). Viewing targets in highly eccentric viewing i.e. with near targets situated far from the midline, provides unequal accommodation demands to the two eyes. With the exception of a single subject in one study, early studies with this viewing arrangement confirmed consensuality within a few tenths of a Dioptre (Ogle, 1937; Rosenberg *et al.*, 1953; Spencer and Wilson, 1954). Marran and Schor (1998) have recently re-examined this issue and provided evidence for a limited ability for non-consensual responses to unequal accommodation demands. These non-consensual responses were of small amplitude, 0.75D for a 3D intra-ocular stimulus difference, and very slow compared to normal accommodation responses. Differences in accommodation between the two eyes take more than ten seconds to occur after introduction of an anisometropic demand (Marran and Schor, 1998; Figure 3.3).

Another issue that might be interesting to look at more carefully is whether accommodation is always consensual early in life where an anisometropia is present, either naturally, or artificially introduced in the context of studying the effects of spectacle-wear on eye growth (e.g. Hung, Crawford and Smith, 1995; Judge and Graham, 1995; Graham and Judge, 1999).

Another question that has been asked about binocularity and accommodation is how the blur signals in the two eyes are combined to form a consensual drive. Flitcroft, Judge and Morley (1992) showed that if stimuli of the same form, but dynamically anisometropic, were presented to the two eyes the accommodation response was the average of demand in each eye. If stimuli of different form were viewed by the two eyes, accommodation rivalry was observed (Flitcroft and Morley, 1997).

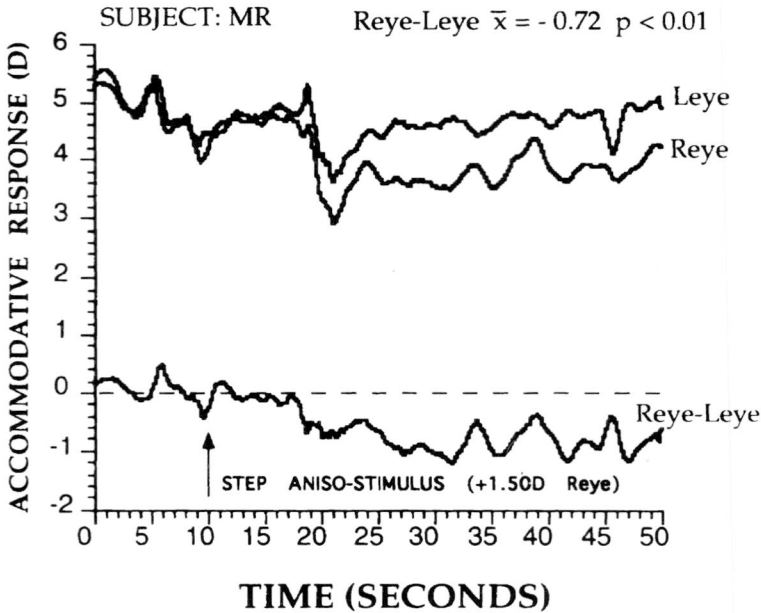

Figure 3.3 Anisometropic accommodation response. (Reprinted from Marran and Schor, 1998, with permission of Elsevier Science).

DYNAMIC ACCOMMODATION RESPONSES

Human accommodation has a latency of about 360 ms (Campbell and Westheimer, 1960; O'Neill and Stark, 1968; Phillips, Shirachi and Stark, 1972; Smithline, 1974; Tucker and Charman, 1979), and a roughly exponential response with a time constant of about 250 ms response to a step change in accommodation demand (Campbell and Westheimer, 1960; Tucker and Charman, 1979). Monkey accommodation latencies are somewhat shorter, averaging 230 ms in three monkeys (Cumming, 1985).

The frequency response of accommodation is usually thought of as low-pass with the 3 dB point at about 0.3 Hz in humans (Campbell and Westheimer, 1960; Krishnan, Phillips and Stark, 1973; Stark, Takahashi and Zames, 1965) and somewhat higher in monkeys (Cumming, 1985; Cumming and Judge, 1986). Responses to smaller amplitude stimuli show better gain but similar phase characteristics to those to larger amplitude stimuli. Intermediate spatial frequencies between 1 and 8 c/degree are most effective in driving dynamic responses with the best responses to spatial frequencies of 3 to 5 c/degree (Mathews and Kruger, 1994; Figure 3.4). Flitcroft (1988a) studied the contrast threshold for accommodation *initiation* as a function of target temporal frequency and found a weakly bandpass characteristic with best frequencies around 5 Hz.

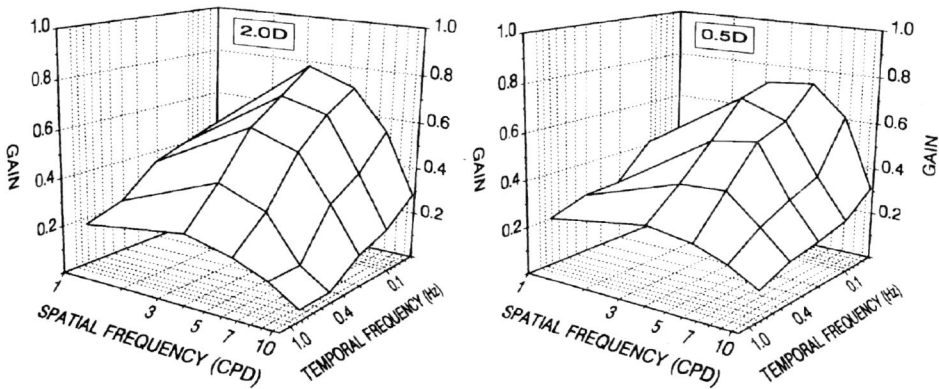

Figure 3.4   Three-dimensional plots of accommodation gain as a function of spatial and temporal frequency, at two amplitudes of target motion. Average of data from four subjects. (Reprinted from Mathews and Kruger, 1994, with permission of Elsevier Science).

FLUCTUATIONS

Campbell, Robson and Westheimer (1959) found that accommodation fluctuated slightly even when subjects were attempting to maintain accommodation on a target. The fluctuations were highly correlated in the two eyes, and the amplitude of the fluctuations increased with increasing levels of accommodation. Fourier analysis showed that the power in these fluctuations fell into two frequency bands: a low-frequency one (0–0.5 Hz) and a separate higher frequency band, around about 2 Hz and varying in frequency somewhat between subjects. Power in the two bands varied in different ways with circumstances: for example, reducing pupil diameter from 7 to 1 mm with an artificial pupil increased the power in the low frequency band but reduced the high frequency fluctuations. Gray, Winn and Gilmartin (1993) replicated the former result but did not find a reduction in the high frequency component with small pupils. In the rhesus monkey open loop accommodation achieved with a servo controlled haploscope rather than a small pupil does, however, result in reduced amplitude of high frequency (1–2.5 Hz) accommodative fluctuations (Flitcroft, 1988b). The same group (Winn et al., 1990) reported a correlation between the high frequency fluctuations and pulse, and this has been confirmed by Collins, Davis and Wood (1995). The implication drawn from these findings was that high-frequency fluctuations are of peripheral origin, rather than of central neurological origin as had previously been assumed. If this is so it is puzzling that macaques, which have considerable higher pulse rates than humans, have high frequency fluctuations of a similar frequency to humans (Flitcroft, 1988b). Indeed, fluctuations in humans at 2Hz would imply a heart rate within the definition of clinical tachycardia. In addition to the potential for pulsatile ocular blood flow to alter zonular tension and hence lens power, the use of an infra-red optometer by Winn et al., 1990 raises the potential for instrumentation artefacts since, unlike visible wavelengths, infra-red light passes though the retinal pigment epithelium to the highly vascularised choroid. Notwithstanding such issues, the somewhat contradictory nature of studies in this area could be reconciled by the presence of both central and haemodynamic contributions to the high frequency component of accommodation responses.

SENSORY MECHANISMS OF ACCOMMODATION

The linear portion of the static accommodation stimulus-response curve is what one would expect if accommodation was governed by a feedback system in which the accommodative error was sensed, passed through some circuit with gain, and used to drive accommodation response. Because the response alters the error, the system is closed-loop. This is what engineers call a proportional control system (Toates, 1972). The internal gain would need to be quite low (10 or so, sometimes less) to match real stimulus-response curves.

Recently, Yamada and Ukai (1997) have raised the question of whether the feed-back loop idea might be fundamentally mistaken, at least for near-to-far changes, by showing that the time course of accommodation responses to a range of near-to-far step changes took the form of an exponential relaxation, independent of step size, toward the dark focus value, which was clipped at the target accommodation level. It will be interesting to see whether this result holds up.

Although the proportional controller model has been useful for thinking about accommodation control, its use can have the effect of concealing a difficulty: how does the visual system measure accommodation errors? How, for example, is one to know whether the lack of high spatial frequencies in the image on the retina is caused by incorrect accommodation or by fog obscuring the object one is looking at? Worse still, how is one to know whether the accommodative error is in one direction or another, because to a first approximation the effects on the image are the same?

There are two main proposals about these issues. The first is that in many circumstances the chromatic aberration of the eye may allow good estimates to be made of the accommodative error. Chromatic aberration causes the three classes of cones to be supplied with information about three slightly different distances of object plane: the image of a real object at one particular distance must therefore be more defocused as 'seen' by one class of cone than another, and it is not difficult to see that in principle the accommodative error could be computed. The second proposal is that the fluctuations in accommodation can be used, in something like the same way, to explore how image contrast co-varies with small changes in accommodation and therefore yield information about extent and direction of accommodative error. This would require the visual system to have information about the time course of the fluctuations, most plausibly from efference copy of accommodation commands (assuming the fluctuations are of central rather than peripheral origin). We are unaware of any evidence for or against the existence of such a signal.

In many natural situations, accommodation can be driven by information, other than blur, about target distance. The two sources of information that have been most thoroughly investigated are binocular disparity and the convergence eye movement associated with it; and the change in size of the retinal image that is a consequence of a real object moving toward or away from the observation. Consideration will be limited to these two factors because they are the only two for which something is known about the neural machinery that processes the cues.

ROLE OF CHROMATIC ABERRATION

Monochromatic illumination does not affect the slope of the static accommodation stimulus-response relationship (Charman and Tucker, 1978). Moreover, if the longitudinal and transverse chromatic aberration of the eye are manipulated, the static stimulus-response function of accommodation is not affected (Bobier, Campbell and Hinch, 1992).

The slope of the stimulus-response relationship is much lower if the target contains only chromatic rather than luminance contrast (Wolfe and Owens, 1981) and it was suggested that this might mean that accommodation was 'colour blind'. Flitcroft (1988b) pointed out that this does not follow, because the most physiologically plausible mechanism that could make use of chromatic aberration will be impaired by an object that contains chromatic but not luminance contrast.

Fincham (1951) showed that many subjects were unable to accommodate in the correct direction in monochromatic illumination. A number of studies have investigated the effect of monochromatic illumination on dynamic accommodation — typically tracking of targets presenting an accommodation demand (in dioptres) varying sinusoidally with time (not the same thing as a target moving sinusoidally in depth because of the reciprocal relationship between accommodation and fixation distance). A series of studies by Kruger and his colleagues have shown that accommodation gain is lower and phase-lag longer in monochromatic light (Kruger and Pola, 1986; Kruger et al., 1995a, 1997). Neutralising longitudinal chromatic aberration (LCA) with a compensatory lens produces somewhat smaller effects than monochromatic illumination, and reversing LCA larger ones (Kruger et al., 1993). Similarly, if subjects viewed a yellow-black grating produced by superimposing red (600 nm) and green (520 nm) gratings, and the effective LCA was set to different levels by altering the distance of the red grating from the subject, an extra 0.5 D of LCA in the normal direction enhanced dynamic accommodation response and 0.25 D in the reverse direction markedly inhibited response (Kruger et al., 1995b).

In the monkey Flitcroft and Judge (1988) reported that dynamic accommodation response was impaired in monochromatic light; further details are given in Flitcroft (1988b).

ROLE OF FLUCTUATIONS

Owens and Wolfe (1985) studied the effect on accommodation response of sinusoidal modulation of stimulus contrast at temporal frequencies between 3 and 40 Hz and for high, middle, and low spatial frequency targets. They found that flicker diminished the accuracy of accommodation, especially with high and low spatial frequency targets, and that the size of the effect was greater for lower frequency flicker. Flitcroft (1991) examined the effect of low temporal frequency more fully and suggested that the reason for flicker having a disruptive effect on accuracy of accommodation was probably that it interfered with the ability of the accommodation system to utilise information derived from the small fluctuations that are a feature of accommodation under many circumstances. In support of this view, he showed that temporal modulation of stimulus contrast with a broad-band filtered noise signal had minimal effects on accommodation response at high and low temporal frequencies, but substantial effects at frequencies between and 1 Hz and 5 Hz.

SIZE-CHANGE

Kruger and Pola (1985, 1986) showed that adding size-change to blur reduced the phase lag of accommodation. McLin, Schor and Kruger (1988) inferred from their studies of the effect of changing size on accommodation and vergence that size-change stimulated accommodation directly and vergence indirectly through the accommodative-vergence cross-link. Judge (1988) showed from Kruger and Pola's data that size-change and blur cues do not interact linearly, but that a simple non-linear model could account for their data.

VERGENCE-ACCOMMODATION

Fincham and Walton (1957) showed that if vergence was manipulated in a haploscopic display incorporating binocular pinhole pupils, 'vergence-accommodation' was induced despite the absence of blur, and indeed despite the distance of the real target remaining constant. Similar results were obtained by Kersten and Legge (1983), using quite different methods. Others have studied the dynamics of vergence-accommodation (Yoshida and Watanabe, 1969; Cumming and Judge, 1986 (in the monkey); Schor and Kotulak, 1986) and have shown that vergence accommodation has a good dynamic response, often superior to that of blur-driven accommodation.

One question, raised by Semmlow (1982), is the relative weight of the contributions of blur and vergence accommodation in normal binocular viewing. A variation of this question is whether there are subjects who rely to a very large extent on vergence-accommodation and make little use of blur. Interestingly, White and Wick (1995) have shown that binocular cues improved the accommodation of some subjects with juvenile macular degeneration.

INTERACTION OF ACCOMMODATION AND VERGENCE

Disparity-driven vergence and blur-driven accommodation can be thought of as two interacting feedback systems (Westheimer, 1963; Hung and Semmlow, 1980; Schor and Kotulak, 1986). The two systems are not independent, but cross-coupled to form what is sometimes called the dual interactive controller. Accommodation to a monocular target (i.e. without disparity input) produces convergence (termed accommodative-convergence) as well as accommodation. Vergence-accommodation supplies the opposite link, from disparity to accommodation. The static gains of the cross-links are not small (Hung, Ciuffreda and Semmlow (1986) reported mean values of about 0.6 or 0.7; Miles, Judge and Optican (1987) values almost always greater than 0.7), so the two feedback loops are strongly coupled. Moreover, when expressed in commensurate units the disparity associated with a given level of vergence is much smaller than the blur associated with the corresponding level of accommodation. In other words, in the dual interactive control model the gain of the disparity controller is much higher than that of the blur controller (Hung and Semmlow, 1980). A model with only such elements would predict that in binocular viewing the control of both accommodation and vergence would be dominated by the disparity input. It is unclear to what extent this is generally true.

A more realistic model will need to incorporate the additional features that both accommodation and convergence have tonic levels that do not depend on sensory input; that the tonic levels of both accommodation and vergence are adaptable; as are the cross-link gains (Judge and Miles, 1985; Miles, Judge and Optican, 1987; Bobier, Campbell and Hinch, 1992; but for a contrary view see Schor and Tsuetaki, 1987); and that there is some uncertainty about the position of the cross-links (Schor and Kotulak, 1986). Such a model has a large number of parameters. Several papers have recently modelled how interactions between accommodation and convergence might affect the accuracy of accommodation (Flitcroft, 1988; Schor, 1999).

ANATOMY AND NEUROPHYSIOLOGY

I shall consider the anatomy and physiology of the control of accommodation in the conventional order of sensory, intermediate, and pre-motor. The role of the Edinger-Westphal nucleus in accommodation is discussed by P.D.R. Gamlin in Chapter 4 of this volume. A recent review of the neural basis of the near response as a whole is given by Mays and Gamlin (1995).

IS THE VISUAL CORTEX NECESSARY?

It is usually assumed that the cortex is necessary for accommodation, but there is little direct evidence for this. Manor *et al.* (1988) have described a 37-year old patient with a bilateral accommodative paresis associated with a spontaneous parieto-occipital haematoma. There was no evidence of transtentorial herniation or upper brain stem pathology. Accommodation returned to normal when the haemotoma resolved.

What kind of cells?

Flitcroft (1990) has shown that double-opponent cells (i.e. ones that are both spatially opponent and chromatically opponent) could provide the appropriate signals for the main (blur) drive to accommodation (Figure 3.5). Double opponent cells, rather than singly, chromatically opponent cells, are better because chromatic information is relatively unaffected at low spatial frequencies, and high spatial frequencies can only provide information about small degrees of defocus (Flitcroft, 1990). As double opponent cells have only been found in the visual cortex (Michael, 1978; Ts'o and Gilbert, 1988) this would imply a necessary role for visual cortex.

The various types of disparity-selective cell in the visual cortex (Barlow, Blakemore and Pettigrew, 1967; Pettigrew, Nikara and Bishop, 1968; Poggio and Fischer, 1977) could supply disparity information to induce vergence-accommodation. Cells sensitive to size change have been seen in extrastriate visual cortex area V5 (Zeki, 1974).

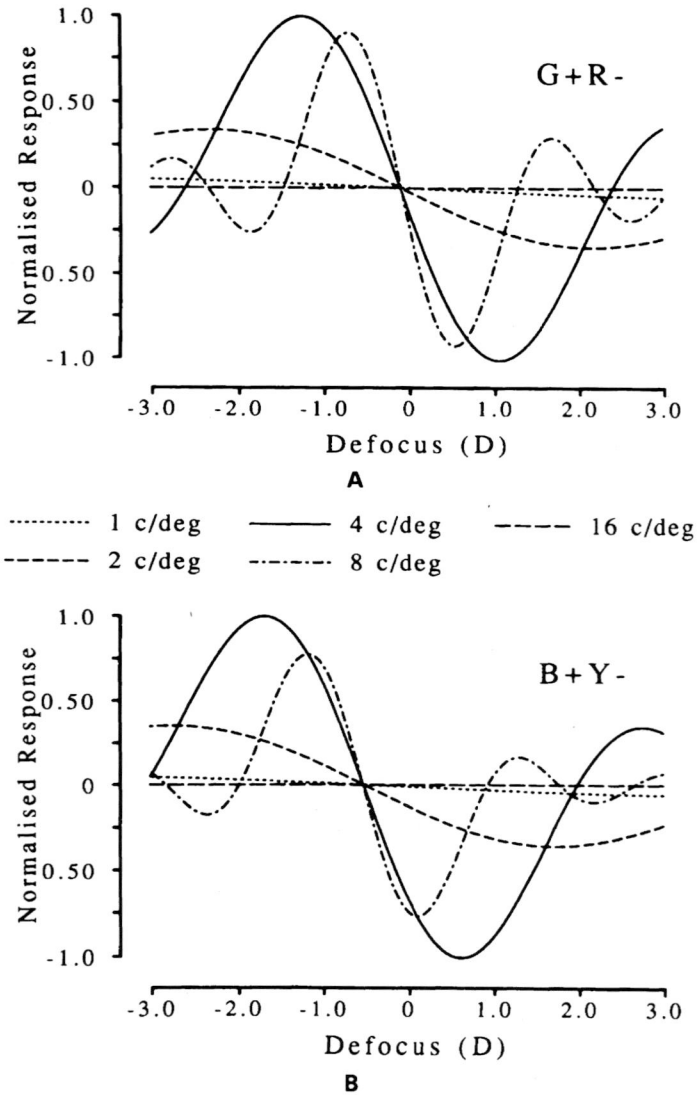

Figure 3.5   (A). Effects of defocus on the normalised response of an idealised double opponent mechanism, with a green-on/red-off centre and a green-off/red-on surround, to grating stimuli of different spatial frequency. The structure of the receptive field was modelled with four Gaussian subunits, and a ratio of size of centre and surround Gausssians that gave rise to spatial band pass filtering with a peak at 4 c/degree. (B). Effect of defocus on the normalised response of an idealised double opponent mechanism with a blue-on/yellow-off centre and blue-off/yellow-on surround, with other parameters as in (A). (Reprinted from Flitcroft, 1990, with permission from Cambridge University Press).

Parietal cortex

Holmes (1918a,b) noted patients with parietal lesions who had defects of depth perception. Jampel (1960) found that electrical stimulation of a somewhat ill-defined area at the junction of the occipital, parietal and temporal lobes produced all three components of the near response (accommodation, convergence and pupil constriction) in lightly anaesthetised macaques. Sakata, Shibutani and Kawano (1980, 1983) reported cells in the inferior parietal lobule of the macaque whose activity was related to distance of fixation, or to ocular tracking of targets moving in depth, and subsequent studies by the same group have found size-change related cells in the same general region (Sakata *et al.*, 1985; Watanabe, Kusunoki and Sakata, 1990). Gnadt and Mays (1995) found that many cells in the lateral interparietal sulcus (LIP) area had responses that were tuned for the position in depth of the target, and that these responses depended both on binocular and monocular cues. More recently, it has been shown that the LIP cells that project to superior colliculus have depth-related responses (Gnadt and Beyer, 1998).

In the cat, the lateral suprasylvian (LS) cortex has been implicated in the control of accommodation. Some cells in the PMLS and to a lesser extent the PLLS subdivision of the LS cortex of chloralose-anaesthetized and pharmacologically immobilised cats increased their firing before spontaneous increases in accommodation (Bando, Yamamoto and Tsukahara, 1984). About 70% of a subset of such cells whose projections were investigated by antidromic activation projected to the superior colliculus or pretectum. Micro-stimulation of the same parts of the LS cortex evoked small increases in accommodation (Bando *et al.*, 1984; Sawa, Maekawa and Ohtsuka, 1992). The area of the LS cortex in which stimulation evokes accommodation projects to the rostral superior colliculus, and to the nucleus of the optic tract, but not to other pretectal nuclei (Maekawa and Ohtsuka, 1993; Ohtsuka and Sato, 1996). Stimulation of PMLS and PLLS can also evoke disjunctive eye movements in the anaesthetised cat (Toda *et al.*, 1991), with some sites evoking both disjunctive movements and accommodation, and others only one response. In PMLS of alert cats, cells were found whose firing rate was correlated with convergence eye movement (Takagi *et al.*, 1992), though presumably accommodation would also be changing, and one cannot exclude the possibility that the activity of these cells was related to accommodation. One worrying feature of the studies of the neural control of accommodation in the cat is the small size of the accommodation responses — on average about 0.5 D, and not uncommonly less than half that. At least some of the responses must have been less than the depth of focus and so of doubtful functional significance.

Evidence compatible with a relationship of some kind to vergence is that Hiraoka and Shimamura (1989) found electromyogram changes in the ciliary muscle with a latency of about 20 ms after PMLS stimulation.

FRONTAL CORTEX

Although for many years it was believed that only saccadic eye movement was represented in the frontal eye fields of the monkey (Robinson and Fuchs, 1968), there is now evidence for a representation of smooth pursuit (Lynch, 1987; MacAvoy, Gottlieb and Bruce, 1991;

Gottlieb, Bruce and MacAvoy, 1993) and also reports of near-response-related neurons (Gamlin and Yoon, 1995; Gamlin, Yoon and Zhang, 1996). There are several reports of projections from monkey frontal cortex to the Edinger-Westphal nucleus or nearby midbrain areas (Hartman-von Monakow, Akert and Kuenzle, 1979; Leichnetz, 1982; Leichnetz, Spencer and Smith, 1984).

SUPERIOR COLLICULUS AND PRETECTUM

Micro-stimulation of the rostral superior colliculus evokes small changes in accommodation in the anaesthetised and paralysed cat (Sawa and Ohtsuka, 1994; Sato and Ohtsuka, 1996). Moreover, accommodation responses were also produced by micro-stimulation of the posteromedial and posterolateral pretectum, including the nucleus of the optic tract, the posterior pretectal nucleus, the nucleus of the posterior commissure and the adjacent commissural fibres (Konno and Ohtsuka, 1997).

Microstimulation of the superior colliculus of the alert monkey has no effect on accommodation when the monkey is looking at a distant target, but relaxes accommodation (1–2 D) and convergence if the animal is looking at a near target (Billitz and Mays, 1997). Stimulation of deeper layers of the colliculus produces relaxation of accommodation without saccades, whereas saccades are also produced when the less deep layers are stimulated.

Some monkeys whose superior colliculus has been lesioned have a characteristic blank stare. Lawler and Cowey (1986) showed by careful testing of the animal's ability to distinguish patterns with a small number of vertical bars that they were diplopic. Moreover, it was argued that the critical lesion for producing this defect was the adjacent pretectum rather than the superior colliculus itself.

Mays *et al.* (1986) and Judge and Cumming (1986) reported cells in the awake monkey pretectum whose activity was correlated with the near-response, and the exact area in which the near-response related cells lie (perhaps the nucleus of the posterior commissure) needs to be better defined.

Buttner-Ennever *et al.* (1996) have advanced arguments on the basis of anatomical connectivity for a role in the near response for the nucleus of the optic tract (NOT) and the pretectal olivary nucleus (PON), which is surrounded by the NOT in the monkey. The finding that neurons in the pretectal olivary nucleus do not alter their firing rate during the near response (Zhang, Clarke and Gamlin, 1996) is evidence against this proposal.

PONS

Nucleus reticularis tegmenti pontis (NRTP) is best known as part of the subcortical pathway for optokinetic eye movements, but studies in monkeys have revealed neurons in NRTP whose activity is related to smooth pursuit or saccadic eye movements, and Gamlin and Clarke (1995) have described roughly equal numbers of cells in the medial NRTP of the monkey that increased firing rate with convergence ($n = 32$) or increased firing rate during divergence ($n = 33$) — in the sense of looking from near to far, rather than absolute divergence of the visual axes (Figure 3.6) Microstimulation at the site of such near or far neurons often produced changes in vergence angle and accommodation, and in many cases saccades as well. One interesting detail was the statement that the

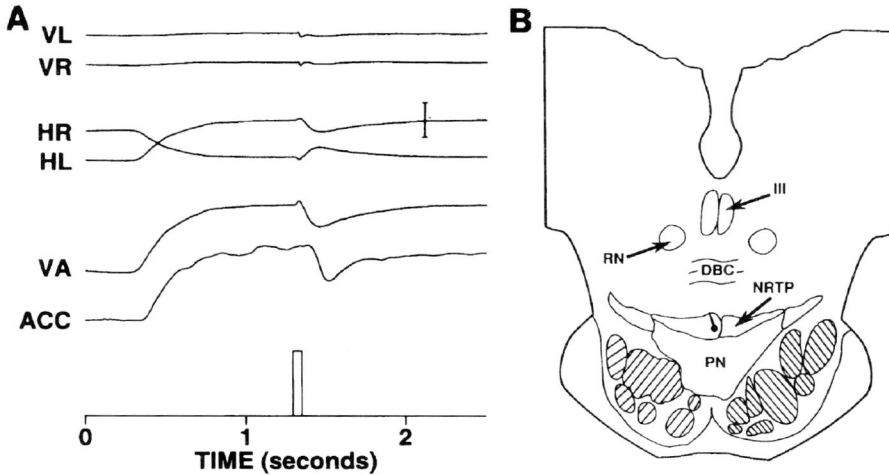

Figure 3.6   (**A**) Stimulation (60 ms, 500 Hz, 25 μA) at the location of a nucleus reticularis tegmenti pontis (NTRP) 'far response' neuron elicits a far response. VL = vertical position of the left eye; VR = Vertical position of the right eye; HL = horizontal position of the left eye; HR = Horizontal position of the right eye; VA = vergence angle; ACC = accommodation. (**B**) schematic diagram of the location of a marking lesion in the left NRTP. The diagram is drawn from a coronal section in the stereotaxic plane through the level of the oculomotor nucleus (III) and NRTP. DBC = decussation of brachium conjunctivium; PN = pontine nuclei; RN = nucleus ruber. (Reprinted from Gamlin and Clarke, 1995, with permission of the American Physiological Society).

activity of both near and far cells appeared correlated with the dark vergence angle of the animal. Gamlin and Clarke (1995) suggest, following Buttner-Ennever and Buttner (1988), that the NRTP cells receive input from the pretectal/superior colliculus area, and perhaps from the frontal eye fields. The monkey NRTP projects to the flocculus, vermis and paravermis of the cerebellum, and also to the fastigial and interpositus nuclei, so there is no known connection outside the cerebellum by which NRTP could influence the near response. One obvious speculation, in line with thinking about the role of the cerebellum in general, would be that such cells may be providing signals to the cerebellum that are necessary for adaptive changes in accommodation or vergence. If so, the finding (Judge, 1987) that monkeys with lesions of flocculus and paraflocculus were able to adapt to prisms (i.e. alter tonic vergence) and to telescopic viewing (i.e., alter their AC/A ratio) suggest that the midline cerebellum might be a better candidate for a role in adaptive aspect of the near response.

Near response related neurons have recently been found in the dorsolateral pontine nucleus (Zhang and Gamlin, 1997).

CEREBELLUM

Hosoba, Bando and Tsukahara (1978) found that stimulation of the contralateral interpositus or fastigial nucleus or ipsilateral interpositus nucleus of the anaesthetised cat

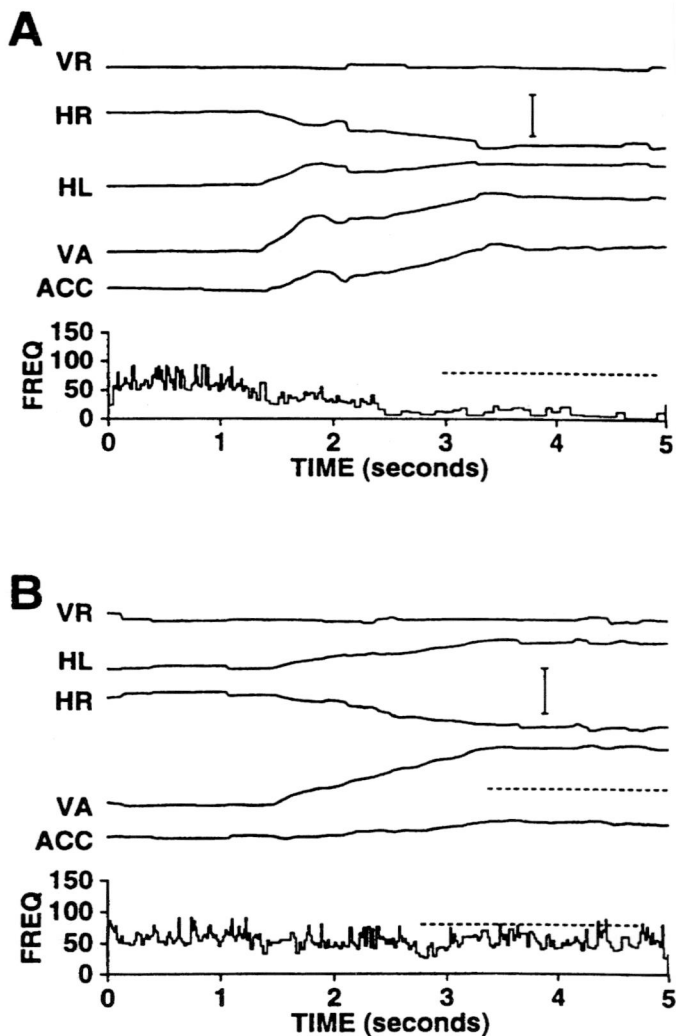

Figure 3.7   Response properties of a 'far' accommodation cell in the monkey interpositus nucleus during normal binocular viewing (**A**) and viewing with accommodation and vergence partially dissociated (**B**). Initial accommodation 1 dioptre, and vergence (VA) commensurate (1 metre-angle). In (**A**) activity decreases as accommodation increases; in (**B**) both activity and accommodation hardly change. Dashed line is at equivalent levels in (**A**)and (**B**). (Reprinted from Zhang and Gamlin, 1998, with permission of the American Physiological Society).

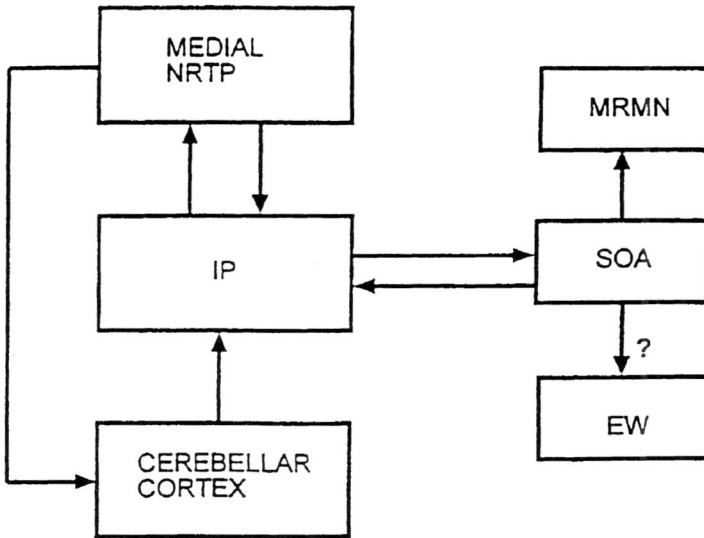

Figure 3.8   Summary diagram of known efferent and afferent connections of posterior interposed nucleus potentially involved in vergence and accommodation. EW = Edinger-Westphal nucleus; MRMN = medial rectus motoneurons; NRTP = nucleus reticularis tegmenti pontis; SOA = supraoculomotor area (midbrain near-response region). (Reprinted from Zhang and Gamlin, 1998, with permission of the American Physiological Society).

elicited accommodation responses with a latency of about 160 ms, and that stimulation of the cerebellar cortex also evoked accommodation, though at a longer latency.

In the monkey, May, Porter and Gamlin (1992) showed that the posterior interposed nucleus and the fastigial nucleus both projected to the supra-oculomotor area in which near response cells are found. Zhang and Gamlin (1998) have described cells in the posterior interposed nucleus of the monkey that decreased their firing rate with divergence and accommodation to far (Figures 3.7 and 3.8). Microstimulation of the area often elicited divergence and accommodation to far.

There seem to be very few reports of accommodation defects associated with cerebellar lesions, and one wonders to what extent this reflects a real absence of effects, and to what extent the issue has escaped attention. Kawasaki *et al.* (1993) have described a patient with slow accommodation release associated with a subtentorial cyst. Accommodation returned to normal when the cyst was removed. Ohtsuka and Sawa (1997) have found that a patient with agenesis of the posterior cerebellar vermis had larger phase lags and lower high frequency gain of accommodation than age matched controls.

MIDBRAIN

Bando *et al.* (1984) described cells in the midbrain of the anaesthetised cat that could be activated both by stimulating the interpositus nucleus and by stimulating the posterior commissure. These cells discharged in correlation with spontaneous fluctuations of accommodation. Some of these cells could be driven antidromically by stimulating the preganglionic fibres to the ciliary ganglion.

In the monkey midbrain, Mays (1984) found cells dorsal and dorsolateral to the oculomotor nucleus whose firing rate was correlated with vergence (or accommodation). By partially dissociating accommodation and convergence, Judge and Cumming (1986) and Zhang, Mays and Gamlin (1992) were able to show that some of these cells responded as though they were driving only vergence, others as though they were driving only accommodation, and others in an intermediate way that could most parsimoniously be accounted for by the assumption that these cells received somewhat mis-matched blur- and disparity-inputs (Zhang, Mays and Gamlin, 1992). Antidromic activation of some of the cells in this midbrain near-response area by stimulation of the medial rectus subdivision of the oculomotor nucleus (Zhang, Gamlin and Mays, 1991) suggests the near-response cells project monosynaptically to motoneurons. Individual cells that were antidromically activated by stimulating the medial rectus subdivision of the oculomotor nucleus were not necessary of the type exclusively related to vergence, but the accommodation coefficients of different cells tended to cancel, so that the overall signal carried by this subset of cells was related predominantly to vergence (Zhang, Gamlin and Mays, 1991; Zhang, Mays and Gamlin, 1992.)

Presumably other midbrain near-response cells project monsynaptically to the Edinger-Westphal nucleus, though the only evidence bearing on this seems to be an observation by Gamlin et al. (1994) in a study of Edinger-Westphal nucleus cells in the monkey. Edinger-Westphal nucleus cells were physiologically identified by the combination of being antidromically activated by third nerve stimulation, but not being ocular motoneurons. What Gamlin et al. (1994) noted was that there were other near-response cells close to these Edinger-Westphal cells that were not identified Edinger-Westphal cells. One of their suggestions, amongst others, was that these might be pre-motor cells projecting to the Edinger-Westphal cells.

REFERENCES

Ball, E.A.W. (1952). A study in consensual accommodation. *American Journal of Optometry and Archives of the American Academy of Optometry*, **29**, 561–574.

Bando, T., Yamamoto, N. and Tsukahara, N. (1984). Cortical neurons related to lens accommodation in posterior lateral suprasylvian area in cats. *Journal of Neurophysiology,* **52**, 879–891.

Bando, T., Tsukuda, K., Yamamoto, N., Maeda, J. and Tsukahara, N. (1984). Physiological identification of midbrain neurons related to lens accommodation in cats. *Journal of Neurophysiology*, **52**, 870–878.

Barlow, H.B., Blakemore, C. and Pettigrew, J.D. (1967). The neural mechanism of binocular depth discrimination. *Journal of Physiology*, **193**, 327–342.

Billitz, M.S. and Mays, L.E. (1997). Effects of microstimulation of the superior colliculus on vergence and accommodation. *Investigative Ophthalmology and Visual Science Supplement*, **38**, S984

Bloom, M. and Berkley, M.A. (1977). Visual acuity and the near point of accommodation in cats. *Vision Research*, **17**, 723–730.

Bobier, W.R., Campbell, M.C. and Hinch, M. (1992). The influence of chromatic aberration on the static accommodative response. *Vision Research*, **32**, 823–832.

Büttner-Ennever, J.A and Büttner, U. (1988). The reticular formation. In: *Neuroanatomy of the Oculomotor System*, edited by J.A. Buttner-Ennever, pp. 119–202. Amsterdam: Elsevier.

Büttner-Ennever, J.A., Cohen, B., Horn, A.K.E. and Reisine, H. (1996). Pretectal projections to the oculo-motor complex of the monkey and their role in eye movements. *Journal of Comparative Neurology*, **366**, 348–359.

Campbell, F.W. (1954). The minimum quantity of light required to elicit the accommodation reflex in man. *Journal of Physiology*, **123**, 357–366.

Campbell, F.W., Robson, J.G. and Westheimer, G. (1959). Fluctuations in accomodation under steady state viewing conditions. *Journal of Physiology*, **145**, 579–594.

Campbell, F.W. and Westheimer, G. (1960). Dynamics of accommodation responses of the human eye. *Journal of Physiology*, **151**, 285–295.

Charman, W.N. and Tucker, J. (1977). Dependence of the accommodation response on the spatial frequency spectrum of the observed object. *Vision Research*, **17**, 129–139.

Charman, W.N. and Tucker, J. (1978). Accommodation and color. *Journal of the Optical Society of America*, **68**, 459–471.

Ciuffreda, K.J. (1991). Accommodation and its anomalies. In: *Vision and Visual Dysfunction*, vol. 1 (ed. W.N. Charman), Macmillan, London, UK, pp. 231–279.

Collins, M., Davis, B, and Wood, J. (1995). Microfluctuations of steady-state accommodation and the cardiopulmonary system *Vision Research*, **35**, 2491–2502.

Cumming, B.G. (1985). The neural control of convergence eye movements and accommodation. *D.Phil. Thesis*, Oxford, UK: University of Oxford.

Cumming, B.G. and Judge, S.J. (1986). Disparity-induced and blur-induced convergence eye movement and accommodation in the monkey. *Journal of Neurophysiology*, **55**, 896–913.

Fincham, E.F. (1951). The accommodation reflex and its stimulus. *British Journal of Ophthalmology*, **35**, 381–393.

Fincham, E.F. and Walton, J. (1957). The reciprocal actions of accommodation and vergence. *Journal of Physiology*, **137**, 488–508.

Fisher, R.F. (1971). The elastic constants of the human lens. *Journal of Physiology*, **212**, 147–180.

Flitcroft, D.I. (1988a). Effects of temporal frequency on the contrast sensitivity of the human accommodation system. *Vision Research*, **28**, 269–278.

Flitcroft, D.I. (1988b). Sensory control of ocular accommodation. *D.Phil. Thesis*, Oxford, UK: University of Oxford.

Flitcroft, D.I. (1990). A neural and computational model for the chromatic control of accommodation. *Visual Neuroscience*, **5**, 547–555.

Flitcroft, D.I. (1991). Accommodation and flicker: evidence of a role for temporal cues in accommodation control? *Ophthalmic and Physiological Optics*, **11**, 81–90.

Flitcroft, D.I. (1998). A model of the contribution of oculomotor and optical factors to emmetropization and myopia. *Vision Research*, **38**, 2869–2879.

Flitcroft, D.I. and Judge, S.J. (1988). The effect of stimulus chromaticity on ocular accommodation in the monkey. *Journal of Physiology*, **398**, 36P.

Flitcroft, D.I. and Morley, J.W. (1997). Accommodation in binocular rivalry. *Vision Research*, **37**, 121–125.

Flitcroft, D.I., Judge, S.J. and Morley, J.W. (1992). Binocular interactions in accommodation control: effects of anisometropic stimuli. *Journal of Neuroscience,* **12**, 188–203.

Gamlin, P.D.R. and Clarke, R.J. (1995). Single-unit activity in the primate nucleus reticularis tegmenti pontis related to vergence and ocular accommodation. *Journal of Neurophysiology*, **73**, 2115–2119.

Gamlin, P.D.R. and Yoon K. (1995). Single-unit activity related to the near-response in area 8 of the primate frontal cortex. *Society of Neuroscience Abstracts*, **21**, 1918.

Gamlin, P.D.R., Zhang, Y., Clendaniel, R.A. and Mays, L.E. (1994). Behavior of identified Edinger-Westphal neurons during ocular accommodation. *Journal of Neurophysiology*, **72**, 2368–2382.

Gamlin, P.D.R., Yoon, K. and Zhang, H. (1996). The role of cerebro-ponto-cerebellar pathways in the control of vergence eye movements. *Eye*, **10**, 167–171.

Glasser, A. and Campbell, M.C. (1998). Presbyopia and the optical changes in the human crystalline lens with age. *Vision Research*, **38**, 209–229.

Gnadt, J.W. and Beyer, J. (1998). Eye movements in depth: what does the monkey's parietal cortex tell the superior colliculus? *NeuroReport*, **9**, 233–238.

Gnadt, J.W. and Mays, L.E. (1995). Neurons in monkey parietal area LIP are tuned for eye-movement parameters in three-dimensional space. *Journal of Neurophysiology*, **73**, 280–297.

Gottlieb, J.P., Bruce, C.J. and MacAvoy, M.G. (1993). Smooth eye movements elicited by microstimulation in the primate frontal eye field. *Journal of Neurophysiology*, **69**, 786–799.

Graham, B. and Judge, S.J. (1999). The effects of spectacle wear in infancy on eye growth and refractive error in the marmoset (*Callithrix jacchus*). *Vision Research*, **39**, 189–206.

Gray, L.S., Winn, B. and Gilmartin, B. (1993). Accommodative microfluctuations and pupil diameter. *Vision Research*, **33**, 2083–2090.

Green, D.G., Powers, M.K. and Banks, M.S. (1980). Depth-of-focus, eye size and visual acuity. *Vision Research*, **20**, 827–835.

Hall, S.E. and Mitchell, D.E. (1991). Grating acuity of cats measured with detection and discrimination tasks. *Behavioral Brain Research*, **44**, 1–9.

Hamasaki, D., Ong, J. and Marg, E. (1956). The amplitude of accommodation in presbyopia. *American Journal of Optometry*, **33**, 3–14.

Hartmann-von Monakow, K., Akert, K. and Kuenzle, H. (1979). Projections of precentral and premotor cortex to the red nucleus and other midbrain areas in *Macaca fascicularis*. *Experimental Brain Research*, **34**, 91–106

Hiraoka, M. and Shimamura, M. (1989). The midbrain reticular formation as an integration center for the near response in the cat. *Neuroscience Research*, **7**, 1–12.

Holmes, G. (1918a). Disturbances of visual orientation. *British Journal of Ophthalmology*, **2**, 449–468.

Holmes, G. (1918b). Disturbances of visual orientation. *British Journal of Ophthalmology*, **2**, 506–516.

Hosoba, M., Bando, T. and Tsukahara, N. (1978). The cerebellar control of accommodation of the eye in the cat. *Brain Research*, **153**, 495–505.

Hughes, A. (1977). The topography of vision in mammals of contrasting life style: comparative optics and retinal organisation. In *Handbook of Sensory Physiology Volume VII/5: The visual system in vertebrates*, edited by F. Crescitelli, pp 613–756. Berlin: Springer.

Hung, G.K. and Semmlow, J.L. (1980). Static behavior of accommodation and vergence: computer simulation of an interactive dual-feedback system. *IEEE Transactions on Biomedical Engineering*, **BME-27**, 439–447.

Hung. G.K., Ciuffreda, K.J. and Semmlow, J.L (1986). Static vergence and accommodation: population norms and orthoptic effects. *Documenta Ophthalmologica*, **62**, 165–179.

Hung, L-F, Crawford, M.L. and Smith, E.L. (1995). Spectacle lenses alter eye growth and the refractive status of young monkeys. *Nature Medicine*, **1**, 761–765.

Jacobson, S.G., Franklin, K.B.J. and McDonald, W.I. (1976). Visual acuity of the cat. *Vision Research*, **16**, 1141–1143.

Jampel, R.S. (1960). Convergence, divergence, pupillary reactions and accommodation of the eyes from faradic stimulation of the macaque brain. *Journal of Comparative Neurology*, **115**, 371–399.

Johnson, C.A. (1976). Effects of luminance and stimulus distance on accommodation and visual resolution. *Journal of Optical Society of America*, **66**, 138–142.

Judge, S.J. (1987). Optically-induced changes in tonic vergence and AC/A ratio in normal monkeys and monkeys with lesions of the flocculus and ventral paraflocculus. *Experimental Brain Research*, **66**, 1–9.

Judge, S.J. (1988). Do target angular size-change and blur cues interact linearly in the control of human accommodation? *Vision Research*, **28**, 263–268.

Judge, S.J. and Cumming, B.G. (1986). Neurons in the monkey midbrain with activity related to vergence eye movement and accommodation. *Journal of Neurophysiology*, **55**, 915–930.

Judge, S.J. and Graham, B. (1995). Differential ocular growth of infant marmoset (*Callithrix jacchus jacchus*) induced by optical anisometropia combined with alternating occlusion. *Journal of Physiology*, **485**, 27P.

Judge, S.J. and Miles, F.A. (1985). Changes in the coupling between accommodation and vergence eye movements induced in human subjects by altering the effective interocular separation. *Perception*, **14**, 617–629.

Kawasaki, T., Kiyosawa, M., Fujino, T. and Tokoro, T. (1993). Slow accommodation release with a cerebellar lesion. *British Journal of Ophthalmology*, **77**, 678.

Kersten, D. and Legge, G.E. (1983). Convergence accommodation. *Journal of Optical Society of America*, **73**, 332–338.

Konno, S. and Ohtsuka, K. (1997). Accommodation and pupilloconstriction areas in the cat midbrain. *Japanese Journal of Ophthalmology*, **41**, 43–48.

Krishnan, V.V., Phillips, S. and Stark L. (1973). Frequency analysis of accommodation, accommodative vergence and disparity vergence. *Vision Research*, **13**, 1545–1554.

Kruger, P.B. and Pola, J. (1985). Changing target size is a stimulus for accommodation. *Journal of the Optical Society of America*, **2**, 1832–1835.

Kruger, P.B. and Pola, J. (1986). Stimuli for accommodation: blur, chromatic aberration and size. *Vision Research*, **26**, 957–971.

Kruger, P.B., Mathews, S., Aggarwala, K.R. and Sanchez, N. (1993). Chromatic aberration and ocular focus: Fincham revisited. *Vision Research*, **33**, 1397–1411.

Kruger, P.B., Nowbotsing, S., Aggarwala, K.R. and Mathews, S. (1995a). Small amounts of chromatic aberration influence dynamic accommodation. *Optomometry and Vision Science*, **72**, 656–666.

Kruger, P.B., Mathews, S., Aggarwala, K.R., Yager, D. and Kruger, E.S. (1995b). Accommodation responds to changing contrast of long, middle and short spectral-waveband components of the retinal image. *Vision Research*, **35**, 2415–2429.

Kruger, P.B., Mathews, S., Katz, M., Aggarwala, K.R. and Nowbotsing, S. (1997). Accommodation without feedback suggest directional signals specify ocular focus. *Vision Research*, **37**, 2511–2526.

Lawler, K.A. and Cowey, A. (1986). The effects of pretectal and superior colliculus lesions on binocular vision. *Experimental Brain Research*, **63**, 402–408.

Leichnetz, G.R. (1982). The medial accessory nucleus of Bechterew: a cell group within the anatomical limits of the rostral oculomotor complex receives a direct prefrontal projection in the monkey. *Journal of Comparative Neurology*, **210**, 147–151.

Leichnetz, G.R., Spencer, R.F. and Smith, D.J. (1984). Cortical projections to nuclei adjacent to the oculomotor complex in the medial diencephalic tegmentum in the monkey. *Journal of Comparative Neurology*, **228**, 359–387.

Lynch, J.C. (1987). Frontal eye field lesions in monkeys disrupt visual pursuit. *Experimental Brain Research*, **68**, 437–441.

MacAvoy, M.G., Gottlieb, J.P. and Bruce C.J. (1991). Smooth-pursuit eye movement representation in the primate frontal eye field. *Cerebral Cortex*, **1**, 95–102.

Maekawa, H. and Ohtsuka, K. (1993). Afferent and efferent connections of the cortical accommodation area in the cat. *Neuroscience Research*, **17**, 315–323.

Manor, R.S., Heilbronn, Y.D., Sherf, I and Ben-Sira, I. (1988). Loss of accommodation produced by peristriate lesion in man? *Journal of Clinical Neuro-Ophthalmology*, **8**, 19–23.

Marran, L. and Schor, C.M. (1998). Lens induced aniso-accommodation. *Vision Research*, **38**, 3601–3619.

Mathews, S. and Kruger, P.B. (1994). Spatiotemporal transfer function of human accommodation. *Vision Research*, **34**, 1965–1980.

May, P.J., Porter, J.D. and Gamlin, P.D.R. (1992). Interconnections between the primate cerebellum and midbrian near-response regions. *Journal of Comparative Neurology*, **315**, 98–116.

Mays, L.E. (1984). Neural control of vergence eye movements: convergence and divergence neurons in midbrain. *Journal of Neurophysiology*, **51**, 1091–1108.

Mays, L.E. and Gamlin, P.D.R. (1995). Neuronal circuitry controlling the near response. *Current Opinion in Neurobiology*, **5**, 763–768.

Mays, L.E., Porter, J.D., Gamlin, P.D.R. and Tello, C.A. (1986). Neural control of vergence eye movements: neurons encoding vergence velocity. *Journal of Neurophysiology*, **56**, 1007–1021.

McBrien, N.A. and Millodot, M. (1986). The effect of refractive error on the accommodative response gradient. *Ophthalmic and Physiological Optics*, **6**, 145–149.

McBrien, N.A. and Millodot, M. (1988). Differences in adaptation of tonic accommodation with refractive state. *Investigative Ophthalmology and Visual Science*, **29**, 460–469.

McLin, L.N., Schor, C.M. and Kruger P.B. (1988). Changing size (looming) as a stimulus to accommodation and vergence. *Vision Research*, **28**, 883–898.

Michael, C.R. (1978). Color vision mechanisms in monkey striate cortex: dual opponent cells with concentric receptive fields. *Journal of Neurophysiology*, **46**, 587–604.

Miles, F.A., Judge S.J. and Optican, L.M. (1987). Optically induced changes in the couplings between vergence and accommodation. *Journal of Neuroscience*, **7**, 2576–2589.

Murphy, C.J., Bellhorn, R.W., Williams, T., Burns, M.S., Schaeffel, F. and Howland, H.C. (1990). Refractive state, ocular anatomy and accommodative range of the sea otter (*Enhydra lutris*). *Vision Research*, **30**, 23–32.

O'Neill, W.P. and Stark, L.(1968). Triple function ocular monitor. *Journal of the Optical Society of America*, **58**, 570–573.

Ogle, K.N. (1937). Relative sizes of the ocular images of the two eyes in asymmetric convergence. *Archives of Ophthalmology*, **22**, 1046–1066.

Ohtsuka, K. and Sato, A. (1996). Descending projections from the cortical accommodation area in the cat. *Investigative Ophthalmology and Visual Science*, **37**, 142–1436.

Ohtsuka, K. and Sawa., M. (1997). Frequency characteristics of accommodation in a patient with agenesis of the posterior vermis and normal subjects. *British Journal of Ophthalmology*, **81**, 476–480.

Owens, D.A. (1980). A comparison of accommodative responsiveness and contrast sensitivity for sinusoidal gratings. *Vision Research*, **20**, 159–167

Owens, D.A. and Wolfe, J.M. (1985). Accommodation for flickering stimuli. *Ophthalmic and Physiological Optics*, **5**, 291–296.

Pettigrew, J.D., Nikara, T. and Bishop, P.O. (1968). Binocular interaction on single units in cat striate cortex: simultaneous simulation by single moving slits with receptive fields in correspondence. *Experimental Brain Research*, **6**, 391–410.

Phillips, S., Shirachi, D. and Stark, L. (1972). Analysis of accommodation response times using histogram information. *American Journal of Optometry and Archives of the American Academy of Optometry*, **49**, 389–401.

Poggio, G.F. and Fischer, B. (1977). Binocular interaction and depth sensitivity of striate and prestriate cortical neurons of the behaving rhesus monkey. *Journal of Neurophysiology*, **40**, 1392–1405.

Robinson, D. and Fuchs, A.F. (1968). Eye movements evoked by stimulation of frontal eye fields. *Journal of Neurophysiology*, **32**, 637–648.

Rosenberg, R., Flax, N., Brodsky, B. and Abelman, L. (1953). Accommodation levels under conditions of asymmetric convergences. *American Journal of Optometry and Archives of the American Academy of Optometry*, **30**, 244–254.

Rosenfield, M., Ciuffreda, K.J., Hung G.K. and Gilmartin, B. (1994). Tonic accommodation: a review. II. Accommodative adaptation and clinical aspects. *Ophthalmic and Physiological Optics*, **14**, 265–277.

Sakata, H., Shibutani, H. and Kawano, K. (1980). Spatial properties of visual fixation neurons in posterior parietal association cortex of the monkey. *Journal of Neurophysiology*, **43**, 1654–1672.

Sakata, H., Shibutani, H. and Kawano, K. (1983). Functional properties of visual tracking neurons in posterior parietal association cortex of the monkey. *Journal of Neurophysiology*, **49**, 1364–1380.

Sakata, H., Shibutani, H., Kawano, K. and Harrington, T.L. (1985). Neural mechanisms of space vision in the parietal association cortex of the monkey. *Vision Research*, **25**, 453–463.

Sato, A. and Ohtsuka, K. (1996). Projection from the accommodation-related area in the superior colliculus of the cat. *Journal of Comparative Neurology*, **367**, 465–476.

Sawa, M. and Ohtsuka, K. (1994). Lens accommodation evoked by microstimulation of the superior colliculus in the cat. *Vision Research*, **34**, 975–981.

Sawa, M., Maekawa, H. and Ohtsuka, K. (1992). Cortical area related to lens accommodation in cat. *Japanese Journal of Ophthalmology*, **36**, 371–379.

Schaeffel, F. and Howland, H.C. (1987). Corneal accommodation in chick and pigeon. *Journal of Comparative Physiology A*, **160**, 375–384.

Schor, C.M. (1999). The influence of interactions between accommodation and convergence on the lag of accommodation. *Ophthalmic and Physiological Optics*, **19**, 134–150.

Schor, C.M. and Kotulak, J.C. (1986). Dynamic interactions between accommodation and convergence are velocity sensitive. *Vision Research*, **26**, 927–942.

Schor, C.M. and Tsuetaki, T.K. (1987). Fatigue of accommodation and vergence modifies their mutual interactions. *Investigative Ophthalmology and Visual Science*, **28**, 1250–1259.

Schor, C.M., Kotulak, J.C. and Tsuetaki, T. (1986). Adaptation of tonic accommodation reduces accommodative lag and is masked in darkness *Investigative Ophthalmology and Visual Science*, **27**, 820–827.

Semmlow, J.L. (1982). Oculomotor response to near stimuli: the near triad. In*: Models of Oculomotor Behavior*, edited by B. Zuber, CRC Press, Bocca Raton, FL, USA, pp. 161–191.

Sivak, J.G. (1980). Accommodation in vertebrates: a contemporary survey. In: *Current Topics in Eye Research* Volume III, edited by J.A. Zadunaisky and H. Davson, pp 281–330. San Diego: Academic Press.

Smith, E.L. 3rd and Harwerth, R.S. (1984). Behavioral measurements of accommodative amplitude in rhesus monkeys. *Vision Research*, **24**, 1821–1827.

Smithline, L.M. (1974). Accommodative responses to blur. *Journal of the Optical Society of America*, **64**, 1512–1526.

Spencer, R.W. and Wilson, K.W. (1954). Accommodative response in asymmetric convergence. *American Journal of Optometry and Archives of the American Academy of Optometry*, **31**, 498–505.

Stark, L., Takahashi, Y. and Zames, G. (1965). Nonlinear servoanalysis of human lens accommodation. *IEEE Transactions on Systems Science and Cybernetics*, **SSC-3**, 75–83.

Takagi, M., Toda, H., Yoshizawa, T., Hara, N., Ando, T., Abe, H. *et al.* (1992). Ocular convergence-related neuronal responses in the lateral suprasylvian area of alert cats. *Neuroscience Research*, **15**, 229–234.

Thibos, L.N. and Levick, W.R (1982). Astigmatic visual deprivation in cat: behavioral, optical and retinophysiological consequences. *Vision Research*, **22**, 43–53.

Thorn, F. Grennan, M. and Heath, D. (1983). Consensual accommodation. *Current Eye Research*, **3**, 711–716.

Toates, F.M. (1972). Accommodative function of the human eye. *Physiological Reviews*, **52**, 828–863.

Toda, H., Takagi, M., Yoshizawa, T. and Bando, T. (1991). Disjunctive eye movement evoked by microstimulation in an extrastriate cortical area of the cat. *Neuroscience Research*, **12**, 300–306.

Troilo, D., Howland, H.C. and Judge, S.J. (1993). Visual optics and retinal cone topography in the common marmoset. *Vision Research*, **33**, 1301–1310.

Ts'o, D.Y. and Gilbert, C.D. (1988). The organisation of chromatic and spatial interactions in the primate striate cortex. *Journal of Neuroscience*, **8**, 1712–1727.

Tucker, J. and Charman, W.N. (1979). Reaction and response times for accommodation. *American Journal of Optometry and Physiological Optics*, **56**, 490–503.

Walls, G.L. (1942). *The vertebrate eye and its adaptive radiation.* Michigan: Cranbrook Institute of Science.

Ward, P.A. (1987). The effect of spatial frequency on steady-state accommodation. *Ophthalmic and Physiological Optics*, **7**, 211–217.

Ward, P.A. and Charman, W.N. (1985). Effect of pupil size on steady-state accommodation. *Vision Research*, **25**, 1317–1326.

Watanabe, Y., Kusunoki, M. and Sakata, H. (1990). Neurons in monkey parietal association cortex sensitive to depth movement. *Nippon Ganka Gakkai Zasshi*, **94**, 1031–1039 (in Japanese).

Westheimer, G. (1963). Amphetamine, barbiturates and accommodative convergence. *Archives of Ophthalmology*, **70**, 830–836.

White, J.M. and Wick, B. (1995). Accommodation in humans with juvenile macular degeneration. *Vision Research*, **35**, 873–880.

Whiteside, T.C.D. (1952). Accommodation of the human eye in a bright and empty visual field. *Journal of Physiology*, **118**, 65P.

Winn, B. (1987). Studies in binocular accommodation. Ph.D Thesis. Glasgow Institute of Technology.

Winn, B., Pugh, J.R., Gilmartin, B. and Owens, H. (1990). Arterial pulse modulates steady-state ocular accommodation. *Current Eye Research*, **9**, 971–975.

Wolfe, J.M. and O'Connell, K.M. (1987). Adaptation of the resting state of accommodation. *Investigative Ophthalmology and Visual Science*, **28**, 992–996.

Wolfe, J.M. and Owens, D.A. (1981). Is accommodation colorblind? Focusing chromatic contours. *Perception*, **10**, 53–62.

Yamada, T and Ukai, K. (1997). Amount of defocus is not used as an error signal in the control system of accommodation dynamics. *Ophthalmic and Physiological Optics*, **17**, 55–60.

Yoshida, T. and Watanabe, A. (1969). Analysis of interaction between accommodation and vergence feedback control systems of human eyes. *Bulletin of the NHK Broadcast Science Research Laboratories*, **3**, 72–80 (in Japanese).

Zeki, S.M. (1974). Cells responding to changing image size and disparity in the cortex of the rhesus monkey. *Journal of Physiology*, **242**, 827–841.

Zhang, H.Y. and Gamlin, P.D.R. (1997). The dorsolateral pontine nucleus of the primate: neurons related to vergence and accommodation. *Society of Neuroscience Abstracts.*, **23**, 1556.

Zhang, H. and Gamlin, P.D.R. (1998). Neurons in the posterior interposed nucleus of the cerebellum related to vergence and accommodation. I. Steady-state characteristics. *Journal of Neurophysiology*, **79**, 1255–1269.

Zhang, Y., Gamlin, P.D.R. and Mays, L.E. (1991). Antidromic identification of midbrain near response cells projecting to the oculomotor nucleus. *Experimental Brain Research*, **84**, 525–528.

Zhang, Y., Mays, L.E. and Gamlin P.D.R. (1992). Characteristics of near response cells projecting to the oculomotor nucleus. *Journal of Neurophysiology*, **67**, 944–960.

Zhang, H., Clarke, R.J. and Gamlin, P.D.R. (1996). Behavior of luminance neurons in the olivary pretectal nucleus during the pupillary near response. *Experimental Brain Research*, **112**, 158–162.

4 Functions of the Edinger-Westphal Nucleus

Paul D. R. Gamlin

Vision Science Research Center, 626 Worrell Building, University of Alabama at Birmingham, Birmingham, Alabama 35294, USA

The Edinger-Westphal nucleus (EW), the parasympathetic component of the oculomotor nuclear complex controls several ocular functions by way of its projection to the postganglionic neurons of the ciliary ganglion. These neurons in turn innervate the iris, the ciliary muscle, and the smooth muscle of choroidal blood vessels. It has been shown in both mammals and birds that the EW controls pupilloconstriction and accommodation. It has also been shown in birds, but not so clearly in mammals, that the EW controls choroidal blood flow. Recent studies in primates and pigeons have revealed much about the physiology and anatomy underlying the control of these intraocular functions by the EW. Electrical microstimulation studies of the EW and the intracranial portion of the third nerve in alert primates have confirmed earlier reports that electrical stimulation of the EW elicits pupilloconstriction and ocular accommodation. These studies have also allowed the characteristics and latencies of these two responses to be more precisely described. Single-unit recordings from preganglionic EW neurons related to either pupil diameter or to ocular accommodation have permitted a detailed description of their relationship to these two ocular functions. Further, in birds, electrical microstimulation and lesion studies of the EW have shown that it modulates choroidal blood flow. Anatomical studies in the pigeon have revealed three distinct subdivisions of the EW that control respectively, pupilloconstriction, ocular accommodation, and choroidal blood flow. These subdivisions are selectively innervated by separate afferent pathways related to these three functions. To date, comparable studies have not been conducted in primates. Future studies of the functions of the vertebrate EW can be expected to build upon the electrophysiological approach that has been so successful in primates and the anatomical approach that has been so successful in pigeons.

KEY WORDS: pupillary light reflex; ocular accommodation; choroidal blood flow; ciliary ganglion

INTRODUCTION

The Edinger-Westphal nucleus (EW) is a distinct nucleus lying immediately dorsal to the somatic subdivisions of the oculomotor complex. It was first described in a developmental study of human neuroanatomical material by Edinger (1885) and, a short time afterward, in a neuropathological study by Westphal (1887). While both authors recognised this

Correspondence: Dr. P.D.R. Gamlin, Vision Science Research Center, 626 Worrell Building, University of Alabama at Birmingham, Birmingham, AL 35294, USA. Tel: +1 205 934-0322; Fax: +1 205 934-5725.

cytoarchitecturally distinct nucleus located dorsal to the oculomotor nucleus, the study by Westphal additionally led to the suggestion that the EW was involved in the innervation of the iris and possibly other intraocular muscles. This suggestion was made because Westphal studied tissue from an individual who, while alive, had been diagnosed with complete external ophthalmoplegia but whose pupils had constricted for near vision and was thus presumed to have an intact pupillomotor nucleus. Westphal found that the neurons of the large-celled somatic oculomotor nucleus had degenerated, and he correlated this loss with the observed external ophthalmoplegia. However, he found sparing of the neurons of the smaller-celled dorsal nucleus (now known as the Edinger-Westphal nucleus), and he correlated these spared neurons with the spared pupillomotor function.

Since these pioneering studies, additional studies have clearly shown in many vertebrate classes that the EW is the preganglionic, parasympathetic component of the oculomotor nuclear complex, and is the central source of parasympathetic innervation of the iris, the ciliary body, and certain additional intraocular muscles and tissues. However, in some species, the cells of the cytoarchitecturally-defined EW are not preganglionic neurons but instead have central projections while preganglionic neurons are actually located outside of the EW. For example, cells in the EW of cats are reported to project to the spinal cord and cerebellum (Sugimoto, Itoh and Mizuno, 1978; Loewy and Saper, 1978; Loewy, Saper and Yamodis, 1978; Burde, Parelman and Luskin, 1982; Roste and Dietrichs, 1988), while the ciliary ganglion receives its central input from a collection of preganglionic cells along the midline of the rostral mesencephalon and in the ventral segmental area (Sugimoto, Itoh and Mizuno, 1977; Toyoshima, Kawana and Sakai, 1980; Kuchiiwa, Kuchiiwa and Nakagawa, 1994). This has led to much confusion in the literature, but this confusion does not extend to studies of primates and birds since in both of these groups of animals there is a close correspondence between the EW and the preganglionic, parasympathetic neurons (e.g. Narayanan and Narayanan, 1976; Akert et al., 1980).

In addition, for studies of the physiology of the neurons of the EW, the alert primate has proved most useful for a number of reasons. Alert primates show robust pupillary and accommodative responses (Gamlin et al., 1994; Gamlin, Zhang and Clarke, 1995) and their neural responses are not affected by anaesthesia as was the case for earlier studies in anaesthetised animals (e.g. Sillito and Zbrozyna, 1970a). Also, for anatomical studies of the sources of input to the EW, the primate has yielded more reliable results than those obtained in other mammals such as cats and rabbits. This is because the preganglionic neurons in primates are predominantly confined to the cytoarchitecturally-defined EW and thus anterograde studies documenting inputs to the EW can more reliably be considered as demonstrating inputs to the preganglionic neurons. However, despite these advantages of the primate, the pigeon has proved itself by far the most amenable species for detailed anatomical studies of the sources of input to the vertebrate EW. This is for a number of reasons. First, retrograde pathway tracing studies (Cowan and Wenger, 1968; Narayanan and Narayanan, 1976; Lyman and Mugnaini, 1980) have confirmed that, in contrast to most mammals, the avian EW can be clearly delineated cytoarchitecturally, is the sole source of preganglionic afferents to the ciliary ganglion, and does not project to the spinal cord. Instead, in the pigeon a cell group that lies immediately adjacent to but is clearly distinct from the EW projects to the spinal cord (Cabot, Reiner and Bogan, 1982). Second, the avian midbrain and pretectal nuclei have undergone considerable hypertrophy, and many are more sharply defined cytoarchitecturally than in mammals (Ariens-Kappers, Huber and Crosby, 1936; Kuhlenbeck, 1939). Third, the pigeon is a highly visual animal with well-developed neural circuits for the control of the pupil, ocular accommodation, and choroidal blood flow (Gamlin and Reiner, 1991). The above characteristics have resulted in significant advances being made in our understanding of the central visual pathways involved in the parasympathetic control of these ocular functions in this species (Gamlin and Reiner, 1991). Since the peripheral mechanisms regulating these ocular functions can be expected to be evolutionarily conserved and similar in most vertebrates, findings in the pigeon are very relevant to mammals and other vertebrate classes.

For the reasons described in this introduction, this review will concentrate on recent results in pigeons and primates. These results are combined to present as cohesive a view as is currently possible on the physiology and anatomy of the vertebrate Edinger-Westphal nucleus. Results from other species will be considered when they contribute to a specific understanding of a particular function of the EW.

CYTOARCHITECTURE AND PROJECTIONS OF THE EDINGER-WESTPHAL NUCLEUS

BIRDS

In birds, the EW is located dorsolateral to the somatic oculomotor nucleus from which it is clearly distinct (Figure 4.1A). The EW, which is also known in birds and reptiles as the accessory oculomotor nucleus, extends caudally for approximately 700 μm from the rostral pole of the oculomotor complex. It lies close to the midline and is bordered on all sides except ventromedially by the central grey. In transverse section, the nucleus appears elliptical, with a maximum diameter of 700 μm, and is surrounded by a neuropil about 50 μm wide. The neurons in this nucleus are of two major classes: neurons with large spherical or elliptical somata, approximately 19–30 μm in mean diameter, and neurons with smaller, fusiform cell bodies, 14–21 μm in mean diameter (Figure 4.1B). There are about 500 of the larger cells located mainly in the lateral portion of the EW and between 700 and 1200 of the smaller cells generally located medially (Gamlin *et al.*, 1984; Reiner *et al.*, 1991). Based on these observations and upon studies of afferents, the avian EW has been divided into two cytoarchitecturally distinct subdivisions, a medial (EWm) and a lateral (EWl) (Gamlin, Reiner and Karten, 1982; Reiner *et al.*, 1983, 1991; Gamlin *et al.*, 1984). The majority of the cells in both subdivisions in chicken give rise to parasympathetic preganglionic fibres to the ciliary ganglion and are retrogradely labelled by horseradish peroxidase (HRP) injections of this ganglion (Narayanan and Narayanan, 1976; Lyman and Mugnaini, 1980). The few cells that are unlabelled by these HRP injections may be local interneurons since they are small and have no described extrinsic projections (Lyman and Mugnaini, 1980).

In the avian ciliary ganglion, there are two classes of neurons, choroidal and ciliary. They are so named because the choroidal neurons innervate the smooth muscle of choroidal blood vessels (Marwitt, Pilar and Weakly, 1971; Pilar and Tuttle, 1982), while the ciliary neurons innervate the ciliary body and the iris sphincter muscle, which in birds

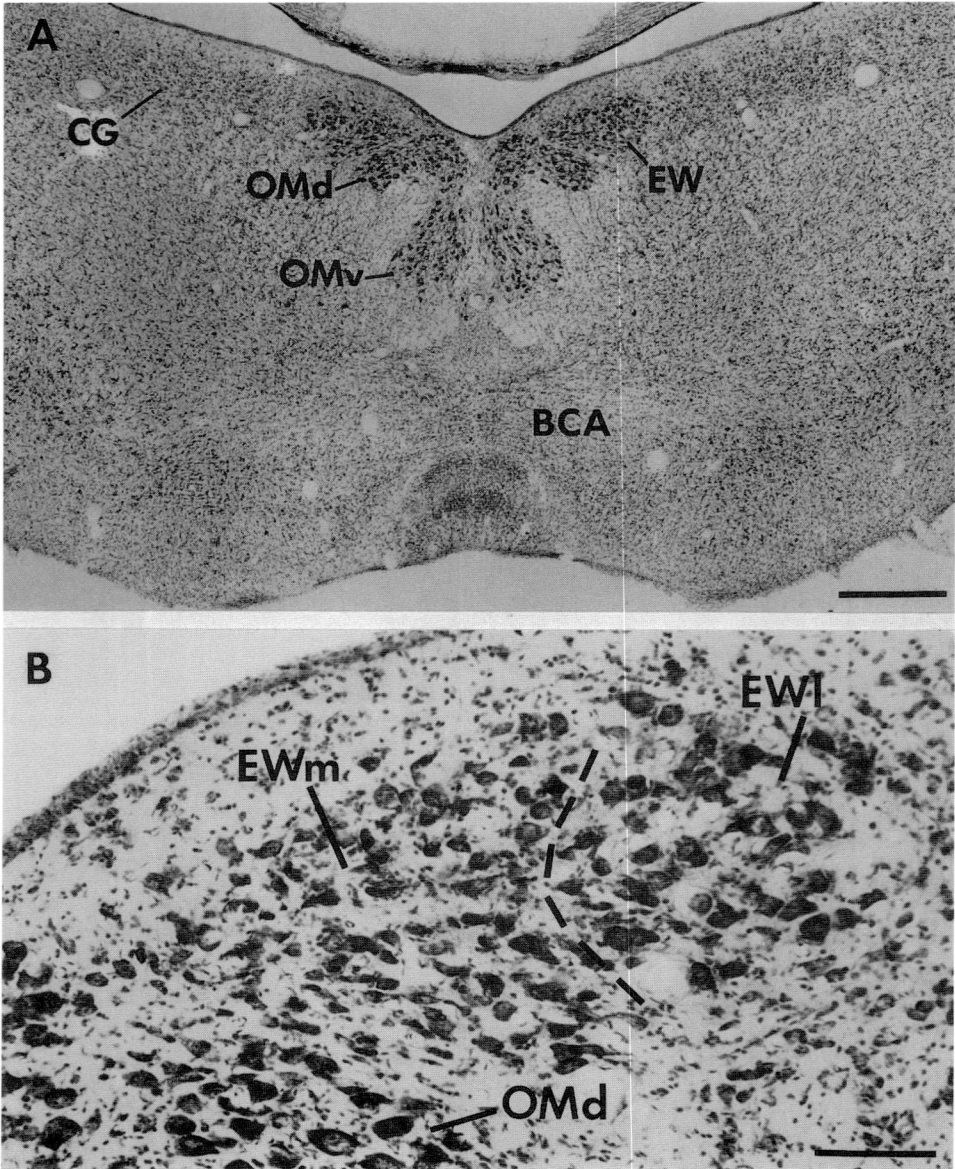

Figure 4.1   (**A**) low-power photomicrograph of the pigeon mesencephalon at the level of the Edinger-Westphal nucleus (EW). (**B**) A higher power photomicrograph of the EW. The nucleus is divided into a medial (EWm) and a lateral (EWl) subdivision. Most of the neurons in the EWm possess fusiform cell bodies of 14–21 μm in average diameter. In contrast, most of the neurons in the EWl possess more spherical cell bodies of 19–30 μm in average diameter. BCA = brachium conjunctivum ascendens; CG = central gray; OMd = Oculomotor nucleus, pars dorsalis; OMV = Oculomotor nucleus, pars ventralis. Scale bars, **A** = 500 μm; **B** = 100 μm. (Reprinted from Gamlin *et al.*, 1984, with permission of Wiley-Liss, Inc.).

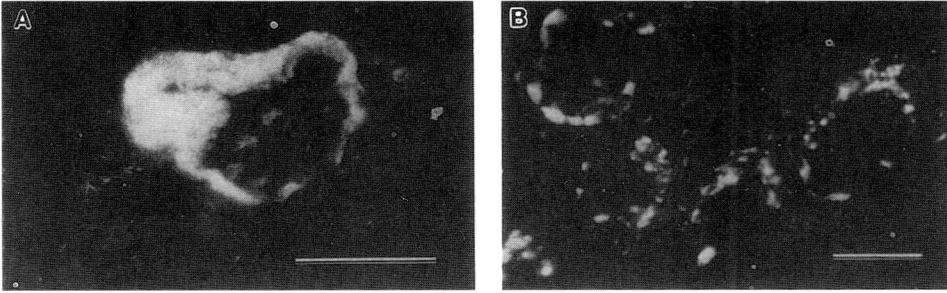

Figure 4.2   Immunohistochemical staining for enkephalin reveals fluorescently-labelled cap-like or calyciform endings (**A**) and boutonal (**B**) endings made by preganglionic neurons on postganglionic neurons within the ciliary ganglion. Scale bars = 20 μm. (Modified from Reiner *et al.*, 1983)

are composed of striated muscles (Pilar and Tuttle, 1982). In addition, there are two classes of terminals in the ciliary ganglion that arise from EW neurons. These terminals contain both acetylcholine and the peptides substance P and enkephalin (Pilar and Tuttle, 1982; Erichsen *et al.*, 1982; Reiner *et al.*, 1991). As shown in Figure 4.2, one type of terminal, the so-called cap-like or calyx endings, envelops a portion of the soma of ciliary neurons. The other type of terminal, the so-called boutonal endings, makes multiple, boutonal synaptic endings on choroidal neurons. Lesions of EWm and EWl differentially affect these terminals in the ciliary ganglion (Reiner *et al.*, 1991). Specifically, lesions of EWl result in a significant decrease in the cap-like terminals in the ciliary ganglion while lesions of EWm result in a significant decrease in the boutonal terminals. From these results it has been concluded that cells in EWl project to the neurons in the ciliary ganglion that innervate the iris sphincter muscle and the ciliary body, while cells in EWm project to the ciliary ganglion neurons that innervate choroidal blood vessels (Reiner *et al.*, 1991). Thus, neurons in the EWl mediate pupilloconstriction and accommodation, while neurons in the EWm modulate choroidal blood flow. Figure 4.3 presents a summary of the EW subdivisions, the two types of terminals and neurons within the ciliary ganglion, and their efferent targets. This figure also shows the sources of afferents to these subdivisions that are described in more detail in each relevant section. This figure thus serves to lay the foundation for a discussion of the role of the EW in the neural control of the pupil, of ocular accommodation, and of choroidal blood flow.

PRIMATES

The primate EW lies dorsal to the somatic subdivisions of the oculomotor nucleus (Figure 4.4). It is composed of relatively large spherical and ovoid cells (approximately 25–40 μm in diameter) and other spindle-shaped neurons (15 μm–30 μm in diameter) (e.g. Warwick, 1954). The preganglionic neurons within the EW were initially identified by a retrograde degeneration study (Warwick, 1954). Subsequently these neurons have been identified by retrograde neuroanatomical tracer studies following injections into the

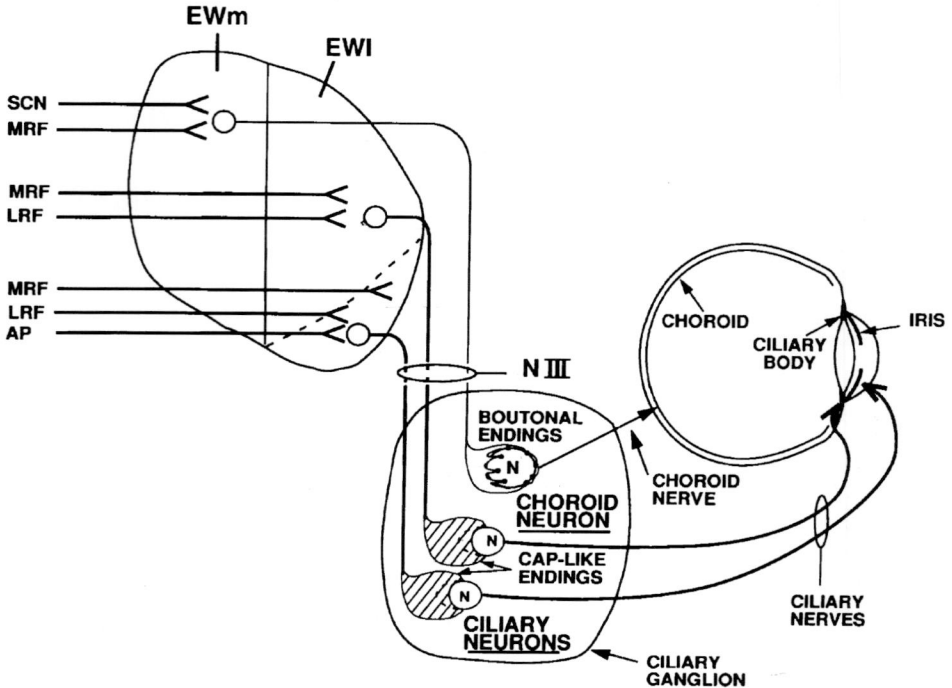

Figure 4.3   A summary diagram showing the efferent projections of the various subdivisions of the Edinger-Westphal nucleus (EW) (shown schematically in horizontal section) to postganglionic neurons within the ciliary ganglion. The targets of the ciliary and choroidal nerves, the iris, the ciliary body, and the choroid, are also identified. In addition, afferents to the specific EW subdivisions that have been described in previous studies are identified (Gamlin, Reiner and Karten, 1982; Gamlin *et al.*, 1984; Gamlin and Reiner, 1991). Abbreviations: AP = Area Pretectalis; EWl = Edinger-Westphal nucleus, pars lateralis; EWm = Edinger-Westphal nucleus, pars medialis; LRF = lateral mesencephalic reticular formation; MRF = medial mesencephalic reticular formation; SCN = suprachiamatic nucleus. (Reprinted from Gamlin and Reiner, 1991, with permission of Wiley-Liss, Inc.).

ciliary ganglion of either HRP (Burde and Loewy, 1980; Clarke, Coimbra and Alessio, 1985b), [^{125}I] wheat germ agglutinin (WGA) (Akert *et al.*, 1980), WGA-HRP (Sun and May, 1993), or fluorescent tracers (Ishikawa, Sekiya and Kondo, 1990). All of these studies reported that labelled, preganglionic neurons are generally the larger, more spherical neurons of the EW and are only slightly smaller than the somatic motoneurons of the oculomotor complex.

    Based on the number of retrogradely labelled cells and the estimated number of parasympathetic axons in the oculomotor nerve, Burde and Loewy (1980) suggested in their study that there were 300–400 preganglionic neurons in EW. However, based on the results of Akert *et al.* (1980), Ishikawa, Sekiya and Kondo (1990), and upon personal observations, the number of preganglionic neurons in the primate EW is more likely to

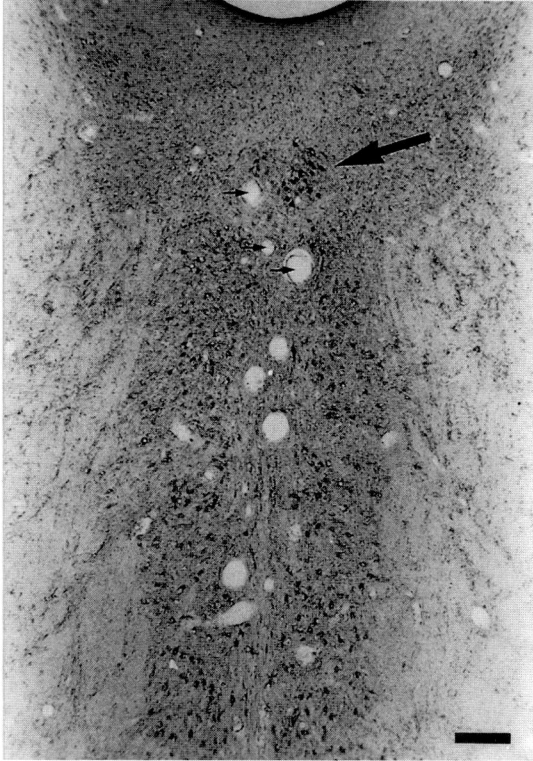

Figure 4.4   A photomicrograph of a coronal section through the Rhesus oculomotor nuclear complex stained for Nissl substance, showing preganglionic Edinger-Westphal nucleus neurons (arrow) labelled with wheat germ agglutinin-horseradish peroxidase (WGA-HRP) as a result of retrograde transneuronal transport following an intravitreal injection of this tracer. Scale bar = 250 μm.

be in the range of 800–1200. This latter estimate is also more consistent with the results reported above for the pigeon.

Throughout much of the length of the oculomotor nucleus, the EW can be seen as the paired medial visceral cell columns of the oculomotor nuclear group. However, anterior to the oculomotor nucleus, this pair of columns merges along the midline to form a contiguous cell group containing cells that are more fusiform than those in the more posterior regions of the EW. As a consequence, this cephalic extension of the EW has often been considered to be separate from the EW and has been termed the anteromedian nucleus by some authors (e.g. Burde and Loewy, 1980; Ishikawa, Sekiya and Kondo, 1990). However, others consider this region a rostral extension of the EW and identify neurons located more ventrally as the anteromedian nucleus (Warwick, 1954; Akert et al., 1980). Based on the retrograde labelling of preganglionic EW neurons, it is clear that the anteriorly located, spindle-shaped, preganglionic neurons that are labelled are

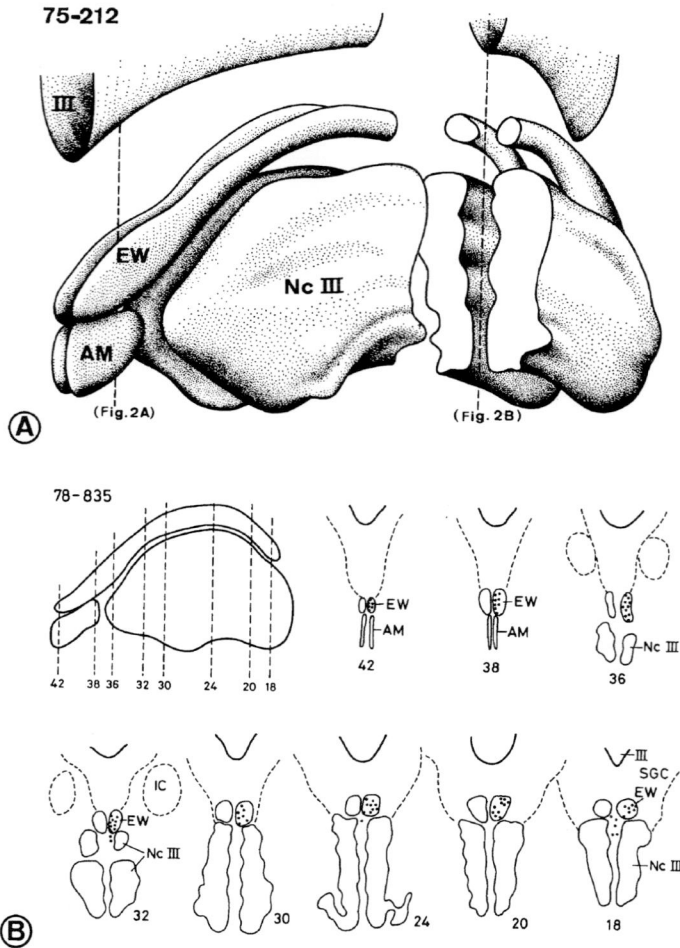

Figure 4.5   (**A**) Three-dimensional reconstruction of the primate oculomotor complex on the basis of serial sections stained for Nissl substance. (**B**). Location of retrogradely labelled cells following injection of [^{125}I]-labelled wheat germ agglutinin into the right ciliary ganglion. Abbreviations: AM = anteromedian nucleus; EW = Edinger-Westphal nucleus; NcIII = oculomotor nucleus; SGC = substantia grisea centralis; III = third ventricle. (Reprinted from Akert *et al.*, 1980, with permission of Elsevier Science NL).

anatomically continuous with labelled neurons within the more caudal regions of the EW and should therefore be included within the boundaries of this nucleus, as was originally suggested by Warwick (1954). This arrangement is shown in Figure 4.5 (Akert *et al.*, 1980) and confirmed by personal observation. As shown in this diagram, the EW in rhesus monkeys is approximately 500 µm in dorsoventral extent, extends anterior to the oculomotor nucleus and caudally to within approximately 500 µm of the most posterior region of the oculomotor nucleus (Akert *et al.*, 1980; Personal Observation). A similar rostrocaudal extent for the preganglionic neurons was also reported by Ishikawa, Sekiya and Kondo (1990), but Burde and Loewy (1980) showed a much more restricted distribution with the cells only extending posteriorly approximately half way through the extent of the oculomotor nucleus.

In addition, these latter two groups of investigators reported a few neurons located between the oculomotor nuclei as an additional source of input to the ciliary ganglion and chose to characterize them as a separate nucleus, Perlia's nucleus (Burde and Loewy, 1980; Ishikawa, Sekiya and Kondo, 1990). From studies in this laboratory and in the marmoset (Clarke, Coimbra and Alessio, 1985b), it is clear however that there are only a very few preganglionic neurons located between the oculomotor nuclei outside of the cytoarchitectural confines of the EW, and it is not clear that this small number of cells constitutes a separate nucleus. Also, in primates there are a few cells located within the boundaries of the Edinger-Westphal nucleus that are not preganglionic neurons and instead project to the cerebellum and spinal cord (Sekiya, Kawamura and Ishikawa, 1984; May, Porter and Gamlin, 1992). Presumably, some of these cells may relay an efference copy signal related to pupilloconstriction or accommodation to the cerebellum that is required by the oculomotor system.

CATS

In cats, a nucleus identified as the EW is located dorsal to the somatic subdivisions of the oculomotor nucleus, but studies indicate that very little of the projection to the ciliary ganglion arises from this nucleus, arising instead from cells in the central grey and ventral tegmental regions (Sugimoto, Itoh and Mizuno, 1977; Toyoshima, Kawana and Sakai, 1980; Kuchiiwa, Kuchiiwa and Nakagawa, 1994). Indeed in the cat, the majority of the neurons in the EW are reported to project not to the ciliary ganglion but to the spinal cord and cerebellum instead (Sugimoto, Itoh and Mizuno, 1978; Loewy, Saper and Yamodis, 1978; Loewy and Saper, 1978; Roste and Dietrichs, 1988).

RABBITS

In rabbits, as in other mammals, the Edinger-Westphal nucleus lies dorsal to the somatic subdivisions of the oculomotor nucleus but, as in cats, it does not contain many preganglionic neurons. Instead, approximately forty preganglionic neurons are reported to lie in the central grey and the tegmental area ventral to the oculomotor nucleus (Johnson and Purves, 1981). As discussed below with respect to birds and primates, this relatively small number of preganglionic neurons is consistent with the observation that in rabbits the EW predominantly subserves the pupillary light reflex (Johnson and Purves, 1983).

OTHER VERTEBRATE CLASSES

Reptiles

The Edinger-Westphal nucleus is well-developed in the monitor lizard (Barbas-Henry and Lohman, 1988). It is an ovoid nucleus lying immediately ventral to the fourth ventricle and dorsal to the somatic oculomotor complex, and it is composed of small to medium-sized bipolar and multipolar neurons. In addition, based on immunohistochemical staining for cholinergic neurons, an EW nucleus has been reported in turtles (Powers and Reiner, 1993). Indeed, according to Barbas-Henry and Lohman (1988), there is a well-defined accessory oculomotor nucleus (Edinger-Westphal nucleus) in all reptilian species studied except for snakes where the nucleus is rudimentary or completely absent presumably due to a decrease in the intrinsic eye muscles.

Fish

The Edinger-Westphal nucleus has been reported in both goldfish and kelp bass (Scherer, 1986; Wathey, 1988; Wathey and Wullimann, 1988). In the kelp bass, the EW lies ventral to the fourth ventricle and dorsolateral to the somatic oculomotor complex with its medial border approximately 340 μm from the midline. It has an extent of approximately 250 μm mediolaterally, 150 μm dorsoventrally, and 250 μm rostrocaudally, and consists of between 60 and 100 medium-sized, multipolar neurons with an average soma size of 20 μm (Wathey, 1988). Since this species lacks a pupillary light reflex, these EW neurons are presumably related to accommodation (Wathey, 1988).

DO ALL EW PREGANGLIONIC NEURONS SYNAPSE IN THE CILIARY GANGLION?

The ciliary ganglion in birds and mammals contains the soma of postganglionic neurons and synaptic contacts from the axons of preganglionic neurons (e.g. Pilar and Tuttle, 1982; May and Warren, 1993). Whilst many studies have reported that all preganglionic neurons synapse in this ganglion in mammals (e.g. Ruskell and Griffiths, 1979; Kuchiiwa, Kuchiiwa and Nakagawa, 1994), there has been some debate as to the existence of a synapse in this ganglion for both pupil-related and accommodation-related preganglionic neurons. For example, based on physiological studies (Westheimer and Blair, 1973) and on retrograde anatomical studies (Jaeger and Benevento, 1980; Parelman, Fay and Burde, 1984) it has been reported that EW neurons do not synapse in the ciliary ganglion but instead have a direct projection to the ciliary muscle. Other reports have disagreed with these studies and there is a growing consensus for the existence of a synapse between the pre- and postganglionic neurons in the primate ciliary ganglion (for a review see Ruskell, 1990). It appears likely that some retrograde tracer experiments appeared to show direct connections between the preganglionic neurons of the Edinger-Westphal nucleus and their eventual peripheral targets because the tracer was taken up by preganglionic fibres en route to the intraocular ganglion cells that are contained within the accessory ciliary ganglia of primates and other mammals. While some of these accessory ganglia are located extraocularly immediately behind the sclera, others are located in the supra-choroid lamina along the intraocular portions of the ciliary nerves from the iris and ciliary body to the scleral canal (Kuchiiwa, Kuchiiwa and Suzuki, 1989; Kuchiiwa, Kuchiiwa and Nakagawa, 1994). Injections in the vicinity of these intraocular cells would thus have involved preganglionic fibres from EW neurons and would have resulted in routine retrograde labelling of these EW neurons.

THE ROLE OF THE EW IN PUPILLOCONSTRICTION

When the intensity of light falling on the retina increases, many vertebrates display a reflexive constriction of the sphincter pupillae muscles of the iris that results in pupilloconstriction. This pupillary light reflex is rapid and well-developed in both birds and mammals and has been extensively studied in both. The anatomical substrate of this reflex was investigated for many years by anatomists who were aware that a number of disorders of the nervous system, e.g., tertiary syphillis, trauma, tumours, etc., could affect the pupillary response to light (Lowenstein and Loewenfeld, 1969). These investigations can be broadly categorised as clinicopathological studies in humans, and lesion, electrical stimulation, and single-unit recording studies in experimental animals.

PUPILLARY DEFICITS RESULTING FROM DAMAGE TO THE EW OR EW EFFERENT PROJECTIONS

Earlier this century, based on experimental lesion studies and clinical studies, it was generally accepted that the final efferent link of the pupillary light reflex consisted of a preganglionic projection from the EW in the midbrain to the ciliary ganglion that, in turn, projected to the sphincter pupillae muscle of the iris (for a complete review of the early literature see Loewenfeld, 1993). Specifically, Bernheimer showed that lesions in the region of the EW resulted in a fixed, dilated pupil ipsilateral to the lesion (Bernheimer, 1909). A later study on primates by Pierson and Carpenter (1974) also showed pupillary deficits following discrete lesions in the area of the anterior Edinger-Westphal nucleus. Similarly, lesions of the EW in birds result in dilated pupils that are unreactive to light and an inability to accommodate (Schaeffel et al., 1990). Pupillary immobility is also associated with third nerve palsies or selective damage to the axons of the preganglionic neurons coursing to the ciliary ganglion, and with damage to the postganglionic fibres which results in Adie's syndrome (Ponsford, Bannister and Paul, 1982; Thompson, 1987).

PUPILLOCONSTRICTION EVOKED BY ELECTRICAL STIMULATION OF THE EW OR EW EFFERENT PROJECTIONS

Further support for the course of the efferent parasympathetic pupillary pathway and the importance of the EW in pupilloconstriction in primates came from stereotaxic, electrical stimulation studies in the vicinity of EW that elicited pupilloconstriction as well as accommodation (Bender and Weinstein, 1943; Jampel and Mindel, 1967; Westheimer and Blair, 1973). Other studies in the cat (Pitts, 1967; Sillito and Zbrozyna, 1970b), marmoset (Clarke, Coimbra and Alessio, 1985a), and chicken (Troilo and Wallman, 1987) have shown that electrical stimulation of the Edinger-Westphal nucleus evokes pupilloconstriction and accommodation in these species. Electrical stimulation of the ciliary ganglion or nerves in cats has also been shown to elicit pupilloconstriction and accommodation (Olmsted, 1944; Marg, Reeves and Wendt, 1954; Ripps, Breinin and Baum, 1961).

Recently, in the alert primate, we have closely examined the pupilloconstriction evoked by electrical microstimulation of the EW or the pupillomotor fibres of the intracranial

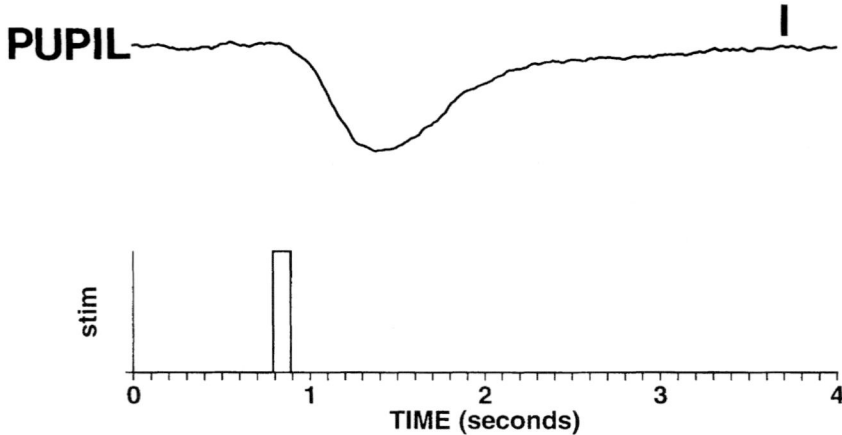

Figure 4.6   A figure showing the pupilloconstriction elicited by electrical microstimulation of the intracranial portion of the third nerve for 100 ms. Note that the pupil begins to constrict approximately 100 ms after the onset of stimulation and reaches maximal constriction at approximately 500 ms after the onset of stimulation. Note the subsequent relatively slow redilation. Scale bar = 0.5 mm.

portion of the oculomotor nerve (Clarke and Gamlin, 1995). A microelectrode was lowered under physiological guidance either to the EW or the oculomotor nerve and microstimulation was carried out over a wide range of parameters. Pupil diameter was measured using an ISCAN video-based pupillometer. As shown in Figure 4.6, in response to a brief stimulus train, the pupil constricts with a latency of approximately 100 ms, and peak pupilloconstriction occurs approximately 300–500 ms after stimulation. Pupil diameter then returns to baseline with a time constant of approximately 600 ms. These responses were only elicited from the area dorsal to the oculomotor nucleus and from a localised region of the oculomotor nerve. Stimulation in the oculomotor nucleus at more ventral sites elicited eye movements as did stimulation at many sites within the oculomotor nerve.

RESPONSES OF EW NEURONS DURING THE PUPILLARY LIGHT REFLEX

In 1970, Sillito and Zbrozyna recorded the activity of preganglionic, pupillomotor neurons in anaesthetised cats. Because of the effects of the chloralose anaesthesia, the pupils were relatively constricted, but still showed a small light reflex. To overcome these effects of anaesthesia, hypothalamic stimulation was used to elicit a "defence reaction" and hence produce pupil dilatation. Using this approach, the authors found that the baseline level of activity of the pupil-related EW neurons was between 6 and 10 spikes/second and was completely inhibited by hypothalamic stimulation. Maximal pupilloconstriction was seen when the pupil-related EW neurons displayed an activity of approximately 8 spikes/second, but these neurons also displayed transient firing rates with light "on" of up to 28 spikes/second. Light "off" was observed to produce a post-excitatory depression lasting

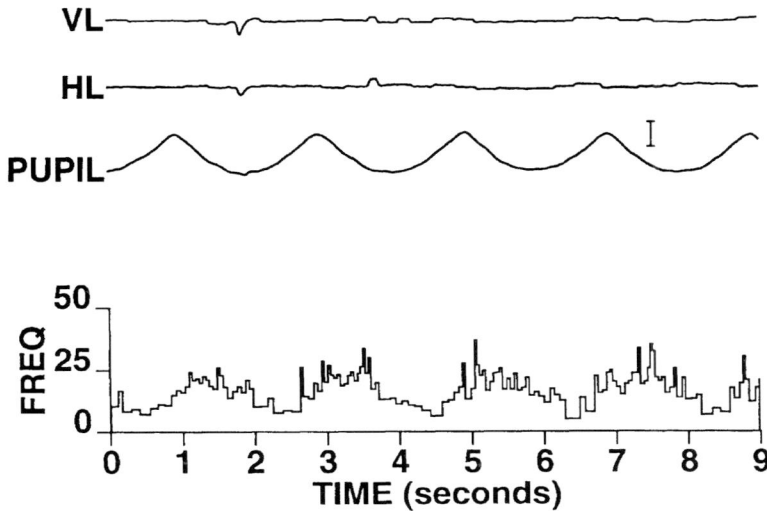

Figure 4.7   The response of a pupil-related neuron of the Edinger-Westphal nucleus during 0.5 Hz sinusoidal modulations in light intensity and the resultant pupillary responses. The activity of the neuron is modulated sinusoidally and also shows a phase advance with respect to the pupilloconstriction. Note that the animal maintained fixation of the target for the entire period of the trial. Abbreviations: HL = Horizontal position of the left eye; VL = Vertical position of the left eye. Scale Bar = 1 mm.

as long as 700 ms. In addition to this study, pupil-related parasympathetic activity has been studied postganglionically in the ciliary nerves of rabbits (Nisida and Okada, 1960; Inoue, 1980), and in the ciliary ganglion of cats (Melnitchenko and Skok, 1970) and rabbits (Johnson and Purves, 1983). The reports of these studies are generally consistent with the results of the study by Sillito and Zbrozyna (1970a).

More recently, in our ongoing studies in alert primates, pupil-related EW neurons have been antidromically identified by electrical stimulation of the intracranial portion of the oculomotor nerve. In all cases antidromic activation was confirmed by collision testing (Schlag and Fuller, 1976). An example of an EW pupillomotor neuron is shown in Figure 4.7. In darkness, the firing rate of the neuron was very low. During sinusoidal modulation of light intensity, the activity of the neuron is modulated sinusoidally and varies from approximately 10 spikes/second at a pupillary diameter of approximately 7 mm to 25 spikes/second for near-maximal pupilloconstriction. This neuron also shows a phase advance with respect to pupilloconstriction. This indicates that, during pupillconstriction, these neural signals show a characteristic pre-emphasis. This will result in the transient increase in muscle force that is required to compensate for the sluggish nature of the iris musculature and its associated tissues. Interestingly, the behaviour of this pupil-related EW neuron is very similar to the luminance neurons of the pretectal olivary nucleus that are presumed to provide EW with input related to the pupillary light reflex (Gamlin, Zhang and Clarke, 1995).

PUPIL-RELATED INPUTS TO THE EW

Once the essential role of the EW in the pupillary light reflex had been established by the early part of this century, investigators began to study the sources of inputs to this nucleus that mediated the reflex. It was soon shown that the central afferent limb of the reflex began with retinal ganglion cell fibres and includes the brachium of the superior colliculus (Karplus and Kreidl, 1913). However, the site of termination of these fibres and their subsequent projections remained unclear until the studies of Magoun and colleagues (Ranson and Magoun, 1933; Magoun and Ranson, 1935a,b, Magoun *et al.*, 1936; Hare, Magoun and Ranson, 1935). These experimenters used localised stimulation and lesioning techniques to follow the trajectory of the reflex pathway and showed for the first time that the pretectum is essential for the integrity of the pupillary light reflex. Since these pioneering studies, most reports have implicated the pretectum in providing the EW with the pupil-related input that mediates the pupillary light reflex. But there has been disagreement as to the precise portion of the mammalian pretectum that projected to the EW (Carpenter and Pierson, 1973; Pierson and Carpenter, 1974; Benevento, Rezak and Santos-Anderson, 1977; Steiger and Büttner-Ennever, 1979; Magnuson, Rezak and Benevento, 1980; Young and Lund, 1994), and some studies have even reported that there was no direct projection from any retinorecipient pretectal nucleus to the EW in the cat (Graybiel and Hartweig, 1974; Berman, 1977), tree-shrew (Weber and Harting, 1980), and rat (Nicholson and Severin, 1981).

Because of the distinctiveness of the avian EW, the avian midbrain, and the pretectal nuclei, our studies in the pigeon have been able to resolve some of these issues. To identify the pupil-related inputs to the EW, it was injected with HRP (Gamlin *et al.*, 1984). Following these injections, only one retinorecipient region of the pretectum was found to contain HRP-labelled cells. This was a dorsomedially situated region termed the area pretectalis (AP) that receives contralateral retinal input. Approximately 100–250 labelled cells that varied in shape from fusiform (15–20 μm in length) to spherical (12–15 μm in diameter) were observed in the AP contralateral to the injection site. Very few labelled neurons were present ipsilaterally. In order to determine the precise portion of the EW to which AP projected, this nucleus was injected with [^{3}H]proline/leucine and autoradiographic techniques were used to show that the projection was to the caudolateral pole of the lateral subdivision of the contralateral EW (Figure 4.8). The terminal field was confined to a discrete region of the caudal EWl approximately 300 μm in rostrocaudal extent and overlying approximately 100 EW neurons. These anatomical results suggest that only a small number of EW cells mediate the pupillary light reflex and are consistent with anatomical studies in the monkey that report that the iris is innervated by only about 3% of ciliary ganglion cells (Warwick, 1954), and with a physiological study in the monkey (Jampel and Mindel, 1967) that reported that few areas in the EW are pupillomotor and that the majority are involved in ocular accommodation.

To establish that this pathway played an essential role in the pupillary light reflex we conducted additional experiments (Gamlin *et al.*, 1984). We showed that unilateral lesions that completely destroyed AP resulted in a fixed, dilated pupil in the eye contralateral to the lesions, while the pupillary light reflex of the eye ipsilateral to the lesioned AP appeared normal. Also, microstimulation of AP was found to elicit a pupillary constriction

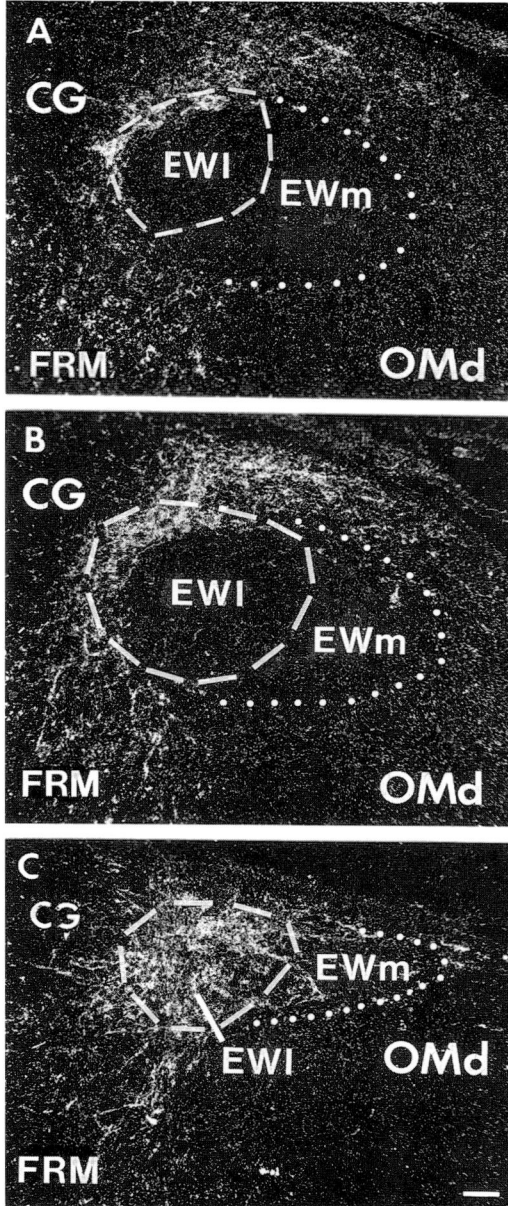

Figure 4.8   A rostrocaudal series (**A-C**) of darkfield photographs of autoradiographic labelling overlying the caudal pole of the Edinger-Westphal nucleus following an injection of tritiated amino acids into the contralateral area pretectalis. Dorsal is to the top and medial is to the right in each photograph. Abbreviations: CG = central grey; EWl = Edinger-Westphal nucleus, pars lateralis; EWm = Edinger-Westphal nucleus, pars medialis; FRM = Medial mesencephalic reticular formation; OMd = Oculomotor nucleus, pars dorsalis. Scale bar = 50 μm. (Reprinted from Gamlin *et al.*, 1984, with permission of Wiley-Liss, Inc.).

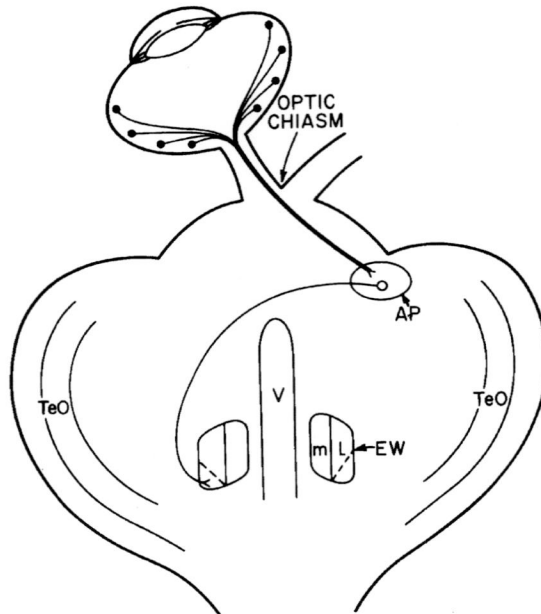

Figure 4.9   Schematic illustration of the central course of the pupillary light reflex in the pigeon. Abbreviations: AP = area pretectalis; EW = Edinger-Westphal nucleus; l = pars lateralis of EW; m = pars medialis of EW; TeO = Optic tectum; V = Ventricle. (Reprinted from Gamlin *et al.*, 1984, with permission of Wiley-Liss, Inc.).

of the contralateral eye, but no pupillary constriction was evident in the ipsilateral eye. Thus our data for pigeon clearly indicate that a retinorecipient nucleus in the pretectum (area pretectalis) plays a major role in the control of pupilloconstriction. These findings are summarised in Figure 4.9.

To directly relate the findings in the pigeon to those in mammals, the mammalian pretectal nucleus that is comparable to AP had to be identified. Studies in the rat, mouse, rabbit, tree-shrew (Scalia, 1972), cat (Berman, 1977), and monkey (Benevento, Rezak and Santos-Anderson, 1977) show that the pretectal olivary nucleus is retinorecipient and occupies a region in the pretectum that is topographically comparable to that occupied by AP in the pigeon. Fibres within AP stain for substance P, catecholamines and enkephalin (Gamlin *et al.*, 1984). The pretectal olivary nucleus (PON) in the rat also contains substance P (Ljungdahl, Hökfelt and Nilsson, 1978), catecholamines (Lindvall *et al.*, 1974), and enkephalin (unpublished observations) and is thus histochemically similar to AP in the pigeon. Thus, based on topographic and histochemical grounds, the PON of mammals appears to be similar to the AP of birds.

In light of these results in the pigeon and the conflicting results regarding the source of pretectal input to the EW in mammals, we have recently investigated the source of the pretectal input to the EW in the rhesus monkey (Gamlin and Clarke, 1995). To identify the afferent pretectal regions, WGA-HRP was injected into the EW under physiological

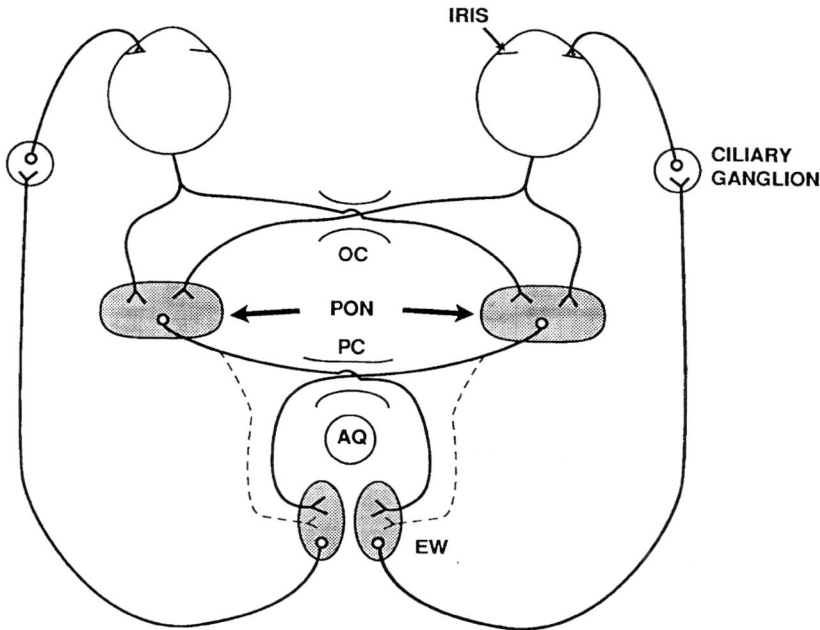

Figure 4.10 Schematic illustration of the subcortical connections mediating the pupillary light reflex in the primate. Abbreviations: AQ = aqueduct; EW = Edinger-Westphal nucleus; OC = optic chiasm; PC = posterior commissure; PON = pretectal olivary nucleus. (Redrawn from Gamlin and Clarke, 1995).

guidance. Intravitreal injection of the same tracer were also made in other animals to define the retinal terminal fields within the pretectum. Following injection of WGA-HRP in the Edinger-Westphal nucleus and appropriate processing, retrogradely labelled cells were found in only one retinorecipient pretectal nucleus, the pretectal olivary nucleus. Almost all labelled cells were located contralateral to the injection site. Intravitreal injection of tracer resulted in anterograde labelling of all the retinorecipient pretectal nuclei including the pretectal olivary nucleus. The retinal terminal field in the pretectal olivary nucleus coincided with the location of the cells that were retrogradely labelled by the injection of tracer into the Edinger-Westphal nucleus and its vicinity. As summarised in Figure 4.10, these results are very similar to those obtained by us in the pigeon and demonstrate that there is a direct projection from the pretectum to the Edinger-Westphal nucleus, that it arises from only one retinorecipient pretectal nucleus, the pretectal olivary nucleus, and that the pretectal olivary nucleus projects predominantly contralaterally to the EW by way of the posterior commissure.

Figure 4.10 indicates that in primates the pretectal projection to EW is predominantly contralateral and not bilateral, and this contrasts with the generally held textbook view (e.g. Thompson, 1987). However, the results of Pasik, Pasik and Bender (1969) are consistent with this proposal. These data show that lesions of the posterior commissure result in bilaterally dilated pupils, which fail to constrict in response to increased illumination, but show a dark reflex. Some clinical evidence in humans (Collier, 1927; Keane and Davis, 1976) also indicates that lesions of the posterior commissure bilaterally abolish the pupillary light reflex, thus supporting the idea that the pretectal projection to the EW is predominantly contralateral in both humans and monkeys. The proposal that the projection from the pretectum to EW is entirely contralateral might also seem incompatible with the consensual pupillary light reflex seen both in normal monkeys and in monkeys with optic tract lesions (Lowenstein, 1954). However, the consensual reflex could be mediated indirectly by way of a projection from the PON to the contralateral PON. Alternatively, since each retina projects bilaterally to the pretectum, the consensual pupillary light reflex could be mediated by this route.

Support for this viewpoint comes from other studies in the monkey that have generally yielded comparable results demonstrating that retrogradely labelled cells in the pretectum are predominantly confined to the contralateral pretectal olivary nucleus after injections of HRP or WGA-HRP into the EW (Steiger and Büttner-Ennever, 1979; Magnuson, Rezak and Benevento, 1980; Büttner-Ennever et al., 1996). However, the results of two recent anterograde studies have raised some questions regarding the details of this proposed pathway. One of these studies investigated the pretectal projection to the EW and suggested that the PON projects not to the EW proper, but immediately lateral to it (Büttner-Ennever et al., 1996). Following intraocular injections of tritiated amino acids, the other study reported transneuronal anterograde labelling over a similar region lateral to the EW proper (Kourouyan and Horton, 1997). Specifically, in both studies, the projection was reported to be to the so-called lateral visceral cell column (Carpenter and Peter, 1970) where, except for a report by Burde and Williams (1989), preganglionic neurons have not been reported. While it is hard to explain these results, it is possible that if the pretectal projection to the EW in primates is as localised as we observed in the pigeon, then the specific region of EW that receives direct pretectal input could have been overlooked in these studies. This would have been particularly likely if the anterograde label in the EW proper was very weak due to the insensitivity of the autoradiographic technique. Alternatively, neurons in the lateral visceral cell column could be interneurons and project to the preganglionic, pupillomotor neurons of the EW. Further study will be required to resolve this issue.

In addition to this pretectal, pupil-related input to the EW, the cerebellum has also been reported to project to the EW (e.g. Thomas et al., 1956; May, Porter and Gamlin, 1992). Electrophysiological and lesion studies in cats (Tsukahara, Kiyohara and Ijichi, 1973; Hultborn, Mori and Tsukahara, 1973, 1978; Ijichi et al., 1977) have reported that this projection may modulate pupillary function. However, as emphasised by Hultborn, Mori and Tsukahara (1978) and as described in the next section, while these papers provide some evidence that the cerebellum modulates the pupillary light reflex, there is even stronger evidence that the cerebellar projection to the EW modulates accommodation.

THE ROLE OF THE EW IN OCULAR ACCOMMODATION

The basic mechanism of accommodation was first described by Helmholtz (1867). Briefly, changes in focus of the lens of the eye are brought about by changes in force from the ciliary muscle which acts in an antagonistic fashion with the fibres of the zonule. These zonular fibres support the lens capsule and act to maintain the lens in a relatively "flattened" state at rest. Ocular accommodation occurs when ciliary muscle contraction results in a reduction in tension in the zonular fibres which produces a "bulging" (increase in convexity) of the lens and a concomitant increase in its refractive power. Accommodative ranges of up to 20 dioptres, i.e. the ability to focus as close as 5 cm, are seen in primates (e.g. Chin *et al.*, 1968; Crawford, Terasawa and Kaufman, 1989) and birds (e.g. Troilo and Wallman, 1987). The parasympathetic innervation of the ciliary muscle dominates the dynamics of these accommodative responses. Increases in parasympathetic innervation act rapidly (within 1 second) to produce substantial positive accommodation of up to 20 dioptres (e.g. Chin *et al.*, 1968; Törnqvist, 1967; Gamlin *et al.*, 1994), while sympathetic innervation acts with the much slower time course of 10–40 s to produce hyperopia (accommodation for far) of at most 1.5 dioptres (e.g. Törnqvist, 1966, 1967; Gilmartin, 1986).

The parasympathetic innervation for the ciliary muscle arises from postganglionic neurons of the ciliary ganglion and from neurons of the accessory ciliary ganglia (e.g. Kuchiiwa, Kuchiiwa and Suzuki, 1989). In turn, these postganglionic neurons are innervated by accommodation-related neurons of the Edinger-Westphal nucleus. Electrical stimulation and single-unit recording in the EW and its efferent pathways in primates and birds have revealed much about the physiology of these connections. Also, anatomical investigations in birds and primates have revealed much about the sources of afferents to the EW that mediate accommodation.

ACCOMMODATION EVOKED BY ELECTRICAL STIMULATION OF THE EW OR EW EFFERENT PROJECTIONS

A number of studies have shown that electrical microstimulation in or immediately adjacent to the EW evokes ocular accommodation in primates (Bender and Weinstein, 1943; Jampel and Mindel, 1967; Westheimer and Blair, 1973; Clarke, Coimbra and Alessio, 1985a; Judge and Cumming, 1986; Crawford, Terasawa and Kaufman, 1989; Gamlin *et al.*, 1994), in cats (Pitts, 1967), and in birds (Troilo and Wallman, 1987). These results of EW stimulation are also consistent with studies in which electrical stimulation of the ciliary ganglion or nerves produced similar increases in accommodation in cats (Olmsted, 1944; Marg, Reeves and Wendt, 1954; Ripps, Breinin and Baum, 1961).

More recently we have examined the effects of electrical microstimulation in the EW of the alert rhesus monkey (Figure 4.11; Gamlin *et al.*, 1994). Accommodation was measured by a continuous recording optometer based on a design of Kruger (1979). To facilitate measurement of the latency of the response, both accommodation (ACC) and accommodative velocity (ACCV) are shown in Figure 4.11A. As can be seen in this figure, the ACCV trace crosses the zero velocity line approximately 75 ms after the beginning of the stimulus train. In Figure 4.11A, some convergence and adduction of the right eye is visible. However, Figure 4.11B shows that, with shorter stimulation times, specific changes in accommodation could be elicited that were not associated with significant changes in eye position. Repeated measures of the latency of the stimulation-induced responses were made at a number of sites within the EW, and the results from two animals

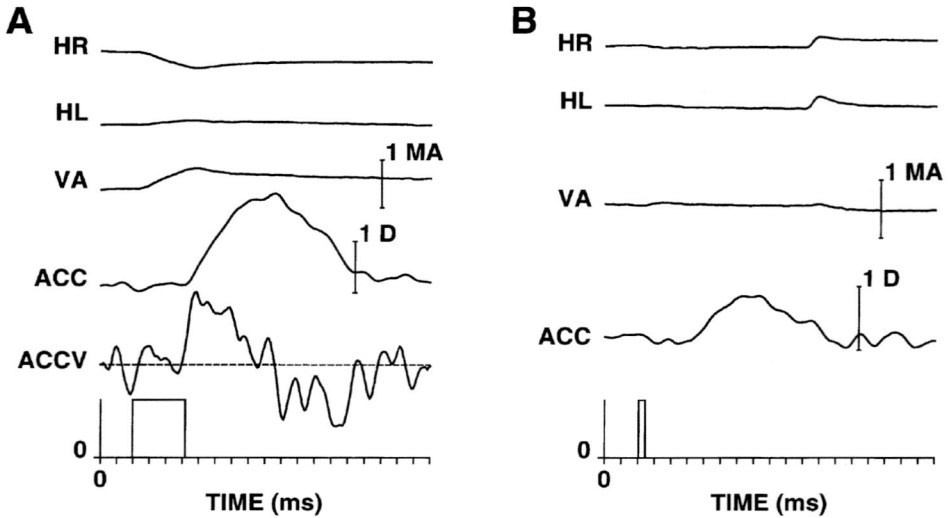

Figure 4.11 Effect of microstimulation of the Edinger-Westphal nucleus on ocular accommodation. (A) shows stimulation (80 ms; 500 Hz; 40 μA) producing an accommodative response with a latency of 75 ms. Accommodative velocity (ACCV) is also shown to facilitate estimation of the latency of the accommodative response. Note that, in addition to accommodation, there is an adduction of the right eye which presumably results from current spread to the nearby medial rectus motoneurons of the "C" subgroup of the right oculomotor nucleus. The stimulation also produces a small amount of convergence that is probably the result of the activation of the axon collaterals of near-response neurons that project to medial rectus motoneurons and presumably also to the Edinger-Westphal nucleus. (B) confirms the specificity of the microstimulation effect by showing that only accommodation is elicited when a stimulation train of shorter duration (10 ms; 500 Hz; 40 μA) is used. Abbreviations: ACC = accommodation; HL = horizontal position of the left eye; HR = horizontal position of the right eye; VA = vergence angle. Scale bar = 1 meter angle and 1 dioptre. (Reprinted from Gamlin et al., 1994, with permission of the American Physiological Society).

were essentially identical, with accommodative responses being evoked with latencies of 75 ms.

RESPONSES OF EW NEURONS DURING OCULAR ACCOMMODATION

To examine the single-unit responses of EW neurons during accommodation, preganglionic neurons were identified by antidromic activation from a stimulating electrode placed in the intracranial portion of the ipsilateral oculomotor nerve just as was done to identify pupil-related preganglionic EW neurons. Also, as was the case for pupil-related preganglionic neurons, antidromic activation was confirmed by collision testing (Schlag and Fuller, 1976). An example of the behaviour of an accommodation-related EW neuron is shown in Figure 4.12 during sinusoidal tracking of a target moving in depth at 0.5 Hz. We found that the behaviour of these neurons during normal binocular viewing was qualitatively the same for all cells. In all cases, their firing rates in darkness and at optical

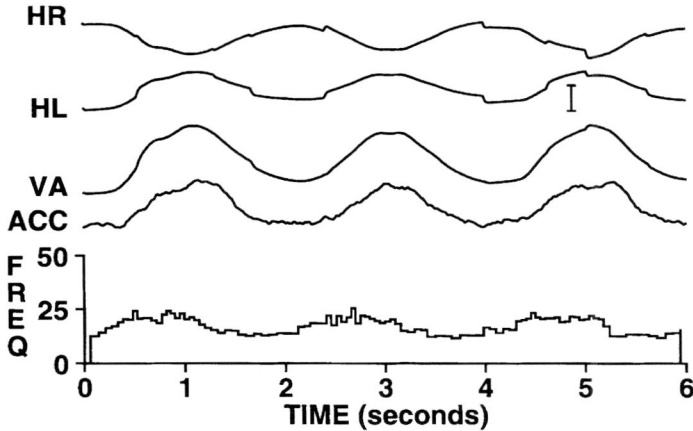

Figure 4.12 Behaviour of a preganglionic EW neuron during sine wave tracking of a target moving in depth. The firing rate modulates between approximately 15 spikes/second and 25 spikes/second for the change in accommodation of 5 dioptres. Note that there is a significant phase lead in the firing rate of this cell with respect to accommodation that cannot be accounted for solely by the latency between the activity of the cell and accommodation. This phase lead results from a substantial component of the firing rate being related to the dynamics of the movement. Abbreviations: ACC = accommodation; HL = horizontal position of the left eye; HR = horizontal position of the right eye; VA = Vergence angle. Scale bar = 2 metre angles and 2 dioptres. (Reprinted from Gamlin *et al.*, 1994, with permission of the American Physiological Society).

infinity were low, but all were active (average activity = 11.6 spikes/second), and all increased their firing rates with increases in accommodation. On average these neurons showed a sensitivity to accommodation of 3.3 (spikes/second)/dioptre. Also as expected, we found that the firing rate of these EW neurons was unaffected by horizontal conjugate eye movements.

In addition, as shown in this figure, the neuron shows a phase lead that cannot be accounted for solely based on the 75 ms delay between neural activity and ocular accommodation that would have been expected based on the results from electrical microstimulation. Instead, this phase lead results from a significant sensitivity to the speed of accommodation that was seen on this cell and on all the accommodation-related EW neurons that were characterised with sine-wave tracking in depth. Overall, accommodation-related preganglionic neurons were related to the rate of change of accommodation with a sensitivity of 1.2 (spikes/second)/(dioptre/second) at 0.5 Hz. This neural activity is presumably required to provide the additional innervation of the ciliary muscle that is needed to generate sufficient transient force to compensate for the characteristics of this muscle and the peripheral accommodative apparatus, which are presumed to be very sluggish (Gamlin *et al.*, 1994).

Interestingly, the firing rates of EW neurons is extremely low when compared with the firing rates of the neurons related to accommodation and convergence that are encountered around the oculomotor nucleus and that are presumed to provide accommodation-related input to the EW. For example, the highest gain for an identified EW neuron was 6.4 (spikes/second)/dioptre while the lowest gain for a near-response cell identified as projecting to the medial rectus subdivision of the oculomotor nucleus was 12.0 (spikes/ second)/dioptre (Zhang, Mays and Gamlin, 1992). Overall, the gain of identified accommodation-related EW neurons is more than six times lower than that of previously reported midbrain near-response neurons (Mays, 1984; Judge and Cumming, 1986; Zhang, Mays and Gamlin, 1992). The functional significance of this is currently unclear, but may be related to the fact that midbrain near-response neurons innervate medial rectus motoneurons with much higher firing rates (Gamlin and Mays, 1992).

ACCOMMODATION-RELATED INPUTS TO THE EW

There have been few anatomical studies of the inputs to the primate EW that might control accommodation. However, as shown in Figure 4.13, we have used anatomical techniques in the rhesus monkey to show that the fastigial nucleus of the cerebellum projects to the EW (May, Porter and Gamlin, 1992). Also, our electrophysiological studies in the alert rhesus monkey have shown that some neurons in this nucleus are related to accommodation and vergence and presumably modulate the activity of EW neurons by way of this projection (Zhang and Gamlin, 1996). In cats, neurons related to accommodation have been reported in both the fastigial and interpositus nucleus (Hosoba, Bando and Tsukahara, 1978; Bando, Ishihara and Tsukahara, 1979), and accommodation-related preganglionic neurons can be orthodromically activated by electrical microstimulation of the interpositus nucleus (Bando et al., 1984). Thus there is strong evidence from studies in both primates and cats that the cerebellum can significantly modulate accommodation by way of its projection to the EW.

Several electrophysiological studies on other subcortical areas of the alert rhesus monkey have characterised the behaviour of neurons that are related to the vergence and accommodative components of the near response. Such neurons have been reported in the reticular formation immediately dorsal and lateral to the oculomotor nucleus (Mays, 1984; Judge and Cumming, 1986; Mays et al., 1986; Zhang, Mays and Gamlin, 1992). Some of these cells appear to control convergence, while others are more closely linked to the control of accommodation (Mays, 1984; Judge and Cumming, 1986; Mays et al., 1986; Zhang, Mays and Gamlin, 1992). Consistent with this suggestion, some convergence-related midbrain cells project monosynaptically to ipsilateral medial rectus motoneurons in the primate (Zhang, Gamlin and Mays, 1991). Comparable monosynaptic projections from accommodation-related midbrain neurons to EW neurons are very likely to exist.

In addition to these near-response cells located immediately adjacent to the oculomotor nucleus, other cells that modulate their activity during the near response have been reported approximately 4–6 mm more dorsal in an ill-defined region that includes, or is close to, the nucleus of the posterior commissure (Judge and Cumming, 1986; Mays et al., 1986). On the basis of cytoarchitecture and embryological considerations, the nucleus of the posterior commissure in primates may be considered comparable to the region of the avian mesencephalon which has been termed the lateral mesencephalic reticular formation (LRF) (Kuhlenbeck, 1937; Kuhlenbeck, 1939).

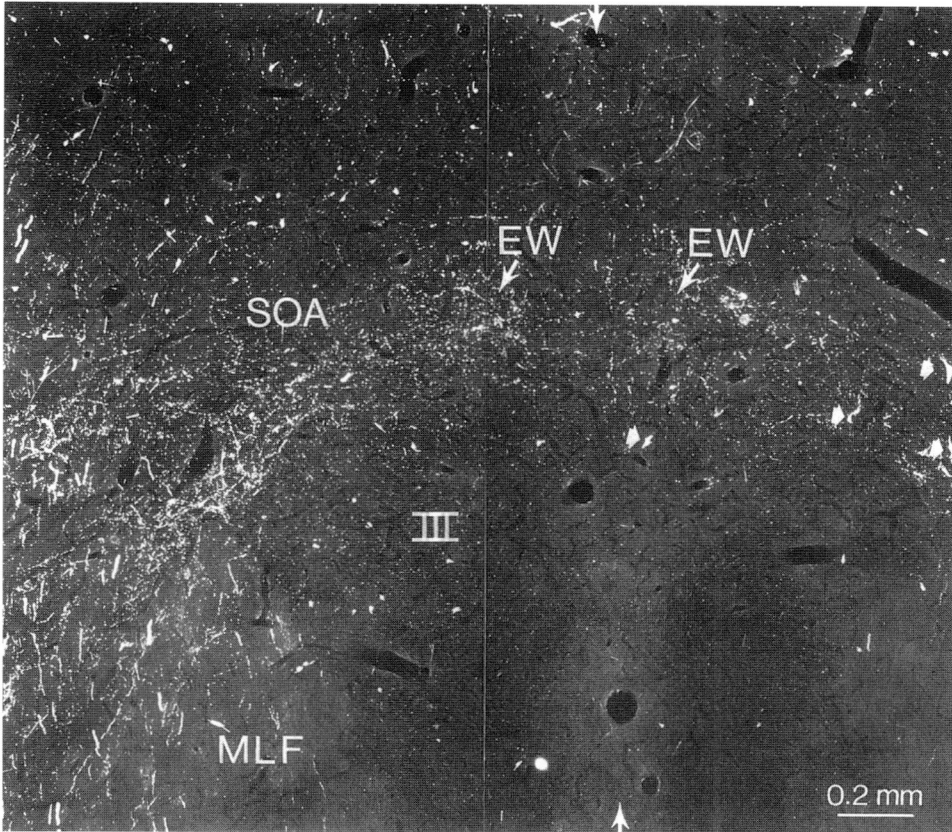

Figure 4.13 Darkfield photomicrograph showing the distribution of terminal labelling in the Edinger-Westphal nucleus (EW) and surrounding regions that resulted from an injection of wheat germ agglutinin-horseradish peroxidase (WGA-HRP) into the right fastigial nucleus. The projection to the contralateral EW is heavier than to the ipsilateral EW. The thin arrows at the top and bottom of the photograph indicate the midline while the stouter arrows indicate retrogradely labelled neurons. Scale bar = 200 μm. (Reprinted from May, Porter and Gamlin, 1992, with permission of Wiley-Liss, Inc.).

In the rhesus monkey, it is possible that these near-response cells in both the midbrain and pretectal region project monosynaptically to the EW. Although this has not been specifically investigated either physiologically or anatomically in this species, there is support for connections of this nature from some anatomical studies in the pigeon (Gamlin and Reiner, 1991). As described below, in this species it has been shown that there are cells in the medial mesencephalic reticular formation (MRF) that are located in a cytoarchitecturally and topographically similar location to cells immediately adjacent to the oculomotor nucleus in primates. These cells project to the entire EW including its accommodative subdivision, the rostromedial EWl. Also, cells in the LRF that are located in a cytoarchitecturally and topographically similar location as cells in the nucleus of the

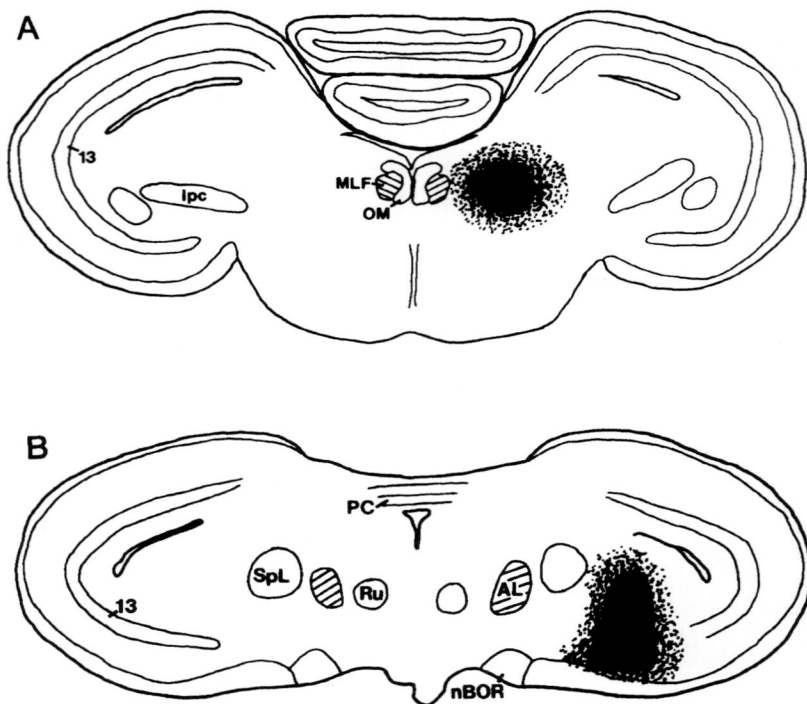

Figure 4.14 (A) Low-power drawing of the mesencephalon showing the location of an injection of tritiated proline/leucine into the medial mesencephalic reticular formation adjacent to the oculomotor nucleus. (B) Low-power drawing of the mesencephalon showing the location of an injection of tritiated proline/leucine into the lateral mesencephalic reticular formation. (Reprinted from Gamlin and Reiner, 1991, with permission of Wiley-Liss, Inc.).

posterior commissure of primates project specifically to the accommodative subdivision of the EW.

To investigate accommodation-related and other inputs to the EW of the pigeon, injections of HRP were placed into it. These injections retrogradely labelled cells in a number of regions including the MRF and LRF. To investigate these apparent sources of input to EW further, a series of anterograde studies were conducted in which animals received ^{3}H proline/leucine injections in and around the MRF. One of these injections, which is shown in Figure 4.14A, resulted in clear anterograde labelling of the entire contralateral EW. Weaker, but nevertheless clear, anterograde labelling of the entire ipsilateral EW was also present. Both contralaterally and ipsilaterally this anterograde label formed a dust-like pattern over the entire nucleus (Figure 4.15). This terminal labelling may reflect a single afferent input related to accommodation, pupil, and choroidal blood flow. Alternatively, it is possible that this pattern of terminal label in the EW results from the involvement of two afferent systems. It appears as though a rostrolateral region of the MRF projects to the EWl and that this input is related to accommodation and to

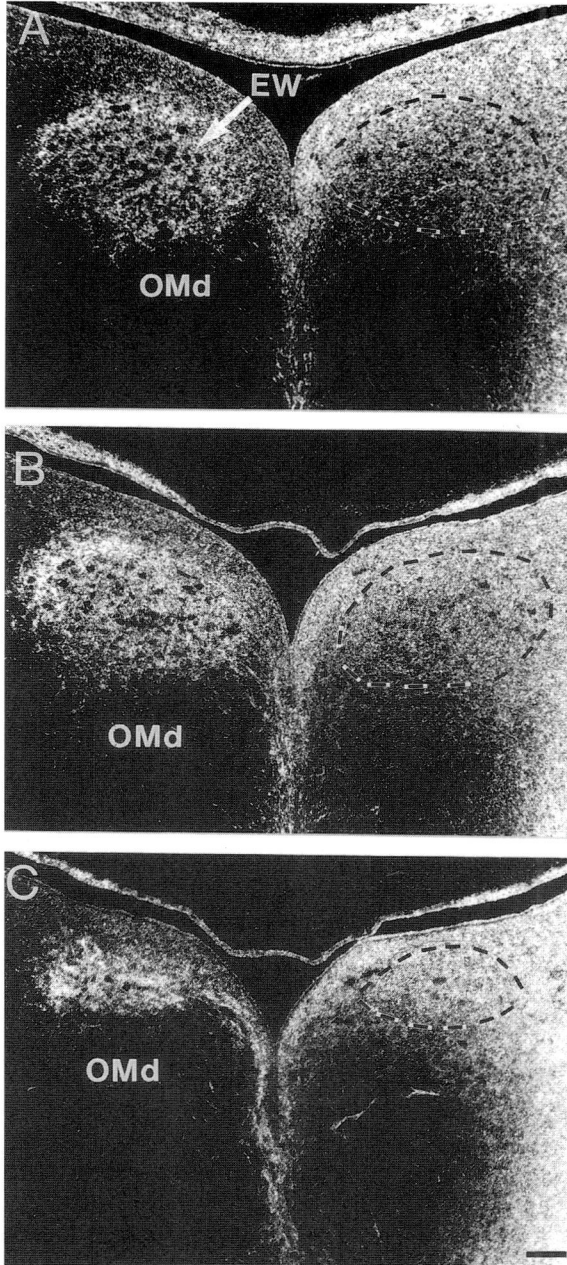

Figure 4.15 A rostrocaudal series (**A-C**) of darkfield photographs showing the distribution of autoradiographic labelling in the Edinger-Westphal nucleus (EW) following the injection of tritiated amino acids into the medial mesencephalic reticular formation that is shown in Figure 4.14A. Dotted lines indicate the boundaries of the EW ipsilateral to the injection site. OMd = Oculomotor nucleus, pars dorsalis. Scale bar = 100 μm. (Reprinted from Gamlin and Reiner, 1991, with permission of Wiley-Liss, Inc.).

pupilloconstriction. In addition, a separate and independent pathway, which appears to arise from more medial reticular regions, projects to all EW subdivisions and possibly modulates the overall transmission through this nucleus. As discussed in a subsequent section, this would be consistent with evidence in mammals for a modulatory influence of the reticular formation on transmission through the EW.

The possibility that the MRF input to the EW plays both a direct and a modulatory role of the near response is also supported by immunohistochemical studies that have revealed that some MRF cells projecting to the EW stain for the presence of the neuropeptide enkephalin, while others do not (Gamlin and Reiner, 1987; Reiner, unpublished observations). Thus there appear to be two functionally distinct subpopulations of cells in the MRF projecting to EW. One population, immediately adjacent to the oculomotor nucleus controls accommodation and pupilloconstriction, the other, more caudally located in the raphe, projects to all the EW subdivisions and modulates transmission through the EW.

To investigate the apparent projection of the LRF to EW, we carried out a series of anterograde studies in which animals received [^{3}H]proline/leucine injections in and around this nucleus. One of these injections, which is shown in Figure 4.14B, resulted in clear anterograde labelling of the contralateral EWl (Figure 4.16). Weaker, but nevertheless clear, anterograde labelling of the ipsilateral EWl was also present (Figure 4.16). The pattern of anterograde label was very different from that seen following MRF injections. Instead of being dust-like, the anterograde label was present over large-calibre fibres that terminated specifically in the lateral region of the EW (Figure 4.16). The terminal field overlay the accommodative subdivision of EWl; it was also closely associated with the smaller, pupilloconstrictor subdivision of the EWl, appearing to overlie some cells in this caudolateral subdivision.

Our anterograde studies showed that cells in the LRF project to the accommodative subdivision of EWl. They also showed a projection field closely associated with the localised region of EWl that we have identified as the pupilloconstrictor subdivision. It therefore seems likely that the input to the EWl from the LRF is to both the accommodative and pupillomotor subdivisions and that this pathway may mediate or modulate both accommodation and pupilloconstriction. The results of the above studies are summarised in Figure 4.3.

These results in pigeons suggest that, if the projections in primates are comparable, there are monosynaptic projection from the near response cells in both the supraoculomotor area and the nucleus of the posterior commissure onto the EW. Projections of this nature would presumably mediate or modulate accommodation and possibly the pupillary near response. In mammals, authors have interpreted the projection from the nucleus of the posterior commissure to much of the EW as being involved in the pupillary light reflex (e.g. Benevento, Rezak and Santos-Anderson, 1977). However, since only a small percentage of the EW cells are involved in the pupillary light reflex in both mammals and birds (Warwick, 1954; Gamlin et al., 1984), it is unlikely that this extensive an input would mediate only the pupillary light reflex. It is instead likely that the nucleus of the posterior commissure input to the EW in mammals is related to both accommodation and to the pupilloconstriction associated with the near response. To resolve the details of these connections, further study of this pathway and of the proposed projection from the supraoculomotor area to the EW will be needed.

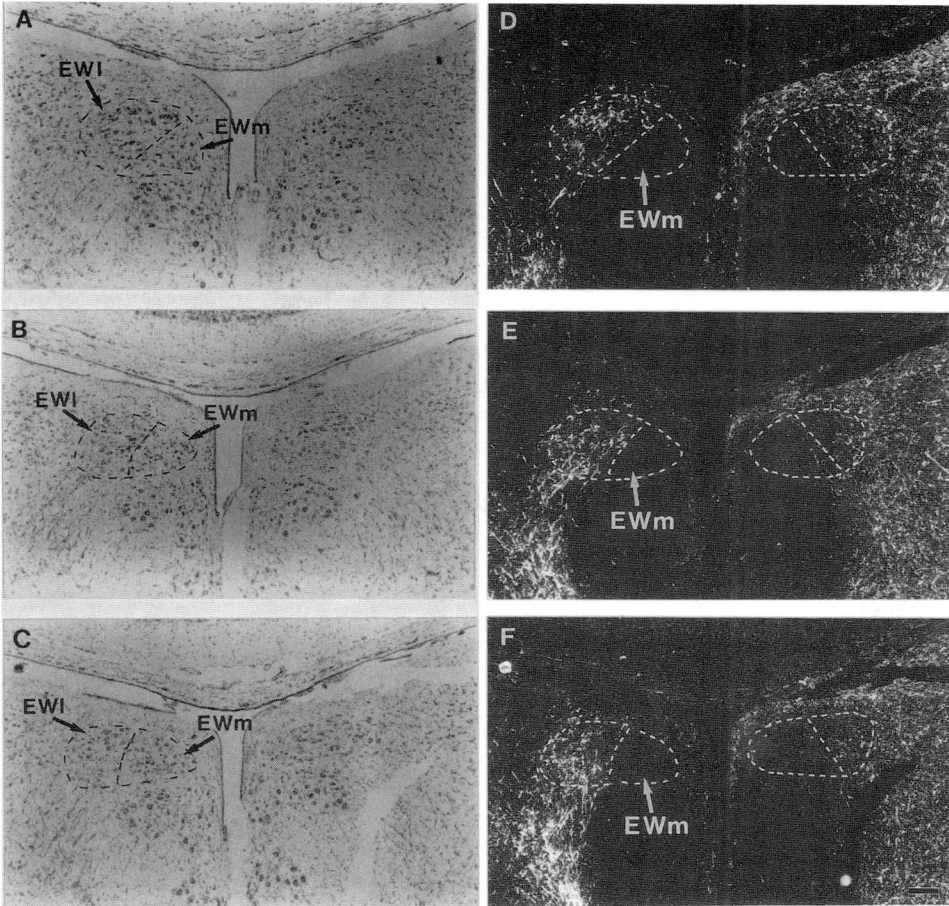

Figure 4.16 A rostrocaudal series of brightfield (**A-C**) and matching darkfield photographs (**D-F**) showing the distribution of autoradiographic labelling in the Edinger-Westphal nucleus (EW) following the injection of tritiated amino acids into the lateral mesencephalic reticular formation that is shown in Figure 4.14B. Scale bar = 200 μm. (Reprinted from Gamlin and Reiner, 1991, with permission of Wiley-Liss, Inc.).

THE ROLE OF THE EW IN REGULATING CHOROIDAL BLOOD FLOW

Increases in illumination of the retina places both metabolic and thermoregulatory demands on this structure (Parver *et al.*, 1982; Parver, Auker and Carpenter, 1983). These demands are particularly pronounced for the outer layers of the retina including the photoreceptor layer which are supplied with blood by the choroid. Thus, the choroid is essential to meeting the metabolic and thermoregulatory demands of the outer layers of the retina. Changes in retinal illumination alter these demands and presumably require a compensatory alteration on the part of choroidal blood flow. Indeed, as described below, a centrally-mediated reflex controlling choroidal blood flow appears to play a role in preventing the deleterious effects of light on photoreceptors under normal ambient light levels. Parver and colleagues (1982, 1983) have shown in monkeys and humans that increases in illumination of a given eye result in increased blood flow within both that eye and the contralateral eye. Dysfunctions of this reflex may play a role in at least some classes of degenerative diseases of the retina. In support of this, retinal photoreceptors have been shown to degenerate when the choroidal blood supply is occluded (Goldor and Gay, 1967), or the innervation to the choroid is compromised (Shih, Fitzgerald and Reiner, 1993).

In both birds and mammals, sympathetic innervation of the choroid arises from the superior cervical ganglion and serves to decrease choroidal blood flow (e.g. Bill and Nilsson, 1985; Gherezghiher, Hey and Koss, 1989). In contrast, the parasympathetic input has a vasodilatory action and serves to increase choroidal blood flow in both mammals (Bill and Sperber, 1990) and birds (Fitzgerald, Vana and Reiner, 1990a). In birds, this parasympathetic innervation arises predominantly from the choroidal neurons of the ciliary ganglion (Pilar and Tuttle, 1982; Meriney and Pilar, 1987; Cuthbertson *et al.*, 1996), with a lesser contribution from the facial nerve by way of the sphenopalatine ganglion (Cuthbertson *et al.*, 1997). In contrast, based on anatomical studies in mammals the parasympathetic innervation of the choroid is reported to arise predominantly from the sphenopalatine ganglion (Ruskell, 1971), but physiological studies in mammals also suggest a contribution from the oculomotor nerve by way of the ciliary ganglion (Gherezghiser, Hey and Koss, 1989; Nakanome *et al.*, 1995).

DEFICITS IN THE CONTROL OF CHOROIDAL BLOOD FLOW THAT RESULT FROM LESIONS OF THE EW OR EW EFFERENT PROJECTIONS

Under normal conditions pigeons display a stable baseline choroidal blood flow that increases significantly with increases in retinal illumination (Fitzgerald *et al.*, 1996). However, following permanent lesions of the EW, baseline choroidal blood flow is decreased by approximately half and does not increase with increases in retinal illumination (Fitzgerald *et al.*, 1996). Importantly, these permanent lesions of EW in the pigeon result in clear retinal pathology as evidenced by Müller cells expressing increased levels of glial fribrillary acidic protein (Fitzgerald, Vana and Reiner, 1990b). In chicks, when the choroidal nerves are sectioned, the decrease in choroidal blood flow is even more significant, decreasing to approximately one fifth of normal, and this results in a histologically verified loss of the cells of the outer retina (Shih, Fitzgerald and Reiner, 1993). To my knowledge, to date, similar experiments have not been conducted in mammals.

CHANGES IN CHOROIDAL BLOOD FLOW EVOKED BY ELECTRICAL STIMULATION OF THE EW OR EW EFFERENT PROJECTIONS

Electrical microstimulation of the Edinger-Westphal nucleus or the ciliary nerves has been shown to evoke increases in choroidal blood flow in the cat (e.g. Gherezghiser, Hey and Koss, 1989; Nakanome *et al.*, 1995) and the rabbit (e.g. Stjernschantz, Alm and Bill,

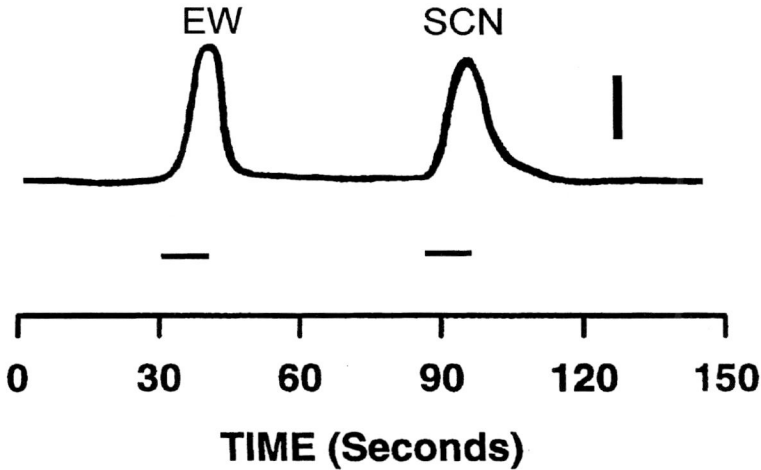

Figure 4.17 Effects on choroidal blood flow of electrical microstimulation of the Edinger-Westphal nucleus (EW) and suprachiasmatic nucleus (SCN). Horizontal solid bars indicate the timing and duration of electrical stimulation. Choroidal blood flow was measured transsclerally at a site below the superior rectus muscle in an animal deeply anaesthetised with ketamine/xylazine. Scale Bar = 50 Blood Flow Units. (Redrawn from Fitzgerald *et al.*, 1996).

1976). It has also been reported that, as a possible consequence of this increase in intraocular blood flow, stimulation of the Edinger-Westphal elicits increases in intraocular pressure in the cat (e.g. Gherezghiser, Hey and Koss, 1990). However, the mechanism underlying this increase in intraocular pressure has not been investigated further.

Studies of the influence of electrical stimulation of the EW on choroidal blood flow have also been conducted in the pigeon (Fitzgerald, Vana and Reiner, 1990a; Fitzgerald *et al.*, 1996). In these studies, a TSI Laserflo blood perfusion monitor using laser doppler flowmetry was used to measure choroidal blood flow. Measurements were made transsclerally from the vascular beds beneath the superior rectus or superior oblique muscle. Since, the blood flow in these particular vascular beds has not been calibrated with an independent method such as microspheres (Lindsberg *et al.*, 1989), it is presented in arbitrary Blood Flow Units, although these can be expected to correspond quite closely to the units of flow of ml/min/100g tissue. Figure 4.17 shows an example of the clear increase in choroidal blood flow that was evoked by electrical microstimulation (100 Hz, 400 µA) of the EW of the pigeon.

BLOOD FLOW-RELATED INPUTS TO EW

To identify the source of the input to EW that was related to the control of choroidal blood flow, we injected HRP into this nucleus. Retrogradely-labelled cells were found in a small retinorecipient, hypothalamic nucleus in the rostral diencephalon termed the suprachiasmatic nucleus (SCN) (Gamlin, Reiner and Karten, 1982). Using

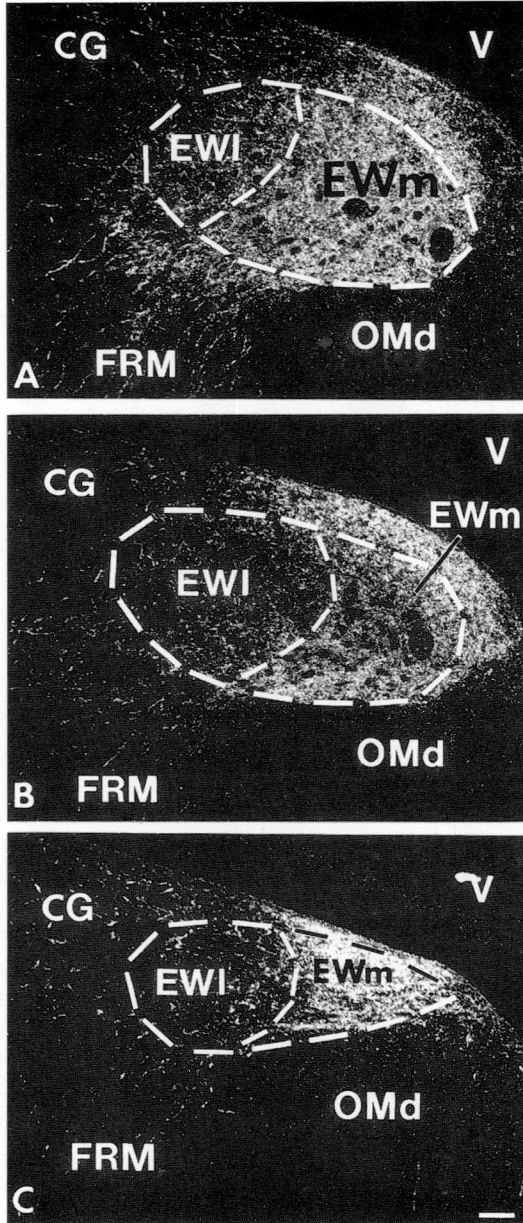

Figure 4.18 A rostrocaudal series (**A-C**) of darkfield photographs of autoradiographic labelling overlying the Edinger-Westphal nucleus, pars medialis (EWm) following an injection of tritiated amino acids into the contralateral suprachiasmatic nucleus. Dorsal is to the top and medial is to the right in each photograph. CG = central grey; EWl = Edinger-Westphal nucleus, pars lateralis; FRM = medial reticular formation; OMd = oculomotor nucleus, pars dorsalis. Scale bar = 50 μm. (Reprinted from Gamlin *et al.*, 1984, with permission of Wiley-Liss, Inc.).

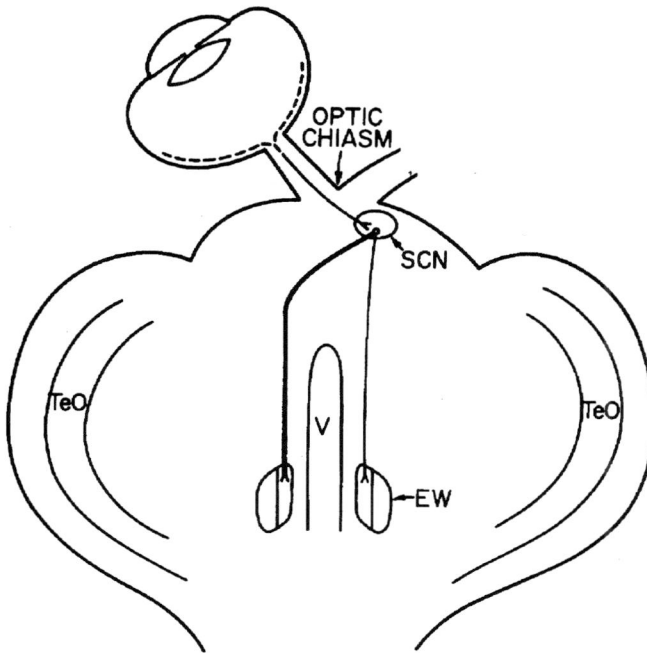

Figure 4.19 Schematic illustration of the projection from the suprachiasmatic nucleus (SCN) to the Edinger-Westphal nucleus (EW) in the pigeon. Abbreviations: TeO = Optic tectum; V = Ventricle.

autoradiographic techniques, we confirmed that the SCN projected to the choroidal subdivision of the EW, the EWm (Figure 4.18). In additional studies, we found that the terminals and fibres of this projection contain substance P (Gamlin, Reiner and Karten, 1982). Also, as shown in Figure 4.17, we have found that electrical microstimulation of the SCN elicits increases in choroidal blood flow (Fitzgerald *et al.*, 1996). In addition, lignocaine blockade of the EW can prevent these increases in blood flow that are observed with electrical stimulation of the SCN (Fitzgerald *et al.*, 1996). Thus, on the basis of anatomical, electrical stimulation, and lesion studies this retino-SCN-EWm pathway has been shown to be a significant component of the parasympathetic control of choroidal blood flow in pigeons and it is summarised in Figure 4.19.

As described above, physiological studies suggest that the mammalian EW may be involved with the control of choroidal blood flow as well as with the control of accommodation and pupilloconstriction. However, to date, no study has described a separate choroidal subdivision of the mammalian EW that is equivalent to the avian EWm. This may well result from the fact that all the cells in the mammalian ciliary ganglion innervate smooth muscle and thus the preganglionic neurons providing their input can be expected to appear similar. Separate subdivisions of the EW containing morphologically different preganglionic neurons may be present in birds because cells in these two subdivisions innervate cells in the ciliary ganglion that project to different muscle types as peripheral targets. Ciliary neurons innervate striated muscle while choroidal neurons innervate smooth muscle (Pilar and Tuttle, 1982). Thus it is possible that cells in a cytoarchitecturally indistinct region of the mammalian EW are involved with the control of choroidal blood flow. However, to my knowledge, this possibility has not been systematically examined anatomically or physiologically in primates.

MODULATION OF INFORMATION FLOW THROUGH THE EW

A number of studies have shown that pupillomotor activity can be modulated by non-visual factors. For example, as a subject becomes fatigued, their pupils will tend to tonically constrict and the subject will show a reduced amplitude of pupilloconstriction in response to the same light stimulus (e.g. Lowenstein and Loewenfeld, 1964). Generally, as described below, most studies have explained these tonic changes in pupil diameter and changes in pupillomotor sensitivity by assuming that there are non-visual modulatory influences on pupillomotor neurons within the EW. To my knowledge, little attention has been paid to possible modulatory influences on accommodation-related or choroidal-related neurons within the EW.

Studies have shown that electrical stimulation of the posterior hypothalamus produces pupil dilatation, presumably because it elicits a "defence reaction" that impinges on neurons of the EW (e.g. Loewenfeld, 1958; Sillito and Zbrozyna, 1970a; Koss, Gherezghiher and Nomura, 1984). Similarly, electrical stimulation of the sciatic nerve elicits pupil dilatation, presumably through pain pathways that modulate EW neuron activity (e.g. Koss, Gherezghiher and Nomura, 1984). Finally, it has been shown that electrical stimulation of the medullary reticular formation or paramedian pontine reticular formation elicits pupil dilatation, presumably by activating ascending adrenergic or serotonergic pathways to the EW (Bonvallet and Zbrozyna, 1963; Loewy, Araujo and Kerr, 1973; Koss, Gherezghiher and Nomura, 1984). Consistent with these suggested inputs, adrenergic fibres have been reported in the EW of the rat (e.g. Dahlström et al., 1964), and enkephalinergic fibres have been reported in the EW of the pigeon (Gamlin and Reiner, 1987). Further, opiate receptors of both the μ- and δ-subtypes have been reported in the pigeon EW (Reiner et al. 1989). Also, pharmacological studies in rats and cats have clearly demonstrated that adrenergic inputs to the EW modulate the activity of pupil-related neurons by selectively activating α_2-adrenoreceptors (Koss, Gherezghiher and Nomura, 1984; Koss, 1986; Heal et al., 1995). Furthermore, micro-injections of either clonidine, an α_2-adrenoceptor agonist, or morphine, an opiate, into the EW of cats produces a well-defined pupil dilatation, presumably by inhibiting the activity of pupil-related EW neurons (Sharpe and Pickworth, 1985). Based on these data, it appears likely that adrenergic, serotonergic, and opiate-containing pathways impinge directly on the EW and that these three neurochemically distinct pathways modulate the activity of pupil-related EW neurons and the flow of information through this nucleus.

ACKNOWLEDGMENTS

I would like to thank Dr. Anton Reiner for his comments on this manuscript. I would like to acknowledge the support of NEI grants R01 EY-07558 and EY-09380, and of NEI CORE grant P30 EY-03039.

REFERENCES

Akert, K., Glicksman, M.A., Lang, W., Grob, P. and Huber, A. (1980). The Edinger-Westphal nucleus in the monkey. A retrograde tracer study. *Brain Research,* **184**, 491–498.

Ariens-Kappers, C.U., Huber, G.C. and Crosby, E.C. (1936). *The Comparative Anatomy of the Nervous System of Vertebrates, including Man.* New York: Macmillan.

Bando, T., Ishihara, A. and Tsukahara, N. (1979). The mode of cerebellar control of lens accommodation. In *Integrative Control Functions of Brain*, Volume 2, edited by M. Ito, N. Tsukahara, K. Kubota and K. Yagi, pp.149–150. Amsterdam, Oxford: Elsevier/North Holland Biomedical Press

Bando, T., Tsukuda, K., Yamamoto, N., Maeda, J. and Tsukahara, N. (1984). Physiological identification of midbrain neurons related to lens accommodation in cats. *Journal of Neurophysiology*, **52**, 870–878.

Barbas-Henry, H.A. and Lohman, A.H. (1988). The motor nuclei and sensory neurons of the IIIrd, IVth, and VIth cranial nerves in the monitor lizard, *Varanus exanthematicus*. *Journal of Comparative Neurology*, **267**, 370–386.

Bender, M.B. and Weinstein, E.A. (1943). Functional representation in the oculomotor and trochlear nuclei. *Archives of Neurology and Psychiatry (Chicago)*, **49**, 98–106.

Benevento, L.A., Rezak, M. and Santos-Anderson, R. (1977). An autoradiographic study of the projections of the pretectum in the rhesus monkey (*Macaca mulatta*): evidence for sensorimotor links to the thalamus and oculomotor nuclei. *Brain Research* **127**,197–218.

Berman, N. (1977). The connections of the pretectum in the cat. *Journal of Comparative Neurology*, **174**, 227–254.

Bernheimer, S. (1909). Weitere experimentelle Studien zur Kenntnis der Lage des Sphincter- und Levator-kerns. *Albrecht von Graefes Archiv für Ophthalmologie*, **70**, 539–562.

Bill, A. and Nilsson, S.F. (1985). Control of ocular blood flow. *Journal of Cardiovascular Pharmacology*, **7**, 96–102.

Bill, A. and Sperber, G.O. (1990). Control of retinal and choroidal blood flow. *Eye*, **4**, 319–325.

Bonvallet, M. and Zbrozyna, A. (1963). Les commandes réticulaires du système autonome et en particulier de l'innervation sympathique et parasympathique de la pupille. *Archives Italienne de Biologie*, **101**, 174–207.

Burde, R.M. and Loewy, A.D. (1980). Central origin of oculomotor parasympathetic neurons in the monkey. *Brain Research*, **198**,434–439.

Burde, R.M., Parelman, J.J. and Luskin, M. (1982). Lack of unity of Edinger-Westphal nucleus projections to the ciliary ganglion and spinal cord: a double-labeling approach. *Brain Research*, **249**,379–382.

Burde, R.M. and Williams, F. (1989). Parasympathetic nuclei. *Brain Research*, **498**, 371–375.

Büttner-Ennever, J.A., Cohen, B., Horn, A.K. and Reisine, H. (1996). Pretectal projections to the oculomotor complex of the monkey and their role in eye movements. *Journal of Comparative Neurology*, **366**, 348–359.

Cabot, J.B., Reiner, A. and Bogan, N. (1982). Avian bulbospinal pathways: Anterograde and retrograde studies of cells of origin, funicular trajectories and laminar terminations. *Progress in Brain Research*, **57**, 79–108.

Carpenter, M.B. and Peter, P. (1970). Accessory oculomotor nuclei in the monkey. *Journal für Hirnforschung*, **12**, 405–418.

Carpenter, M.B. and Pierson, R.J. (1973). Pretectal region and the pupillary light reflex. An anatomical analysis in the monkey. *Journal of Comparative Neurology* **149**, 271–300.

Chin, N.B., Ishikawa, S., Lappin, H., Davidowitz, J. and Breinin, G.M. (1968). Accommodation in monkeys induced by midbrain stimulation. *Investigative Ophthalmology*, **7**, 386–396.

Clarke, R.J. and Gamlin, P.D.R. (1995). Latency and dynamics of pupilloconstriction determined by microstimulation of the Edinger-Westphal nucleus and oculomotor nerve in the primate. *Society for Neuroscience Abstracts*, **21**, 1918.

Clarke, R.J., Coimbra, C.J. and Alessio, M.L. (1985a). Oculomotor areas involved in the parasympathetic control of accommodation and pupil size in the marmoset (*Callithrix jacchus*). *Brazilian Journal of Medical and Biological Research*, **18**, 373–379.

Clarke, R.J., Coimbra, C.J. and Alessio, M.L. (1985b). Distribution of parasympathetic motoneurones in the oculomotor complex innervating the ciliary ganglion in the marmoset (*Callithrix jacchus*). *Acta Anatomica*, **121**, 53–58.

Collier, J. (1927). Nuclear ophthalmoplegia, with special reference to retraction of the lids and ptosis and to lesions of the posterior commissure. *Brain*, **50**, 488–498.

Cowan, W.M. and Wenger, E. (1968). Degeneration in the nucleus of origin of the preganglionic fibres to the chick ciliary ganglion following early removal of the optic vesicle. *Journal of Experimental Zoology*, **168**, 105–124.

Crawford, K., Terasawa, E. and Kaufman, P.L. (1989). Reproducible stimulation of ciliary muscle contraction in the cynomologous monkey via a permanent indwelling midbrain electrode. *Brain Research*, **503**, 265–272.

Cuthbertson, S., White, J., Fitzgerald, M.E., Shih, Y.F. and Reiner, A. (1996). Distribution within the choroid of cholinergic nerve fibers from the ciliary ganglion in pigeons. *Vision Research*, **36**, 775–786.

Cuthbertson, S.L., Jackson, B., Toledo, C.B., Fitzgerald, M.E., Shih, Y.F., Zagvazdin, Y. and Reiner, A. (1997). Innervation of orbital and choroidal blood vessels by the pterygopalatine ganglion in pigeons. *Journal of Comparative Neurology*, **386**, 422–442.

Dahlström, A., Fuxe, K., Hillarp, N. and Malmfors, T. (1964). Adrenergic mechanisms in the pupillary light reflex pathway. *Acta Physiologica Scandinavica*, **62**, 119–124.

Edinger, L. (1885). Ueber den Verlauf der centralen Hirnnervenbahnen mit Demonstration von Präparaten. *Archiv für Psychiatrie und Nervenkrankheim*, **16**, 858–859.

Erichsen, J.T., Karten, H.J., Eldred, W.D. and Brecha, N.C. (1982). Localization of substance P-like and enkephalin-like immunoreactivity within preganglionic terminals of the avian ciliary ganglion: light and electron microscopy. *Journal of Neuroscience*, **2**, 994–1003.

Fitzgerald, M.E., Vana, B.A. and Reiner, A. (1990a). Control of choroidal blood flow by the nucleus of Edinger-Westphal in pigeons: a laser Doppler study. *Investigative Ophthalmology and Visual Science*, **31**, 2483–2492.

Fitzgerald, M.E., Vana, B.A. and Reiner, A. (1990b). Evidence for retinal pathology following interruption of neural regulation of choroidal blood flow: Müller cells express GFAP following lesions of the nucleus of Edinger-Westphal in pigeons. *Current Eye Research*, **9**, 583–598.

Fitzgerald, M.E., Gamlin, P.D., Zagvazdin, Y. and Reiner, A. (1996). Central neural circuits for the light-mediated reflexive control of choroidal blood flow in the pigeon eye: a laser Doppler study. *Visual Neuroscience*, **13**, 655–669.

Gamlin, P.D.R. and Clarke, R.J. (1995). The pupillary light reflex pathway of the primate. *Journal of the American Optical Association*, **66**, 415–418.

Gamlin, P.D.R. and Mays, L.E. (1992). Dynamic properties of medial rectus motoneurons during vergence eye movements. *Journal of Neurophysiology*, **67**, 64–74.

Gamlin, P.D.R. and Reiner, A. (1987). The avian Edinger-Westphal nucleus: Sources of input controlling accommodation, pupilloconstriction and choroidal blood flow. *Society for Neuroscience Abstracts*, **13**, 173.

Gamlin, P.D.R. and Reiner, A. (1991). The Edinger-Westphal nucleus: Sources of input influencing accommodation, pupilloconstriction and choroidal blood flow. *Journal of Comparative Neurology*, **306**, 425–438.

Gamlin, P.D.R., Reiner, A. and Karten, H.J. (1982). Substance P-containing neurons of the avian suprachiasmatic nucleus project directly to the nucleus of Edinger-Westphal. *Proceedings of the National Academy of Sciences USA*, **79**, 3891–3895.

Gamlin, P.D.R., Reiner, A., Erichsen, J.T., Cohen, D.H. and Karten, H.J. (1984). The neural substrate for the pupillary light reflex in the pigeon (*Columba livia*). *Journal of Comparative Neurology*, **226**, 523–543.

Gamlin, P.D.R., Zhang, Y., Clendaniel, R.A. and Mays, L.E. (1994). Behavior of identified Edinger-Westphal neurons during ocular accommodation. *Journal of Neurophysiology*, **72**, 2368–2382.

Gamlin, P.D.R., Zhang, H. and Clarke, R.J. (1995). Luminance neurons in the pretectal olivary nucleus mediate the pupillary light reflex in the rhesus monkey. *Experimental Brain Research*, **106**, 169–176.

Gherezghiser, T., Hey, J. and Koss, M. (1989). Cholinergic control of intraocular pressure. *Investigative Ophthalmology and Visual Science (Suppl.)*, **30**, 20.

Gherezghiser, T., Hey, J.A. and Koss, M.C. (1990). Parasympathetic nervous control of intraocular pressure. *Experimental Eye Research*, **50**, 457–462.

Gilmartin, B. (1986). A review of the role of sympathetic innervation of the ciliary muscle in ocular accommodation. *Ophthalmic and Physiological Optics*, **6**, 23–37.

Goldor, H. and Gay, A.J. (1967). Chorioretinal vascular occlusions with latex microspheres (a long-term study). *Investigative Ophthalmology*, **6**, 51–58.

Graybiel, A.M. and Hartwieg, E.A. (1974). Some afferent connections of the oculomotor complex in the cat: an experimental study with tracer techniques. *Brain Research*, **81**, 543–551.

Hare, W.K., Magoun, H.W. and Ranson, S.W. (1935). Pathways for pupillary constriction: Location of synapses in the path for the pupillary light reflex and of constrictor fibers of cortical origin. *Archives of Neurology and Psychiatry*, **34**, 1188–1194.

Heal, D.J., Prow, M.R., Butler, S.A. and Buckett, W.R. (1995). Mediation of mydriasis in conscious rats by central postsynaptic α_2-adrenoceptors. *Pharmacology, Biochemistry and Behavior*, **50**, 219–224.

Helmholtz, H. (1867). *Handbuch der Physiologischen Optik*. Leipzeig, Voss.

Hosoba, M., Bando, T. and Tsukahara, N. (1978). The cerebellar control of accommodation of the eye in the cat. *Brain Research*, **153**, 495–505.

Hultborn, H., Mori, K. and Tsukahara, N. (1973). The neuronal pathway subserving the pupillary light reflex and its facilitation from cerebellar nuclei. *Brain Research*, **63**, 357–361.

Hultborn, H., Mori, K. and Tsukahara, N. (1978). Cerebellar influence on parasympathetic neurones innervating intra-ocular muscles. *Brain Research*, **159**, 269–278.

Inoue, T. (1980). The response of rabbit ciliary nerve to luminance intensity. *Brain Research*, **201**, 206–209.

Ijichi, J., Kiyohara, T., Hosoba, M. and Tsukahara, N. (1977). The cerebellar control of the pupillary light reflex. *Brain Research*, **128**, 69–79.

Ishikawa, S., Sekiya, H. and Kondo, Y. (1990). The center for controlling the near reflex in the midbrain of the monkey: a double labelling study. *Brain Research*, **519**, 217–222.

Jaeger, R.J. and Benevento, L.A. (1980). A horseradish peroxidase study of the innervation of the internal structures of the eye. *Investigative Ophthalmology and Visual Science* **19**, 575–583.

Jampel, R.S. and Mindel, J. (1967). The nucleus for accommodation in the midbrain of the macaque. The effect of accommodation, pupillary constriction, and extraocular muscle contraction produced by stimulation of the oculomotor nucleus on the intraocular pressure. *Investigative Ophthalmology*, **6**, 40–50.

Johnson, DA. and Purves, D. (1981). Post-natal reduction of neural unit size in the rabbit ciliary ganglion. *Journal of Physiology*, **318**, 143–159.

Johnson, D.A. and Purves, D. (1983). Tonic and reflex synaptic activity recorded in ciliary ganglion cells of anaesthetized rabbits. *Journal of Physiology*, **339**, 599–613.

Judge S.J. and Cumming B.G. (1986). Neurons in the monkey midbrain with activity related to vergence eye movement and accommodation. *Journal of Neurophysiology*, **55**, 915–930.

Karplus, J.P. and Kreidl, A. (1913). Ueber die Bahn des Pupillarreflexes. *Pflugers Archiv für die Gestame Physiologie de Menschen und der Tiere*, **149**,115–155.

Keane, J.R. and Davis, R.L. (1976). Pretectal syndrome with metastatic malignant melanoma to the posterior commissure. *American Journal of Ophthalmology*, **82**, 910–914.

Koss, M.C., Gherezghiher, T. and Nomura, A. (1984). CNS adrenergic inhibition of parasympathetic oculomotor tone. *Journal of the Autonomic Nervous System*, **10**, 55–68.

Koss, M. C. (1986). Pupillary dilation as an index of central nervous system alpha 2–adrenoceptor activation. *Journal of Pharmacological Methods*, **15**, 1–19.

Kourouyan, H and Horton, J.C. (1997). Transneuronal retinal input to the primate Edinger-Westphal nucleus. *Journal of Comparative Neurology*, **381**, 68–80.

Kruger, P. (1979). Infrared recording retinoscope for monitoring accommodation. *American Journal of Optometry and Physiological Optics*, **56**, 116–123.

Kuchiiwa, S., Kuchiiwa, T. and Suzuki, T. (1989). Comparative anatomy of the accessory ciliary ganglion in mammals. *Anatomy and Embryology*, **180**, 199–205.

Kuchiiwa, S., Kuchiiwa, T. and Nakagawa, S. (1994). Localization of preganglionic neurons of the accessory ciliary ganglion in the midbrain: HRP and WGA-HRP studies in the cat. *Journal of Comparative Neurology*, **340**, 577–591.

Kuhlenbeck, H. (1937). The ontogenetic development of the diencephalic centers in the bird's brain (chicken) and comparison with the reptilian and mammalian diencephalon. *Journal of Comparative Neurology*, **66**, 23–75.

Kuhlenbeck, H. (1939). The development and structure of the pretectal cell masses in the chick. *Journal of Comparative Neurology*, **71**, 361–387.

Lindsberg, P.J., O'Neill, J.T., Paakkari, I., Hallenbeck, J. and Feuerstein, G. (1989). Validation of laser-doppler flowmetry in measurements of spinal cord blood flow. *American Journal of Physiology*, **257**, 674–680.

Lindvall, O., Bjorklund, A., Nobin, A. and Stenevi, A. (1974). The adrenergic innervation of the rat thalamus as revealed by the glyoxylic acid fluorescence method. *Journal of Comparative Neurology*, **154**, 317–348.

Ljungdahl, A., Hökfelt, T. and Nilsson, G. (1978). Distribution of Substance P-like immunoreactivity in the central nervous system of the rat — I. Cell bodies and nerve terminals. *Neuroscience*, **3**, 861–943.

Loewenfeld, I.E. (1958). Mechanisms of reflex dilatation of the pupil. Historical review and experimental analysis. *Documenta Ophthalmologica*, **12**, 185–448.

Loewenfeld, I.E. (1993). *The Pupil: Anatomy Physiology, and Clinical Applications*. Ames: Iowa State University Press.

Loewy, A.D., Araujo, J.C. and Kerr, F.W.L. (1973). Pupillodilator pathways in the brain stem of the cat: Anatomical and electrophysiological identification of a central autonomic pathway. *Brain Research*, **60**, 65–91.

Loewy, A.D. and Saper, C.B. (1978). Edinger-Westphal nucleus: projections to the brainstem and spinal cord in the cat. *Brain Research*, **150**, 1–27.

Loewy, A.D., Saper, C.B. and Yamodis, N.D. (1978). Re-evaluation of the efferent projections of the Edinger-Westphal nucleus in the cat. *Brain Research*, **141**, 153–159.

Lowenstein, O. (1954). Clinical pupillary symptoms in lesions of the optic nerve, optic chiasm and optic tract. *Archives of Ophthalmology*, **52**, 385–403.

Lowenstein, O. and Loewenfeld, I.E. (1964). The sleep-waking cycle and pupillary activity. *Annals of the New York Academy of Sciences*, **117**, 142–156.

Lowenstein, O. and Loewenfeld, I.E. (1969). The Pupil. In *The Eye, Vol. 3*, edited by H. Davson, pp. 255–337. New York: Academic Press.

Lyman, D. and Mugnaini, E. (1980). The avian accessory oculomotor nucleus. *Society for Neuroscience Abstracts*, **6**, 479.

Magnuson, D.J., Rezak, M. and Benevento, L.A. (1980). Some afferent connections to the oculomotor complex in the macaque monkey. *Society for Neuroscience Abstracts*, **6**, 478.

Magoun, H.W. and Ranson, S.W. (1935a). The central path of the light reflex: A study of the effect of lesions. *Archives of Ophthalmology*, **13**, 791–811.

Magoun, H.W. and Ranson, S.W. (1935b). The afferent path of the light reflex: A review of the literature. *Archives of Ophthalmology*, **13**, 862–874.

Magoun, H.W., Atlas, D., Hare, W.K. and Ranson, S.W. (1936). The afferent path of the pupillary light reflex in the monkey. *Brain* **59**, 234–249.

Marg, E., Reeves, J.L. and Wendt, W.E. (1954). Accommodative response of the eye to electrical stimulation of the ciliary ganglion in cats. *American Journal of Optometry*, **31**, 127–136.

Marwitt, R., Pilar, G. and Weakly, J.N. (1971). Characterization of two ganglion cell populations in avian ciliary ganglion. *Brain Research*, **25**, 317–334.

Mays, L.E. (1984). Neural control of vergence eye movements: convergence and divergence neurons in the midbrain. *Journal of Neurophysiology*, **51**, 1091–1108.

May, P.J. and Warren, S. (1993). Ultrastructure of the macaque ciliary ganglion. *Journal of Neurocytology*, **22**, 1073–1095.

Mays L.E., Porter J.D., Gamlin P.D.R and Tello C.A. (1986). Neural control of vergence eye movements: neurons encoding vergence velocity. *Journal of Neurophysiology*, **56**, 1007–1021.

May, P.J., Porter, J.D. and Gamlin, P.D.R. (1992). Interconnections between the primate cerebellum and midbrain near-response regions *Journal of Comparative Neurology*, **315**, 98–116.

Melnitchenko, L.V. and Skok, V.I. (1970). Natural electrical activity in mammalian parasympathetic ganglion neurons. *Brain Research*, **23**, 277–279.

Meriney, S.D. and Pilar, G.J. (1987). Cholinergic innervation of the smooth muscle cells in the choroid coat of the chick eye and its development. *Journal of Neuroscience*, **7**, 3827–3839.

Nakanome, Y., Karita, K., Izumi, H. and Tamai, M. (1995). Two types of vasodilation in cat choroid elicited by electrical stimulation of the short ciliary nerve. *Experimental Eye Research*, **60**, 37–42.

Narayanan, C.H. and Narayanan, Y. (1976). An experimental inquiry into the central source of preganglionic fibers to the chick ciliary ganglion. *Journal of Comparative Neurology*, **166**, 101–110.

Nicholson, J. and Severin, C.M. (1981). Afferent projections of the Edinger-Westphal nucleus in the rat. *Anatomical Record*, **199**, 183.

Nisida, I. and Okada, H. (1960). The activity of the pupilloconstrictory centers. *Japanese Journal of Physiology*, **10**, 64–72.

Olmsted, J.M.D. (1944). The role of the autonomic nervous system in accommodation for far and near vision. *Journal of Nervous and Mental Disease*, **99**, 794–798.

Parelman, J.J., Fay, M.T. and Burde, R.M. (1984). Confirmatory evidence for a direct parasympathetic pathway to internal eye structures. *Transactions of the American Ophthalmological. Society*, **82**, 371–380.

Parver, L.M., Auker, C.R., Carpenter, D.O. and Doyle, T. (1982). Choroidal blood flow II. Reflexive control in the monkey. *Archives of Ophthalmology*, **100**, 1327–1330.

Parver, L.M., Auker, C.R. and Carpenter, D.O. (1983). Choroidal blood flow. III. Reflexive control in human eyes. *Archives of Ophthalmology*, **101**, 1604–1606.

Pasik, P., Pasik, T. and Bender, M.B. (1969). Autonomic dysfunctions in monkeys with pretectal lesions. *Federation Proceedings*, **28**, 3175.

Pierson, R.J. and Carpenter, M.B. (1974). Anatomical analysis of pupillary reflex pathways in the Rhesus monkey. *Journal of Comparative Neurology*, **158**, 121–144.

Pilar, G. and Tuttle, J.B. (1982). A simple neuronal system with a range of uses: The avian ciliary ganglion. In *Progress in Cholinergic Biology: Model Cholinergic synapses*, edited by A. Goldberg and I. Hanin, pp. 213–247. New York: Raven Press.

Pitts, D.G. (1967). Electrical stimulation of oculomotor nucleus: The effects of stimulus voltage and anesthesia. *American Journal of Optometry*, **44**, 505–516.

Ponsford, J.R., Bannister, R. and Paul, E.A. (1982). Methacholine pupillary responses in third nerve palsy and Adie's syndrome. *Brain*, **105**, 583–597.

Powers, A.S. and Reiner, A. (1993). The distribution of cholinergic neurons in the central nervous system of the turtle. *Brain, Behavior and Evolution*, **41**, 326–345.

Ranson, S.W. and H.W. Magoun (1933). The central path of the pupilloconstrictor reflex in response to light. *Archives of Neurology and Psychiatry (Chicago)*, **30**, 1193–1204.

Reiner, A., Karten, H.J., Gamlin, P.D.R. and Erichsen, J.T. (1983). Parasympathetic ocular control: Functional subdivisions and circuitry of the avian nucleus of Edinger-Westphal. *Trends in Neurosciences*, **6**, 140–145.

Reiner, A., Brauth, S.E., Kitt, C.A. and Quirion, R. (1989). Distribution of μ, δ, and κ opiate receptor types in the forebrain and midbrain of pigeons. *Journal of Comparative Neurology*, **280**, 359–382,

Reiner, A., Erichsen, J.T., Cabot, J.B., Evinger, C., Fitzgerald, M.E. and Karten, H.J. (1991). Neurotransmitter organization of the nucleus of Edinger-Westphal and its projection to the avian ciliary ganglion. *Visual Neuroscience*, **6**, 451–472.

Ripps, H., Breinin, G.M. and Baum, J.L. (1961). Accommodation in the cat. *Transactions of the American Ophthalmological Society*, **59**, 176–193.

Roste, G.K. and Dietrichs, E. (1988). Cerebellar cortical and nuclear afferents from the Edinger-Westphal nucleus in the cat. *Anatomy and Embryology*, **178**, 59–65.

Ruskell, G.L. (1971). Facial parasympathetic innervation of the choroidal blood vessels in monkeys. *Experimental Eye Research*, **12**, 166–172.

Ruskell, G.L. (1990). Accommodation and the nerve pathway to the ciliary muscle: a review. *Ophthalmological and Physiological Optics*, **10**, 239–242.

Ruskell, G.L. and Griffiths, T. (1979). Peripheral nerve pathway to the ciliary muscle. *Experimental Eye Research*, **28**, 277–284.

Scalia, F. (1972). The termination of retinal axons in the pretectal region of mammals. *Journal of Comparative Neurology*, **145**, 223–257.

Schaeffel, F., Troilo, D., Wallman, J. and Howland, H.C. (1990). Developing eyes that lack accommodation grow to compensate for imposed defocus. *Visual Neuroscience*, **4**, 177–183.

Scherer, S.S. (1986). Reinnervation of the extraocular muscles in goldfish is nonselective. *Journal of Neuroscience*, **6**, 764–773.

Schlag, J.D. and Fuller, J.H. (1976). Determination of antidromic excitation by the collision test: problems of interpretation. *Brain Research*, **112**, 283–298.

Sekiya, H., Kawamura, K. and Ishikawa, S. (1984). Projections from the Edinger-Westphal complex of monkeys as studied by means of retrograde axonal transport of horseradish peroxidase. *Archives Italienne de Biologie*, **122**, 311–319.

Sharpe, L.G. and Pickworth, W.B. (1985). Opposite pupillary size effects in the cat and dog after microinjections of morphine, normorphine and clonidine in the Edinger-Westphal nucleus. *Brain Research Bulletin*, **15**, 329–333.

Shih, Y.F., Fitzgerald, M.E. and Reiner, A. (1993). Effect of choroidal and ciliary nerve transection on choroidal blood flow, retinal health, and ocular enlargement. *Visual Neuroscience*, **10**, 969–979.

Sillito, A.M. and Zbrozyna, A.W. (1970a). The activity characteristics of the preganglionic pupilloconstrictor neurones. *Journal of Physiology (London)*, **211**, 767–779.

Sillito, A.M. and Zbrozyna, A.W. (1970b). The localization of pupilloconstrictor function within the midbrain of the cat. *Journal of Physiology (London)*, **211**, 461–477.

Steiger, H-J. and Büttner-Ennever, J.A. (1979). Oculomotor nucleus afferents in the monkey demonstrated with horseradish peroxidase. *Brain Research*, **160**, 1–15.

Stjernschantz, J., Alm, A. and Bill, A. (1976). Effects of intracranial oculomotor nerve stimulation on ocular blood flow in rabbits: Modification by indomethacin. *Experimental Eye Research*, **23**, 461–469.

Sugimoto, T., Itoh, K. and Mizuno, N. (1977). Localization of neurons giving rise to the oculomotor parasympathetic outflow: A HRP study in cat. *Neuroscience Letters*, **7**, 301–305.

Sugimoto, T., Itoh, K. and Mizuno, N. (1978). Direct projections from the Edinger-Westphal nucleus to the cerebellum and spinal cord in the cat: An HRP study. *Neuroscience Letters*, **9**, 17–22.

Sun, W. and May, P.J. (1993). Organization of the extraocular and preganglionic motoneurons supplying the orbit in the lesser Galago. *Anatomical Record*, **237**, 89–103.

Thomas, D.M., Kaufman, R.P., Sprague, J.M. and Chambers, W.W. (1956). Experimental studies of the vermal cerebellar projections in the brain stem of the cat (Fastigiobulbar tract). *Journal of Anatomy*, **90**, 371–385.

Thompson, H.S. (1987). The Pupil. In *Adler's Physiology of the Eye: Clinical Application*, 8th edn., edited by R.A. Moses and W.M. Hart, Jr., St. Louis: C.V. Mosby Co.

Törnqvist, G. (1966). Effect of cervical sympathetic stimulation on accommodation in monkeys. *Acta Physiologica Scandinavica*, **67**, 363–372.

Tornqvist, G. (1967). The relative importance of the parasympathetic and sympathetic nervous systems for accommodation in monkeys. *Investigative Ophthalmology*, **6**, 612–617.

Toyoshima, K., Kawana, E. and Sakai, H. (1980). On the neuronal origin of the afferents to the ciliary ganglion in cat. *Brain Research*, **185**, 67–76.

Troilo, D. and Wallman, J. (1987). Changes in corneal curvature during accommodation in chicks. *Vision Research*, **27**, 241–247.

Tsukahara, N., Kiyohara, T. and Ijichi, Y. (1973). The mode of cerebellar control of pupillary light reflex. *Brain Research*, **60**, 244–248.

Warwick, R. (1954). The ocular parasympathetic nerve supply and its mesencephalic sources. *Journal of Anatomy*, **88**, 71–93.

Wathey, J.C. (1988). Identification of the teleost Edinger-Westphal nucleus by retrograde horseradish peroxidase labeling and by electrophysiological criteria. *Journal of Comparative Physiology A*, **162**, 511–524.

Wathey, J.C. and Wullimann, M.F. (1988). A double-label study of efferent projections from the Edinger-Westphal nucleus in goldfish and kelp bass. *Neuroscience Letters*, **93**, 121–126.

Weber, J.T and Harting, J.K. (1980). The efferent projections of the pretectal complex: An autoradiographic and horseradish peroxidase analysis. *Brain Research*, **194**, 1–28.

Westheimer, G. and Blair, S.M. (1973). The parasympathetic pathways to internal eye muscles. *Investigative Ophthalmology*, **12**, 193–197.

Westphal, C. (1887). Ueber einen Fall von chronischer progressiver Lähmung der Augenmuskeln (ophthalmoplegia externa) nebst Beschreibung von Ganglienzellengruppen in Bereiche des Oculomotoriuskerns. *Archiv für Psychiatrie und Nervenkrankheim*, **18**, 846–871.

Young, M.J. and Lund, R.D. (1994). The anatomical substrates subserving the pupillary light reflex in rats: origin of the consensual pupillary response. *Neuroscience*, **62**, 481–496.

Zhang, H.Y. and Gamlin, P.D.R. (1996). Single-unit activity within the posterior fastigial nucleus during vergence and accommodation in the alert primate. *Society for Neuroscience Abstracts*, **22**, 2034.

Zhang, Y., Gamlin, P.D.R. and Mays, L.E. (1991). Antidromic identification of midbrain near response cells projecting to the oculomotor nucleus. *Experimental Brain Research*, **84**, 525–528.

Zhang, Y., Mays, L.E. and Gamlin, P.D.R. (1992). Characteristics of near response cells projecting to the oculomotor nucleus. *Journal of Neurophysiology*, **67**, 944–960.

5 Evidence for the Intrinsic Innervation of Retinal Vessels: Anatomical Substrate of Autoregulation in the Retina?

John Greenwood[1], Philip L. Penfold[2] and Jan M. Provis[2,3]

[1]*Department of Clinical Ophthalmology, Institute of Ophthalmology, University College London, Bath Street, London, EC1V 9EL, UK*
[2]*Departments of Clinical Ophthalmology, University of Sydney, Sydney Eye Hospital, New South Wales 2001, and* [3]*Anatomy and Histology, University of Sydney, New South Wales 2006, Australia*

The factors controlling blood flow to the retina are believed to be complex and multifarious and include processes that are both extrinsic and intrinsic to the retina. Although autonomic innervation of the central retinal artery may provide a limited degree of control over retinal perfusion, it is autoregulatory processes that are thought to be the main regulator of retinal vascular tone and blood flow. As no meaningful extrinsic innervation of the retinal vasculature beyond the lamina cribrosa has been demonstrated, the control of retinal vascular tone is most likely due to the local release of vasoactive products. Unlike the retinal vasculature, which forms one aspect of the blood-retinal barrier, blood flow in the choroid is regulated predominantly through autonomic activity. In this review we survey the limited literature relating to the control of retinal blood flow and provide new evidence for the existence of intrinsic innervation of retinal vessels as one possible basis for autoregulation. The spatial arrangement of substance-P and NADPH diaphorase positive processes and terminals are described in detail. The relevance of intrinsic retinal vascular innervation to the control of vascular tone and the impact that such factors may have upon the function of the blood-retinal barrier are discussed.

KEY WORDS: blood-retinal barrier; blood-flow; nitric oxide; retina; substance-P; retinal vasculature.

INTRODUCTION

The regulation of blood flow to the retina is a critical factor in the maintenance of normal retinal function and any failure of the processes regulating retinal perfusion is likely to

Correspondence: J. Greenwood, Department of Clinical Ophthalmology, Institute of Ophthalmology, University College London, Bath Street, London, EC1V 9EL, UK; Tel: +44 171 608 6858; Fax: +44 171 608 6810; E-mail: j.greenwood@ucl.ac.uk have a profound effect on retinal homeostasis. In particular, dysfunctional retinal blood flow has been associated with a variety of retinal disorders including glaucoma and hypertensive retinopathy. The retina receives its blood supply from two separate and distinct vascular beds. In general, the choroidal vasculature supplies the *outer*, avascular region of the retina (the photoreceptor layer) by diffusion across the retinal pigmented epithelium (RPE). The retinal vasculature is normally confined to the inner retina as far as the boundary between the inner nuclear layer and outer plexiform layer, and supplies these *inner* layers as well as 10% of the oxygen requirements of the photoreceptors during dark adaptation (Ahmed *et al.*, 1993).

The vessels comprising these two anatomically discrete vasculatures differ extensively in their phenotype (Greenwood, 1992; Greenwood *et al.*, 1995) as well as in the level of autonomic innervation and autoregulation of blood flow within them (Bill and Nilsson, 1985; Bill and Sperber, 1990; Haefliger *et al.*, 1994; Brown and Jampol, 1996; Funk, 1997). It is generally believed that blood flow to the choroid is regulated predominantly through autonomic activity, whilst that to the retina is almost entirely through autoregulation. Despite the importance of the retinal blood supply the precise mechanisms controlling retinal perfusion remain poorly understood. Recently, however, evidence is emerging that retinal vascular tone may also be controlled to some degree by direct intrinsic innervation. Regulation of retinal blood flow, including both vascular innervation and autoregulatory mechanisms, should not be considered in isolation but must be examined in the context of the unique nature and importance of the retinal vasculature and the considerable retinal metabolic demands. In this review we will summarise the current understanding of the control of blood flow to the retina and speculate on the possible involvement of retinal vascular innervation in controlling this process.

BLOOD SUPPLY TO THE RETINA

Before examining the factors controlling retinal perfusion, it is important that the organisation of the blood supply to the retina should be understood. The retina is supplied by the central retinal artery where, in the human, it arises from the ophthalmic artery approximately 1 cm behind the eye. With the central retinal vein it enters the optic nerve and passes forward in the centre of the optic nerve, through a gap in the lamina cribrosa and emerges centrally through the papilla. As described in more detail below, this region forms an important point of demarcation between the different predominant factors controlling flow through the central retinal artery and its branches and that of the retinal vasculature. From here, the central retinal artery branches into the superior and inferior branches which then subdivide into nasal and temporal arteries. The superior and inferior temporal arteries curve above and below the macula and foveal regions. In approximately 20% of people there exists a cilioretinal artery (arising from a posterior ciliary artery) which forms a small anastomotic connection between the choroidal and retinal circulation and explains why macula function may be preserved in central retinal artery occlusion. At present, it is unclear whether the innervation of the branches of the posterior ciliary artery that form these connections extend into the retina itself or whether, like those of the central retinal artery, they stop at the point of entry (Ye, Laties and Stone, 1990).

Each of the large retinal arterial branches projects through the nerve fibre layer beneath the inner limiting membrane. There are four major branches each supplying a quadrant of the retina and, as no overlap exists between them, these vessels are considered to be functional end-arteries. Throughout the retina there exist two main levels of capillary networks (although this pattern does vary in some areas). The inner plexus, which is situated at the level of the ganglion cell layer, and the outer plexus, which is at the level of the deep aspect of the inner nuclear layer. The outer retina (outer plexiform and photoreceptor layers) and a 500 μm diameter region centred on the fovea are avascular. The density of capillaries varies, being most dense at the macula, decreasing in density towards the peripheral retina where neuronal populations are also less dense. Capillary free zones also surround arterioles and arteries.

The outer retina and the entire thickness of the foveal retina derive much of their nutrition from the choroidal circulation, which lies beneath the RPE extending from the optic nerve margins to the periphery of the retina where it is continuous with the vascular plexus of the ciliary body. This vascular bed is supplied predominantly by the two long posterior ciliary arteries and the short posterior ciliary arteries, with some contribution arising from anastomoses with the anterior ciliary arteries. The capillary bed of the choroidal circulation is the choriocapillaris, a wide-bore (20–40 μm), fenestrated capillary network which abuts Bruch's membrane, a basement membrane-like layer which intervenes between the choriocapillaris and the RPE.

The primate choroid, unlike the retinal vasculature, is richly innervated, particularly at the posterior pole, in the vicinity of the fovea (Flugel et al., 1994; Flugel-Koch, Kaufman and Lutjen-Drecoll, 1994). Evidence supports the involvement of parasympathetic fibres from the pterygopalatine ganglion (Ruskell, 1971) and sympathetic innervation from the superior cervical ganglion (Nuzzi, Gugleilmone and Grignolo, 1995; Klooster et al., 1996). Stimulation of sympathetic nerves from the cervical sympathetic ganglion brings about a frequency-dependent vasoconstriction of choroidal vessels and is thought to prevent choroidal overperfusion (Bill and Sperber, 1990). Conversely, stimulation of the parasympathetic facial nerve results in increased blood flow (Stjernschantz and Bill, 1980).

THE BLOOD-RETINAL BARRIER

The retinal vascular endothelium is highly specialised, forming the *inner* blood-retinal barrier (BRB), the *outer* BRB being formed by the RPE. This selective cellular barrier regulates the passage of molecules and cells, through specific receptors and nutrient transport systems, both into and out of the retina, helping to maintain homeostasis. The endothelial cells of the retinal vasculature are identical in most respects to the endothelia that form the blood-brain barrier. The morphological correlates of this vascular barrier are the tight junctions between endothelial cells which are capable of excluding ions, as well as a lack of pores, fenestrations and vesicular activity. In addition to this physical barrier, both brain and retinal endothelium express a unique repertoire of surface receptors and may thus respond to external signals differently from those of non-CNS endothelium. It is important to consider, therefore, that the release of vasoactive factors involved in the regulation of local blood flow, as described below, may also affect other functions of the vascular endothelium in particular its capacity to maintain a patent barrier to circulating molecules.

Retinal endothelial cells produce a thick basal lamina within which there also exist numerous pericytes, these being more abundant than in the brain. These pericytes may have important properties relating to the vascular tone, especially in the capillary bed, as they have been shown to possess contractile properties (Kelley *et al.*, 1987; Hirschi and D'Amore, 1996). Surrounding the basement membrane of the retinal vessels is the glia limitans, comprising the processes of both astrocytes and microglia (Stone and Dreher, 1987; Provis *et al.*, 1995). Astrocytes produce vasoactive substances and are also thought to produce factors which induce the tight, barrier phenotype within the vascular endothelium (Janzer and Raff, 1987; Raub, Luentzel and Sawada, 1992). The inclusion of microglia in the glial limitans similarly may suggest that microglia have a role in regulating barrier function (Diaz, Penfold and Provis, 1998).

In contrast to retinal vessels, the choriocapillaris is composed of permeable, fenestrated endothelia (particularly where they abut Bruch's membrane) which do not posses any barrier properties. Between this leaky vascular bed and the neural retina lies the posterior barrier formed by a monolayer of RPE joined together by tight apical junctions, providing a selectively permeable barrier between the choroid and the neurosensory retina.

RETINAL BLOOD FLOW

The retina is extremely metabolically active, having the highest oxygen consumption per weight of any human tissue. Both the choroidal circulation (outer third) and the retinal circulation (inner two thirds) supply the retina with its metabolic requirements. Ninety-eight per cent of blood to the eye passes through the uveal tract (choroid, ciliary body and iris), of which 85% is through the choroid. The choroidal circulation has a high flow rate (150 mm/s), low oxygen extraction (only 5–10% of its oxygen being extracted) and may also provide thermoregulation. The retina on the other hand receives approximately 5% of the total blood flow to the eye and, unlike the choroid, has a low flow rate (25 mm/s) and high oxygen exchange. It has been suggested that such differences are due to the spatial relationship of the respective vascular beds to the retina. Thus, the distance for nutrients to reach the outer retina from the choroid is relatively large and requires a high concentration gradient to overcome these distances and to drive the carrier-facilitated diffusion transport processes at the posterior BRB. In particular, in order to meet the metabolic requirements of photoreceptors, oxygen must diffuse across the entire thickness of the RPE plus the length of the outer segments and inner segments (around 100 μm at the fovea) to reach mitochondria concentrated in the ellipsoid. On the other hand the retinal circulation has only to supply the parenchymal cells in their immediate vicinity and as such does not require such high blood-to-tissue concentration gradients to be maintained.

An important characteristic of the retinal vasculature is its phenomenal capacity to maintain constant retinal blood flow in the face of considerable alterations in both perfusion pressure and intraocular pressure (Bill and Nilsson, 1985). The mechanisms responsible for this exquisite control of blood flow to the retina are not entirely clear although it is recognised that autoregulatory processes are the primary mechanism behind the control of retinal vascular tone.

AUTOREGULATION

Since retinal vessels have no conventional sympathetic innervation (Ye, Laties and Stone, 1990), blood flow in response to raised blood pressure is largely dependent upon autoregulation. In healthy humans and experimental animals retinal blood flow is exceptionally adept at autoregulating flow in response to changes in pressure and metabolic demand. This remarkable regulatory capacity is thought to be achieved through the local vascular response to released vasoactive metabolites including eicosanoids, vasogenic amines, autocoids, peptides and nitric oxide (NO) as well as oxygen and carbon dioxide tensions, pH, intravascular pressure (myogenic response) and sheer stress. The variability of the source and site of action further compound the potential complexity of response brought about by such a diverse array of stimuli. Thus, vasoactive agents may be blood-borne or released from vascular and perivascular cells, including local intrinsic neurones, and may act either directly or indirectly, upon smooth muscle or pericytes and endothelium. It is likely that the complex interaction of these vasoactive metabolites results in a balance between vasoconstriction and vasodilatation and that vessel tone depends on their relative actions. Unlike the retina, the choroid is believed to be largely devoid of autoregulatory mechanisms partly because the supply is too far removed from the local metabolic triggers. In addition, any substances released by the local tissue which may initiate autoregulatory responses would have to overcome the outer BRB to exert their effects upon the choriocapillaris.

Many of the vasoactive metabolites reported to be involved in autoregulation are also inflammatory agents and have a variety of additional actions upon vascular endothelium including the induction of inflammatory molecules and vascular leakage. This can be illustrated by the purine vasodilator, adenosine, which has recently has been proposed as an important mediator in controlling retinal arterial tone (Braunagel, Xiao and Chiou, 1988; Gidday and Park, 1993; Crosson, DeBenedetto and Gidday, 1994). However, as with many vasoactive metabolites it may also lead to breakdown of the blood-retinal barrier and haemorrhage (Sen and Campochiaro, 1989; Campochiaro and Sen, 1989). The literature relating to autoregulation of retinal vascular blood flow is outside the main scope of this review but has been explored in more detail in a number of recent reviews (Haefliger et al., 1994; Brown and Jampol, 1996; Funk, 1997)

INNERVATION OF RETINAL VASCULATURE

Although the retinal vasculature lacks any conventional autonomic innervation, there is growing evidence for the existence of central innervation of retinal vessels which may influence retinal blood flow. Indeed, some of the vasoactive metabolites described above, can be generated by neurones and act as classical neurotransmitters. In addition, it has been recognised for many years that there is autonomic innervation of the central retinal artery and that, as the major supplier of blood flow to the retinal vasculature, such vasomotor control can influence retinal perfusion.

THE CENTRAL RETINAL ARTERY

As a branch of the internal carotid, the central retinal artery is innervated by both sympathetic and parasympathetic neurones as far as the lamina cribrosa. Beyond this point however, the presence of peripheral innervation of the vasculature has not been found to any meaningful degree. Using a histofluorometric technique to visualise catecholamines, Laties (1967) demonstrated in new world monkeys that the central retinal artery is innervated with a fine and plentiful plexus of adrenergic nerve fibres, which were also found to be present in abundance in the choroidal vessels of the same animals. Adrenergic innervation of the central retinal vein was also observed, although to a lesser extent than the artery. However, within the globe the arteriolar branches of the central retinal artery lose their adrenergic innervation although it was reported that in some animals there remained some evidence within the region of the optic disc. Beyond this point the bifurcating arterioles and capillary bed is also devoid of adrenergic innervation whilst the small branching arterioles behind the lamina cribrosa retain their innervation. This situation is slightly different from that in the brain where the plexus of adrenergic nerve fibres of the great vessels continues, albeit very much reduced, over the parenchymal arteries (Owman, Edvinsson and Nielsen, 1974).

In the rabbit, substance P (SP)-like immunoreactivity has also been detected in the central retinal artery, but not in the central retinal vein or retinal vessels (Kumagai *et al.*, 1988) leading to the proposal that the central retinal artery is innervated by peripheral nerves whereas the vessels of the retina are innervated by the central nervous system. There is also good evidence for peptidergic innervation of the central retinal artery of the rat and monkey (Ye, Laties and Stone, 1990) where beaded nerve fibres immunoreactive to the vasoactive peptides, calcitonin gene-related peptide (CGRP), neuropeptide Y (NPY), SP and vasoactive intestinal peptide (VIP) were found to be present in the perivascular space. Both CGRP and SP immunoreactivity co-localised to the same nerve fibres. More recently, both aminergic and cholinergic vasomotor innervation of the human central retinal artery and vein have also been identified (Komai *et al.*, 1995). Similarly, immunohistochemical identification of nitric oxide synthase immunoreactivity (NOS-IR) in nerve fibres within the adventitia and media of the central retinal artery of the dog has been observed (Toda, Kitamura and Okamura., 1994) suggesting that NO liberated from these potential vasodilator nerves may be capable of altering arterial muscle tone and blood flow to the retina. Thus, in addition to a role in autoregulation through local endothelial cell synthesis, NO may also be released from neurones.

The strong evidence for innervation of the central retinal artery would suggest that autonomic innervation provides a degree of control over perfusion of the retina. The relative importance of this control in the overall regulation of retinal blood flow, however, is not entirely clear.

EVIDENCE FOR RETINAL VASCULAR INNERVATION

Although retinal arteries appear to lack extrinsic innervation, the presence of autonomic receptors on their walls implies the existence of direct innervation by retinal neurones. There appear to be two candidate substances for neuromodulation of the retinal vasculature; the potent vasoactive metabolite/neurotransmitter NO, which has been strongly implicated in retinal vasodilatation (Nilsson, 1996), and SP, which is implicated in the regulation of cerebral blood flow (Edvinsson, 1985). While it is likely that alterations in NO production by the retinal vascular endothelium or other perivascular cells (Donati *et al.*, 1995) regulates retinal blood flow, it has also been suggested that endothelial NO-mediated dilatation may also be activated by SP (Kitamura *et al.*, 1993).

Nitroxidergic processes derived from intrinsic retinal cells have been described in human and rat retina (Roufail, Stringer and Rees, 1995; Penfold and Provis, 1998) and are also present in cat retina (Provis and Penfold, unpublished observation). In the human retina, three classes of NADPH-diaphorase positive cells have been described previously, based on soma diameter and dendritic stratification and named ND1-3 (Provis and Mitrofanis, 1990). For convenience we refer to nitroxidergic cells with processes contacting the retinal vasculature, as NDv cells (Figure 5.1). NDv cells are relatively large and characteristically have an intensely positive, coarse process arising from the aspect of the cell body adjacent to a retinal vessel. Additional processes radiate from the soma and presumably contact other retinal cells. The processes of NDv cells are relatively short, reaching vessels within 50 μm, or so. At the vascular basement membrane these processes appear to become broader and flatter, as well as more intensely positive, at the point of contact. Ultrastructurally there appear to be two sizes of NOS-IR fibres associated with the basement membrane of the retinal vessels in humans (Figure 5.2). Whether these correspond to terminals of the 'coarse' and 'fine' fibres is not clear. We have seen the large NOS-IR processes opposed to the glia limitans (Figure 5.2A) while the small NOS-IR processes have been observed within the perivascular space of the retinal vessels (Figure 5.2B).

Given the important role NO plays in the relaxation of blood vessels in other parts of the body as well as, it appears, in the choroid (Flugel *et al*, 1994; Flugel-Koch, Kaufman and Lutjen-Drecoll, 1994) it seems likely that NOS-IR neuronal processes associated with the retinal vessels are involved in modulation of retinal blood flow, although the mechanism of this involvement is not known.

Some of the strongest evidence to emerge for retinal vascular innervation has come from studies investigating the distribution of SP-immunoreactive (SP-IR) retinal neurones and their association with retinal vessels. Immunoreactivity of retinal neurons for SP has been investigated in a range of vertebrates (for a review see Kolb *et al.*, 1995) and several studies have reported SP-IR in neurons of the ganglion cell (GCL) and inner nuclear layers (INL) of the primate retina (Brecha *et al.*, 1982; Marshak, 1989; Li and Lam, 1990; Provis, Yip and Penfold, 1992; Cuenca, De Juan and Kolb, 1995; Cuenca and Kolb, 1998; Penfold and Provis, 1998). In general, SP is localised in amacrine cell populations and in some species, including primates, is also contained in a population of ganglion cells (Kolb *et al.*, 1995; Cuenca, De Juan and Kolb, 1995; Cuenca and Kolb, 1998). Since the retina has abundant amacrine-derived nerve processes containing neurotransmitters and/or neuropeptides, a local mechanism coupled to retinal activity might contribute to the local regulation of retinal blood flow. Indeed, in those few studies in which this possibility has been investigated, the presence of varicose, SP-IR, dendritic terminals associated with the retinal vasculature in primate retinae has been noted (Verstappen *et al.*, 1986; Provis, Yip and Penfold, 1992; Cuenca, De Juan and Kolb, 1995), although their precise structural features and functional significance remain to be determined.

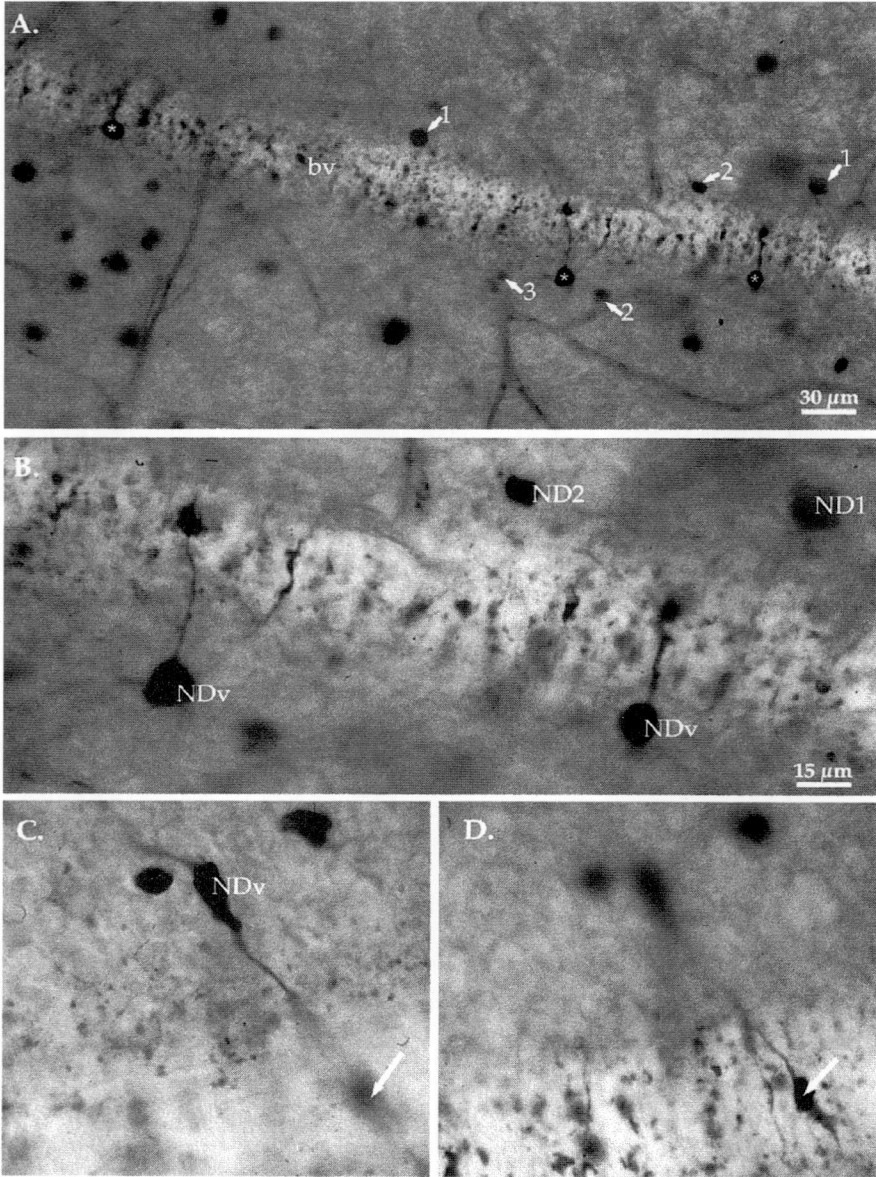

Figure 5.1   Micrographs showing NADPH-diaphorase reactivity in normal adult human retina. (A) Low power showing four types of reactive somas in the inner part of the inner nuclear layer. Cells indicated 1–3 represent the three classes of NADPH-diaphorase positive cells — ND1, ND2, and ND3 — described previously (Provis and Mitrofanis, 1990). A fourth type with processes contacting the retinal vessel which traverses the image is marked with asterisks (*) and represents NDv cells (see **B-D**). (B) The two NDv cells to the right of the field in (**A**) at higher magnification, showing the expansions of their processes at presumed points of contact with the retinal vessel. The same cells in different planes of focus, at the soma (**C**) and level of vascular associations (**D**). Note also the patches of NADPH-diaphorase reactivity scattered along the vascular wall (**A-D**). (**B-D**) are at the same magnification.

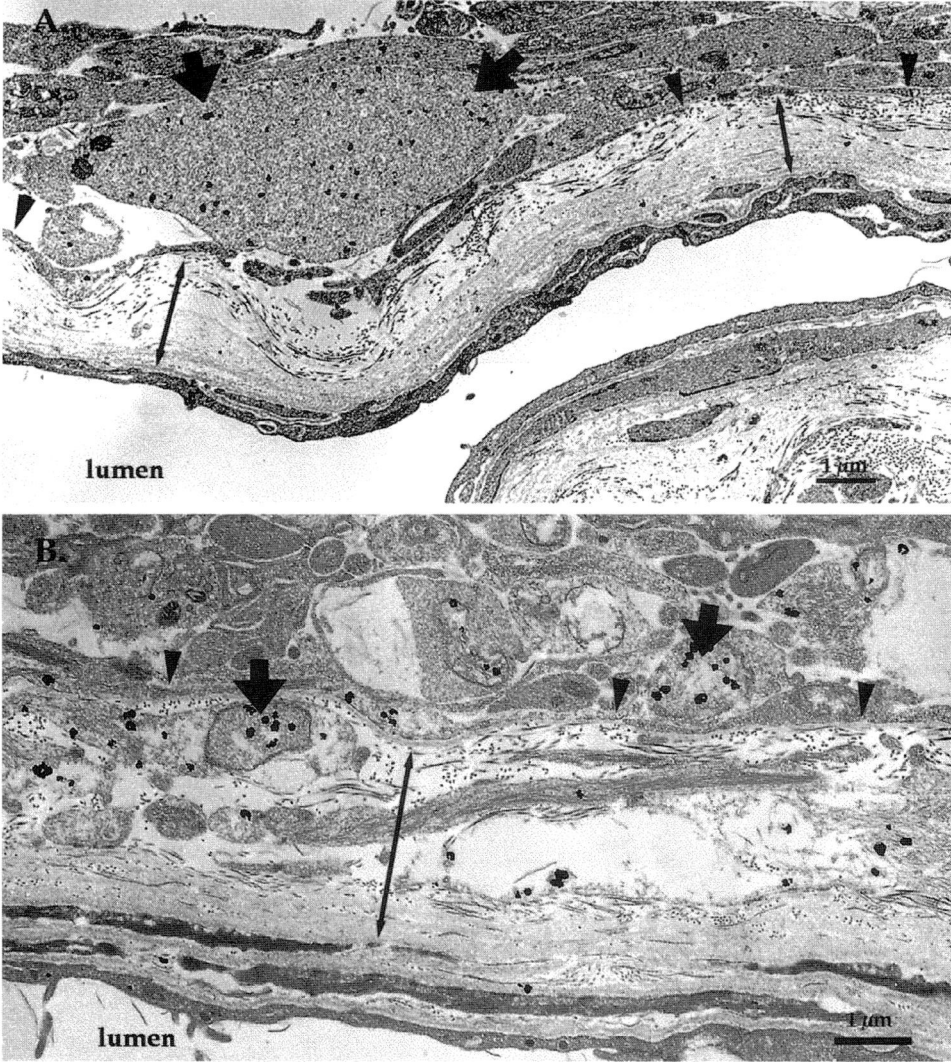

Figure 5.2   Electron micrographs showing nitric oxide synthase-immunoreactive (NOS-IR) processes associated with vessels in normal adult human retina. The basement membrane of the vessels is indicated by the double headed arrows, the glia limitans by the arrowheads. (**A**) A large NOS-IR process (broad arrows) is in close contact with the glia limitans but was not seen to directly interact with the basement membrane of the vascular endothelium in the perivascular space. (**B**) Two small NOS-IR processes (broad arrows); the process on the right is located within the neuropil, in contact with the glia limitans. The process on the left is clearly located within the perivascular space.

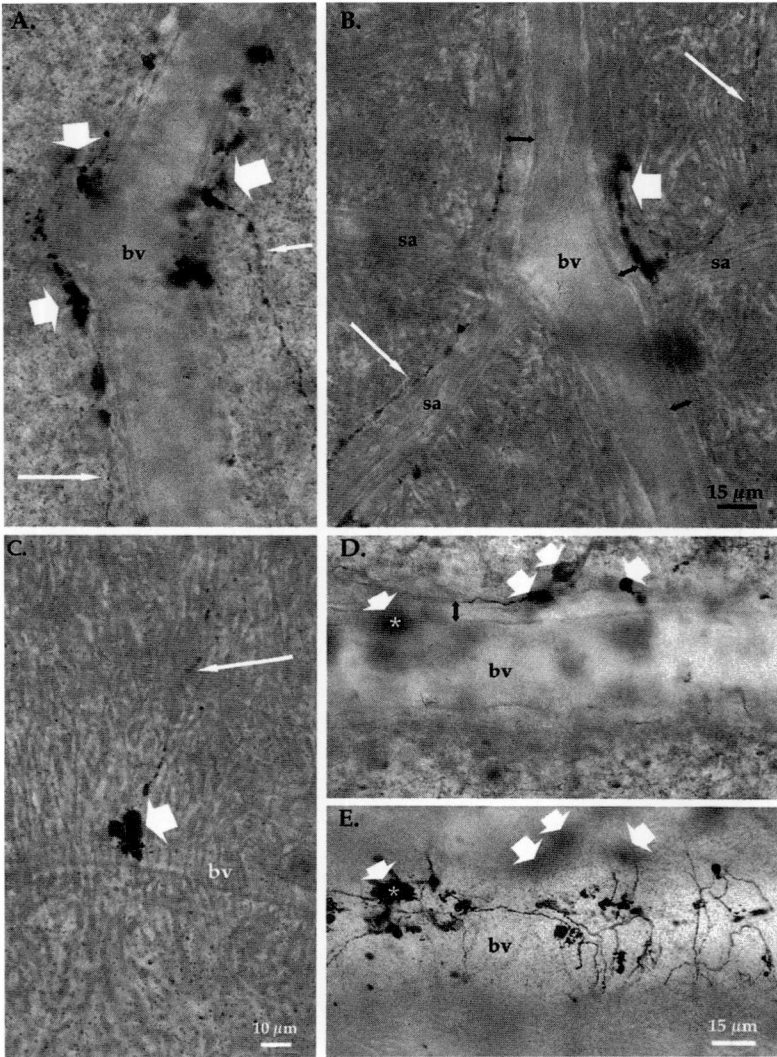

Figure 5.3   Substance P-immunoreactive (SP-IR) terminals associated with blood vessels in whole mounts of human retina. (**A**) Several terminal clusters of SP-IR varicosities (broad arrows), associated with a medium calibre vessel (bv) in the ganglion cell layer (GCL) of human retina. Individual clusters appear to be derived from different nerve fibres (thin arrows) arising from different cells. (**B**) A medium calibre vessel in the GCL with three side arm branches and related SP-IR cell processes (thin arrows) and a single terminal cluster of SP-IR varicosites (broad arrow). The SP-immunoreactive fibres and the SP-IR varicosity appear to be separated from the vascular endothelium by the perivascular space (double-headed arrow). (**C**) A single SP-IR nerve fibre (thin arrow) ending in a cluster of SP-IR varicosites associated with a fine calibre blood vessel (bv) in the inner retina. (**D**) and (**E**) A single large calibre vessel in the GCL of human retina taken in two focal planes. In (**D**) the plane of focus is deep (about midway through the vessel at the point of maximum diameter) showing three associated SP-IR terminals (broad arrows) in the plane of focus and a fourth out of the plane of focus. The gap between the retinal parenchyma and the vascular endothelium (perivascular space) is indicated by the double headed arrow. In (**E**) the plane of focus is on the superficial (inner) aspect of the vessel, showing a plexus of SP-IR fibres with associated varicosities. The features indicated in (**D**) are also indicated in (**E**), for comparison. (**A**) and (**B**) are at the same magnification. (**D**) and (**E**) are at the same magnification.

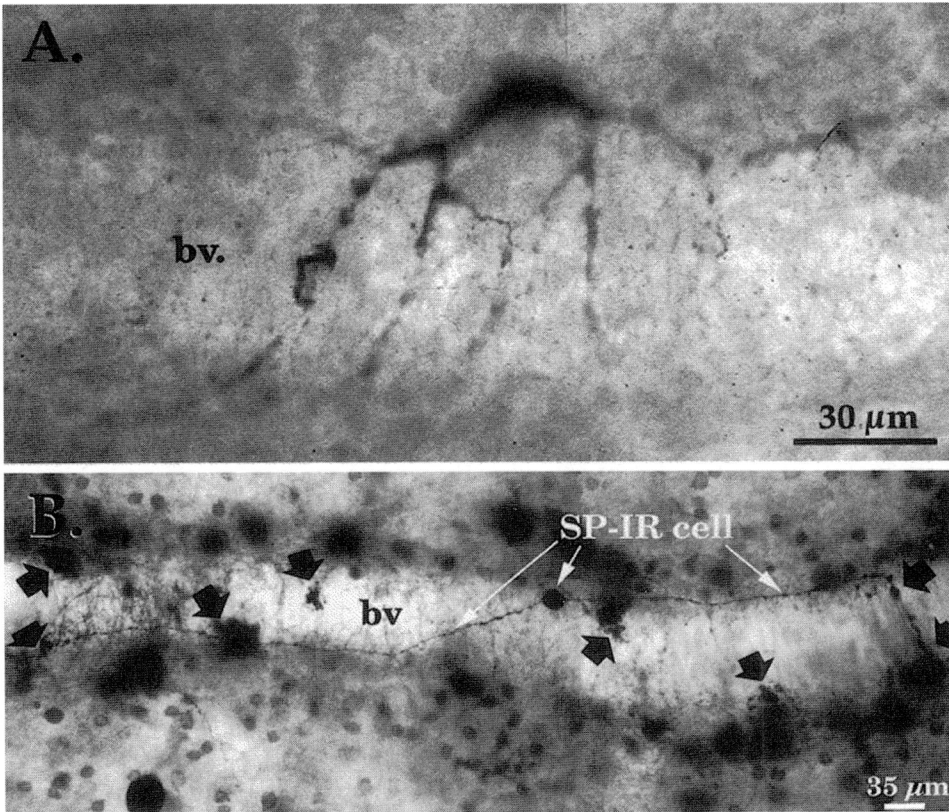

Figure 5.4   Substance P-immunoreactive (SP-IR) vessel-associated cells in whole mounts of human retina. (A) An example of a cell with a very large soma lying adjacent to a large-calibre vessel in the ganglion cell layer. The processes of the cell invest the blood vessel (bv) although no SP-IR varicosities are present. (B) A rare example of a SP-IR soma and processes which course along the length of the vessel and terminate in clusters of SP-IR varicosities (towards the left and right extremities of the micrograph). Other varicosities are also indicated (broad arrows). Note that not all of the SP-IR plexus is derived from this cell.

For sources detailing the morphology and stratification of SP immunoreactive cells and processes in human retina the reader is referred to Cuenca et al., (Cuenca, De Juan and Kolb, 1995; Cuenca and Kolb, 1998). However, two of us (JMP and PLP) have been particularly interested in defining the features of the interaction between SP-IR cell processes and the retinal vasculature.

In whole mounts of human retina the superficial aspect of the major retinal vessels is enmeshed in a network of SP-IR nerve fibres, many of which terminate in clusters of SP-IR varicosities, particularly at the branch points of large vessels (Figure 5.3). In only a few circumstances was it possible to identify the cell somas from which these fibres originate (Figure 5.4) Using double immunolabelling for SP and glial fibrillary acidic protein (GFAP) to label astrocytes, these varicosities are evidently integrated with the glia limitans of the retinal vessels (Figure 5.5). Ultrastructural localisation of SP confirmed

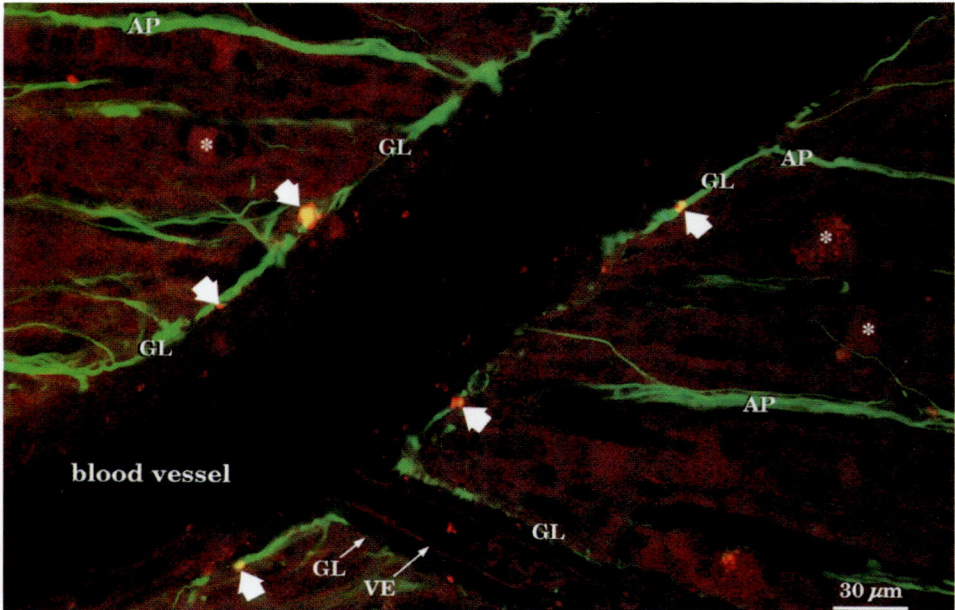

Figure 5.5   An optical section taken using dual-scanning confocal microscopy of a major retinal vessel in a whole mount of adult human retina, double immunolabelled with anti-glial fibrillary acidic protein (GFAP; green) and anti-substance P (SP; red). The plane of focus is the ganglion cell / nerve fibre layer interface. Astrocyte processes (AP) can be seen running approximately parallel to the nerve fibres and investing the retinal vessels, forming the perivascular glia limitans (GL). Some SP-immunoreactive somata can be seen (asterisks). In addition, SP-immunoreactive varicosities can be seen, apparently embedded in the glia limitans (broad arrows). At no time do the SP-immunoreactive varicosites appear to make contact with the vascular endothelium (VE) which is separated from the glia limitans by the perivascular space.

that immunoreactive varicosites do not terminate directly on the vascular endothelium or contractile elements, but are integral to the glia limitans and are separated from the vascular endothelium by the perivascular space (Figure 5.6).

Visualisation of the SP-IR varicosities, by light, confocal and electron microscopy, indicate that SP-IR terminals are separated from the vascular endothelium, and the juxtaposed pericytes and/or smooth muscle cells, by the collagen-containing perivascular space (Figures 5.3–5.6). In addition, both confocal and electron microscopy suggest that SP-IR varicosities are associated with the perivascular glia limitans of the retinal vessels (Figures 5.5 and 5.6).

While these observations provide no data concerning the function of SP-IR varicosities associated with the retinal vasculature, consideration of the known functional roles of SP in the peripheral and central nervous system suggests at least two possibilities. The first is modulation of cytokines, or other inflammatory-associated substances, in microglial cells associated with the glia limitans and/or perivascular space (Martin *et al.*, 1993) which, in turn, may regulate barrier function. The second possibility, more germane to

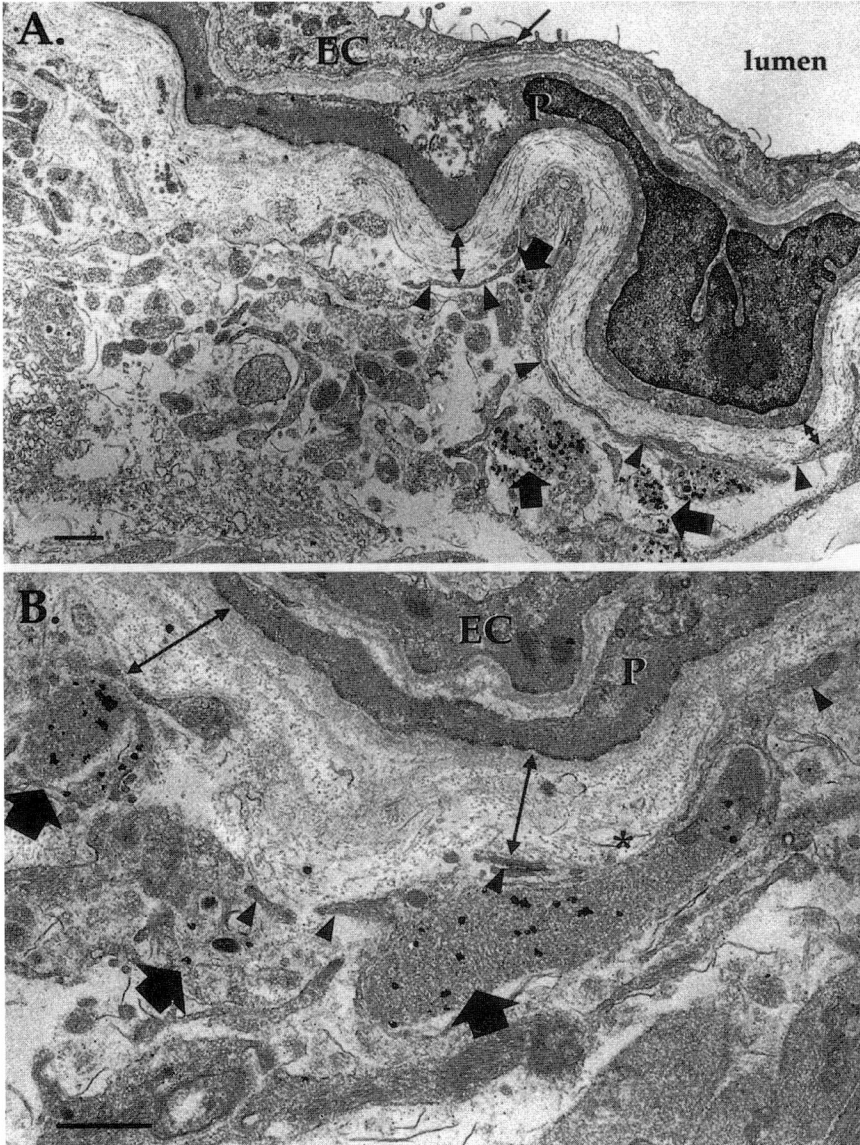

Figure 5.6   Electron micrographs showing the relationship of substance P (SP)-immunoreactive terminals to the retinal blood vessels and perivascular glia limitans. (A) Part of the vessel wall and lumen showing an endothelial cell (EC) and pericyte (P) separated from the parenchyma by the glial limitans (arrowheads) and perivascular space (double-headed arrows). An inter-endothelial cell junction is indicated by the thin arrow. SP-immunoreactive terminals, labelled with electron dense, silver-enhanced immunogold particles, are indicated (broad arrows) adjacent to the glia limitans, within the retinal parenchyma. (B) SP terminals are partially separated from the perivascular space by the glia limitans. The large terminal to the right of the micrograph is partially invested by cell processes (presumably astrocytes) and in some locations directly in contact with the collagenous matrix of the perivascular space (asterisk). Scale bars = 1 μm.

this review, is for a role in autoregulation of retinal blood flow. While activation of autonomic centres has the effect of directing major changes in blood flow, neuropeptides (along with NO) have been proposed to have a role in the rapid modification of vessel tone through autoregulation in response to local demands (Edvinsson, 1985; Said, 1987; Vincent, *et al.*, 1992). In this context SP, particularly in association with the more potent CGRP, can act as a potent vasodilator (McCulloch, *et al.*, 1986; Vincent, *et al.*, 1992). The SP-IR terminals described here which associate with the glia limitans of human retina appear to be derived from intrinsic retinal cells and may represent an anatomical basis for one aspect of autoregulation of the retinal vasculature.

Finally, a recent study has also suggested a role for peptidergic neurons in control of blood flow. Within the human retina thyrotropin-like immunoreactivity has been observed in ganglion cells (Fernandez-Trujillo, Prada and Verastegui, 1996), these authors suggesting that peptides released from terminals associated with blood vessels may be involved in regulating vessel diameter and flow. The data, however, remains speculative and precise association needs to be proved before firm conclusions can be made.

The evidence, therefore, for intrinsic innervation of retinal vessels is now very compelling and provides an anatomical platform for one component of autoregulation of retinal blood flow.

CONCLUSIONS

While there is considerable evidence in support of the existence of neural mechanisms to regulate choroidal blood flow, including specialisation of these mechanisms to modulate flow to the fovea, evidence for neural substrates which mediate regulation of retinal blood flow is less readily available. There is good evidence in the literature indicating innervation of the central retinal artery, but it appears that there is no extrinsic innervation to the retinal vessels themselves. We have reviewed the limited literature describing candidate neurons intrinsic to the retina with processes which appear to make specific contact with the basement membrane of the retinal vessels and provided additional information concerning the nature of those associations.

Whether this data is sufficient to confirm the presence of intrinsic neuronal regulation of retinal blood flow remains to be seen. Certainly, further work is required to integrate physiological and cell biological studies to demonstrate the ability of amacrine cells processes to bring about alterations in vascular tone, either through direct contraction of the endothelial cell or through perivascular-associated cells such as pericytes. The retinal vasculature is essentially devoid of an internal elastic lamina and muscularis raising the intriguing possibility that the tone of the capillary network is being controlled by pericytes. This is supported by the observation that pericytes express neurotransmitter receptors (Ferrari-Dileo, Davis and Anderson, 1991) are contractile (Kelley *et al.*, 1987; Hirschi and D'Amore, 1996) and are present in greater numbers in the retina than in other organs, including the brain. The heterogeneity of the retinal vasculature, and indeed of the retinal endothelium (Yu *et al.*, 1997), is another area of interest that may be of great significance in the local regulation of vascular flow and permeability.

REFERENCES

Ahmed, J., Braun, R.D., Dunn, R. and Linsenmeier, R.A. (1993). Oxygen distribution in macaque retina. *Investigative Ophthalmology and Visual Science*, **34**, 516–521.

Bill, A. and Nilsson, S.F.E. (1985). Control of ocular blood flow. *Journal of Cardiovascular Pharmacology*, **7** (Suppl), S96–S102.

Bill, A. and Sperber, G.O. (1990). Control of retinal and choroidal blood flow. *Eye*, **4**, 319–325.

Braunagel, S.C., Xiao, J.G. and Chiou, G.C.Y. (1988). The potential role of adenosine in regulating blood-flow in the eye. *Journal of Ocular Pharmacology*, **4**, 61–73.

Brecha, N.C., Hendrickson, A., Floren, I. and Karten, H.J. (1982). Localization of substance P-like immunoreactivity within the monkey retina. *Investigative Ophthalmology and Visual Science*, **23**, 147–153.

Brown, S.M. and Jampol, L.M. (1996). New concepts of regulation of retinal vessel tone. *Archives of Ophthalmology*, **114**, 199–204.

Campochiaro, P.A. and Sen, H.A. (1989). Adenosine and its agonists cause retinal vasodilation and hemorrhages. Implications for ischemic retinopathies. *Archives of Ophthalmology*, **107**, 412–416.

Crosson, C.E., DeBenedetto, R. and Gidday, J.M. (1994). Functional evidence for retinal adenosine receptors. *Journal of Ocular Pharmacology*, **10**, 499–507.

Cuenca, C. and Kolb, H. (1998). Circuitry and role of substance P-immunoreactive neurons in the primate retina. *Journal of Comparative Neurology*, **393**, 439–456.

Cuenca, C., De Juan, J. and Kolb, H. (1995). Substance P-immunoreactive neurons in the human retina. *Journal of Comparative Neurology*, **356**, 491–504.

Diaz, C.M., Penfold., P.L. and Provis, J.M. (1998). Modulation of the resistance of a human endothelial cell line by human retinal glia. *Australian and New Zealand Journal of Ophthalmology*, **26** (Suppl.), S62.

Donati, G., Pournaras, C.J., Munoz, J.-L., Poitry, S., Poitry-Yamate, C.L. and Tsacopoulos, M. (1995). Nitric oxide controls arteriolar tone in the retina of the miniature pig. *Investigative Ophthalmology and Visual Science*, **36**, 2228–2237.

Edvinsson, L. (1985). Functional role of perivascular peptides in the control of cerebral circulation. *Trends in the Neurosciences*, **8**, 126–131.

Fernandez-Trujillo, F.J., Prada, A. and Verastegui, C. (1996). Thyrotropin-like immunoreactivity in human retina: immunoreactive co-localization in ganglion cells and perivascular fibres. *Neurochemistry International*, **28**, 381–384.

Ferrari-Dileo, G., Davis, E.B. and Anderson, D.R. (1991). Cholinergic binding sites in pericytes isolated from retinal capillaries. *Blood Vessels*, **28**, 542–546.

Flugel, C., Tamm, E.R., Mayer, B. and Lutjen-Drecoll, E. (1994). Species differences in choroidal vasodilative innervation: evidence for specific intrinsic nitridergic and VIP-positive neruons in the human eye. *Investigative Ophthalmology and Visual Science*, **35**, 592–599.

Flugel-Koch, C., Kaufman, P. and Lutjen-Drecoll, E. (1994). Association of a choroidal ganglion cell plexus with the fovea centralis. *Investigative Ophthalmology and Visual Science*, **35**, 4268–4272.

Funk, R.H.W. (1997). Blood supply of the retina. *Ophthalmic Research*, **29**, 320–325.

Gidday, J.M. and Park, T.S. (1993). Adenosine-mediated autoregulation of retinal arteriolar tone in the piglet. *Investigative Ophthalmology and Visual Science*, **34**, 2713–2719.

Greenwood, J. (1992). Experimental manipulation of the blood-brain and blood-retinal barriers. In: *Physiology and pharmacology of the blood-brain barrier*, edited by M.W.B. Bradbury. *Handbook of Experimental Pharmacology*, **103**, 459–486.

Greenwood, J., Bamforth, S., Wang, Y. and Devine, L. (1995). The blood-retinal barrier in immune-mediated diseases of the retina. In: *New concepts of a blood-brain barrier*, edited by J. Greenwood, D. Begley and M. Segal, pp 315–326. London: Plenum Press.

Haefliger, I.O., Meyer, P., Flammer, J. and Luscher, T.F. (1994). The vascular endothelium as a regulator of the ocular circulation: a new concept in ophthalmology? *Survey of Ophthalmology*, **39**, 123–132.

Hirschi, K.K. and D'Amore, P. (1996). Pericytes in the microvasculature. *Cardiovascular Research*, **32**, 687–698.

Janzer, R.C. and Raff, M.C. (1987). Astrocytes induce blood-brain barrier properties in endothelial cells. *Nature*, **352**, 253–257.

Kelley, C., D'Amore, P., Hechtmann, H.B. and Shepro, D. (1987). Microvascular pericyte contractility in vitro: Comparison with other cells of the vascular wall. *Journal of Cell Biology*, **104**, 483–490.

Kitamura, Y., Okamura, T., Kani, K and Toda, N. (1993). Nitric oxide-mediated retinal arteriolar and arterial dilatation induced by substance P. *Investigative Ophthalmology and Visual Science*, **34**, 2859–2865.

Klooster, J., Beckers, H.J.M, ten Tusscher, M.P.M., Vrensen, G.F.J.M., van der Want, J.J.L. and Lamers, W.P.M.A. (1996). Sympathetic innervation of the rat choroid: an autoradiographic tracing and immunohistochemical study. *Ophthalmic Research*, **28**, 36–43.

Kolb, H., Fernandez, E., Ammermuller, J. and Cuenca, N. (1995). Substance P: a neurotransmitter of amacrine and ganglion cells in the vertebrate retina. *Histology and Histopathology*, **10**, 947–68.

Komai, K., Miyazaki, S., Onoe, S., Shimo-Oku, M. and Hishida, S. (1995). Vasomotor nerves of vessels in the human optic nerve. *Acta Ophthalmologica Scandinavica*, **73**, 512–516.

Kumagai, N., Yuda, K., Kadota, T., Goris, R.C. and Kishida, R. (1988). Substance P-like immunoreactivity in the central retinal artery of the rabbit. *Experimental Eye Research*, **46**, 591–596.

Laties, A.M. (1967). Central retinal artery innervation. *Archives of Ophthalmology*, **77**, 405–409.

Li, H.-B. and Lam, D.M.-K. (1990). Localization of neuropeptide-immunoreactive neurons in the human retina. *Brain Research*, **522**, 30–36.

Marshak, D.W. (1989). Peptidergic neurons of the macaque monkey retina. *Neuroscience Research*, **10**, S117–S130.

Martin, F.C., Anton., P.A., Gornbein, J.A. and Merrill, J.E. (1993). Production of interleukin-1 by microglia in response to substance P: Role for non-classical NK-1 receptor. *Journal of Neuroimmunology*, **42**, 53–60.

McCulloch, J., Uddman, R., Kingman, T.A. and Edvinsson, L. (1986). Calcitonin gene-related peptide: functional role in cerebrovascular regulation. *Proceeding of the National Academy of Sciences USA.*, **83**, 5731–5735.

Nilsson, S.F.E. (1996). Nitric oxide as a mediator of parasympathetic vasodilation in ocular and extraocular tissues in the rabbit. *Investigative Ophthalmology and Visual Science*, **37**, 2110–2119.

Nuzzi, R., Gugleilmone, R. and Grignolo, F.M. (1995). Fluorescence histochemical demonstration of adrenergic terminations in the human choroid. *European Journal of Ophthalmology*, **5**, 251–258.

Owman, C., Edvinsson, L. and Nielsen, K.C. (1974). Autonomic neuroreceptor mechanisms in brain vessels. *Blood Vessels*, **11**, 2–31.

Penfold, P.L. and Provis, J.M. (1998). The neurovascular interface in human retina. *Experimental Eye Research.*, **67** (Suppl), 94.

Provis, J.M. and Mitrofanis, J. (1990). NADPH-diaphorase neurones of human retinae have a uniform topographical distribution. *Visual Neuroscience*, **4**, 619–623.

Provis, J.M., Yip, J. and Penfold, P.L. (1992). Substance P in the human retina. *Proceedings of the Australian Neuroscience Society*, **3**, 62.

Provis, J.M., Penfold, P.L., Edwards, A.J. and van Driel, D. (1995). Human retinal microglia: Expression of immune markers and relationship to the glia limitans. *Glia*, **14**, 243–256.

Raub, T.J., Luentzel, S.L. and Sawada, G.A. (1992). Permeability of bovine brain microvessel endothelial cells in vitro; barrier tightening by a factor released from astroglioma cells. *Experimental Cell Research*, **199**, 330–340.

Roufail, E., Stringer, M. and Rees, S. (1995). Nitric oxide synthase immunoreactivity and NADPH diaphorase staining are co-localised in neurons closely associated with the vasculature in rat and human retina. *Brain Research*, **684**, 36–46.

Ruskell, G.L. (1971). Facial parasympatheitc innervation of the choroid blood vessels in monkeys. *Experimental Eye Research*, **12**, 166–172.

Said, S.I. (1987) Neuropeptides in cardiovascular regulation. In: *Brain Peptides and Catcholamines in Cardiovascular Regulation*, edited by J.P. Buckley and C.M. Ferrario, pp. 93–107. New York: Raven Press.

Sen, H.A. and Campochiaro, P.A. (1989). Intravitreal injection of adenosine or its agonists causes breakdown of the blood-retinal barrier. *Archives of Ophthalmology*, **107**, 1364–1367.

Stjernschantz, J. and Bill, A. (1980). Vasomotor effects of facial nerve stimulation: noncholinergic vasodilation in the eye. *Acta Physiologica Scandinavica*, **109**, 45–50.

Stone, J. and Dreher, Z. (1987). Relationship between astrocytes, ganglion cells and vasculature of the retina. *Journal of Comparative Neurology*, **255**, 35–49.

Toda, N., Kitamura, Y. and Okamura, T. (1994). Role of nitroxidergic nerve in dog retinal arterioles *in vivo* and arteries *in vitro*. *American Journal of Physiology*, **266**, H1985–H1992.

Verstappen, A., Toussaint, D., Lotstra, F. and Van der Haeghen, J.J. (1986). Club endings containing substance P-like immunoreactivity in human retinal vessels: whole mount preparation. *Neurochemistry International*, **8**, 377–380.

Vincent, M.B., White, L.R., Elsas, T., Qvigstad, G. and Sjaastad, O. (1992). Substance P augments the rate of vasodilation induced by calcitonin gene-related peptide in porcine ophthalmic artery *in vitro*. *Neuropeptides*, **22**, 137–141.

Ye, X., Laties, A. and Stone, R. (1990). Peptidergic innervation of the retinal vasculature and optic nerve head. *Investigative Ophthalmology and Visual Science*, **31**, 1731–1737.

Yu, P.K., Yu, D-Y., Alder, V.A., Seydel, U., Su, E-N. and Cringle, S.J. (1997). Heterogeneous endothelial cell structure along the porcine retinal microvasculature. *Experimental Eye Research*, **65**, 379–389.

6 Purinergic Signalling in the Eye

Jesús Pintor

Department of Biochemistry and Molecular Biology IV, University School of Optics and Optometry, University Complutense of Madrid, Arcos de Jalón s/n, 28037 Madrid, Spain

There is a clear therapeutic potential of adenosine and ATP acting through P1 and P2 receptors in the different eye structures. Although there is still a long way to follow until the complete receptor identification and effects are described, nucleosides and nucleotides are interesting new pharmacological tools which might be suitable for the treatment of some ocular pathologies. The actions of adenosine, ATP and other nucleotides in the eye structures are discussed and possible implications in pathological processes and future developments are also indicated.

KEY WORDS: ATP; adenosine; P2 receptor; retina; cornea; lens; ciliary body.

BACKGROUND

Purine compounds such as adenosine 5′-triphosphate (ATP) and adenosine (Figure 6.1) play important roles in energetic metabolism, synthesis of nucleic acids and enzymatic regulation in living organisms. Nevertheless, the functions of these compounds are not restricted to intracellular actions but also to extracellular tasks. It is accepted that purines are stored by neurons and other cells together with other chemical messengers such as hormones and neurotransmitters. Stimulation of these cells evoke the release of purine compounds into the extracellular media, producing different effects in tissues due to their binding to specific purine receptors present in cell membranes. This chapter describes the actions of adenosine, ATP and other nucleotides in the eye structures. Possible implications in pathological processes and future developments are also indicated.

PURINERGIC TRANSMISSION. HISTORICAL PERSPECTIVE

Drury and Szent-Gyorgyi studied the extracellular actions of adenosine and adenine

ADENOSINE

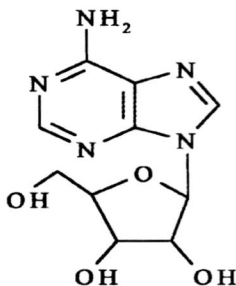

ADENOSINE 5´ TRIPHOSPHATE (ATP)

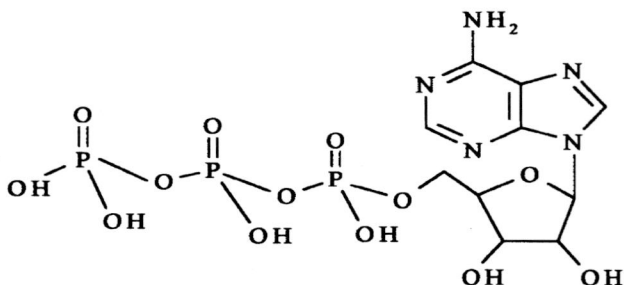

Figure 6.1   Chemical structures adenosine and ATP. These are the two most represented natural purinergic agonists.

nucleotides first in mammalian heart in 1929. In the following years there were several investigations into the actions of purines on the systemic and pulmonary vascular systems (Bennett and Drury, 1931; Wedd, 1931; Gaddum and Holz, 1933; Green and Stoner, 1950). In the 1950s, Holton and Holton proposed that ATP might be the substance released following antidromic stimulation of sensory nerves that was responsible for vasodilatation (Holton and Holton, 1953, 1954; Holton, 1959). In the following decade adenosine was proposed as the local regulator of blood flow after hypoxia in heart and other vascular beds (Berne, 1963), but it was in 1981 when the vasodilator role of ATP acting on receptors present in vascular endothelial cells was demonstrated (De Mey and Vanhoutte, 1981).

The presence of non-adrenergic, non-cholinergic nerves in the gastrointestinal tract was indicated perhaps as long ago as the end of the last century (Courtache and Guyon, 1897; Langley, 1898, 1900; Bayliss and Starling, 1900). In 1963 the existence of autonomic nerves supplying the gastrointestinal tract that were neither adrenergic nor cholinergic was reinvestigated (Burnstock *et al.*, 1963; Martison and Muren, 1963), and later it was suggested that this non-adrenergic, non-cholinergic nerves could supply other organs such as lung, bladder, trachea, parts of the vascular system and the eye (Burnstock, 1969). It was in the nineteen seventies when Burnstock proposed that the principal active substance released from the non-adrenergic non-cholinergic (NANC) nerves was the purine nucleotide ATP that, NANC nerves were re-named as "purinergic" nerves (Burnstock, 1972). Although there was considerable confusion in the published literature on the actions of adenosine nucleotides and nucleosides, a forward step was made when in 1978 it was realised that the receptors for purines could be subdivided into two subclasses: P1 (adenosine receptors) and P2 (ATP and ADP receptors) (Burnstock, 1978). P1 receptors were sensitive to adenosine, coupled to adenylate cyclase and were antagonised by methylxanthines. P2 receptors were sensitive to ATP and ADP, were not antagonised by methylxanthines, did not act via adenylate cyclase but they might activate prostaglandin synthesis (Burnstock, 1978). Since that moment studies on the receptors for adenosine and adenine nucleotides have revealed more information on the nature, distribution and physiological roles of purines.

PURINE RECEPTORS

Receptors for adenosine and adenine nucleotides are nowadays regarded as being independent. For both types of purine receptors, pharmacology studies using synthetic agonists and antagonists *in vitro* or *in vivo* was the only way to study these receptors until the advances in molecular biology and gene expression systems opened new means to search for receptor amino acid sequence, structure and pharmacology. The following paragraphs summarise the most relevant features of adenosine and ATP receptors.

ADENOSINE RECEPTORS

Adenosine produces different effects in many systems and tissues (Daly, 1982; Williams, 1989). Adenosine may be considered as an alarm signal since it is released under hyperactivity conditions or stress. Moreover, adenosine is the final product of extracellular degradation of adenine mono- and dinucleotides.

In the nervous system this nucleoside is an inhibitory modulatory molecule. It can reduce the firing frequency of neurons and reduce the neurotransmitter release, thus inhibiting synaptic transmission (Fredholm *et al.*, 1993). This depressant action of neural activity could explain the sedative and anticonvulsant actions described for adenosine. Furthermore, these mechanisms would explain the stimulant properties of xanthines (caffeine and theophylline) due to the blockade of some adenosine receptors (Williams, 1987).

The effects of adenosine in the periphery are varied. In adipose tissue adenosine is implicated in inhibition of lipolysis and in the potentiation of glucose transport induced by insulin (Olson and Pearson, 1990). In the kidney, adenosine produces antidiuresis and inhibition of the renin-angiotensin system as well as inhibition of transmitter release from renal nerve terminals (Spielman, Arend and Forrest, 1987).

Probably one of the most characteristic actions of adenosine occurs in the vascular system. During inflammation or hypoxia adenosine reduces the metabolic demand and increases nutrient supply. Under hypoxic conditions the concentration of adenosine in interstitial fluid is increased, producing dilatation in vascular beds and a better oxygenation of the tissues (Bruns, 1981). Apart from the peripheral vasodilator role of adenosine it produces direct effects in the heart, which are the basis for the use of adenosine in cardiac therapeutics (Liang, 1992).

Before the application of recombinant DNA techniques, adenosine receptor characterisation was based in pharmacological analysis with agonists and antagonists. Adenosine receptor classification has been clarified by molecular biology techniques: four types of receptors, named A_1, A_{2A}, A_{2B} and A_3, have been found. This classification attends to structural and pharmacological criteria (Fredholm et al., 1994).

Structurally, all adenosine receptors are formed by a single subunit between 36 and 45 kDa and belong to the family of G protein-coupled receptors (GPCR) with seven transmembrane domains (Palmer and Stiles, 1995; Jacobson, 1995). Adenosine receptors contain serine and threonine residues in the third intracellular loop, near the carboxy-terminal, and these are susceptible to phosphorylation that could produce desensitisation of the receptor after long exposure to adenosine. The A_{2A} receptor, for example, contains abundant putative phosphorylation sites and shows rapid desensitisation when exposed to adenosine. On the other hand A_1 receptors do not possess so many phosphorylation sites, and desensitisation occurs slowly in the presence of the agonist (Ramkumar et al., 1991). All the receptors for adenosine are metabotropic and are coupled positively or negatively to adenylate cyclase (Fredholm et al., 1994), nevertheless, other effectors have been described to be coupled to these receptors, such as phospholipase C (Gerwins and Fredholm, 1992), K^+ channels (Trussel and Jackson, 1985) and Ca^{2+} channels (Dolphin, Forda and Scott, 1986). The A_3 receptor seems to be coupled to phospholipase C in addition to adenylate cyclase and it is not sensitive to neutral methylxanthines (Ramkumar et al., 1993). Table 6.1 summarises the molecular properties and pharmacological of the four adenosine receptors.

P2 RECEPTORS

Studying the different behaviour of ATP and its analogues on diverse biological systems, Burnstock and Kennedy (1985) proposed the subdivision of P2 receptors into P2X and P2Y receptors. This division was established on the basis of the agonistic potency of ATP, α,β-methylene ATP (α,β-MeATP) and 2-methylthio ATP (2MeSATP). P2X receptors are more sensitive to α,β-MeATP, 2MeSATP being weaker agonist than ATP. On the other hand 2MeSATP is a better agonist on P2Y receptors, ATP and α,β-MeATP being weaker agonists. P2X receptors are desensitised by the presence of α,β-MeATP whereas P2Y are not affected by this process. Gordon (1986) described other P2 receptors with different

TABLE 6.1
Purinergic receptors for adenosine

Subtype	A_1	A_{2A}	A_{2B}	A_3
Amino Acids	326	410–412	332	320
Mol. weight (kDa)	36–37	45	36	36
Agonists	R-PIA>NECA>s-PIA	NECA>R-PIA>CPA, s-PIA	NECA>R-PIA>s-PIA	NECA>R-PIA>CGS21680
Selective agonist	CPA	CGS21680	—	APNEA
Effective adenosine concentration	3–30 nM	1–20 nM	5–20 µM	>1 µM
Antagonists	XAC, DPCPX (0.5–2 nM)	KF17837, XAC (20–100 nM)	XAC, DPCPX (20–100 nM)	BW-A522, I-ABOPX (1–20 nM)
G protein-coupling	G_i, G_o	G_q	G_q	G_i
Effectors	$\downarrow$cAMP, $\uparrow$IP$_3$/Ca^{2+}, $\uparrow$K$^+$, $\downarrow$Ca^{2+}	$\uparrow$cAMP	$\uparrow$cAMP	$\downarrow$cAMP, $\uparrow$IP$_3$/Ca^{2+}
Distibution	Brain (cortex, hippocampus and cerebellum), testis, adipose tissue, heart, kidney	Brain (striatum, nucleus accumbens, olfactory tubercule), heart, neutrophils	Intestine, urinary bladder, aorta, brain	Testis (rat), peripheral tissues (human and sheep)

APNEA, N-[2-(4-aminophenyl)ethyladenosine]; BW-A522 = I-ABOPX, 3-(3-iodo-4-aminobenzyl)-8-(4-oxyacetate)phenyl-1-propylxanthine; CGS21680, 2-[p-(2-carbonylethyl)-phenylethylamino]-5′-N-ethylcarboxamidoadenosine; CPA, N^6-cyclopentyladenosine; DPCPX, 1,3-dipropyl-8-cyclopentylxanthine; KF17837, 1,3-dialkyl-7-methyl-8-(3,4,5-trimethoxystyryl)xanthine; NECA, N-ethylcarboxamidoadenosine; R-PIA, R-N^6-phenylisopropyl adenosine; s-PIA, s-N^6-phenylisopropyl adenosine; XAC, Xanthine amine congener.

pharmacological features from P2X/P2Y. The P2T receptor is present in platelets, sensitive to ADP and insensitive to ATP. The P2Z receptor is stimulated by the tetrabasic form of ATP, ATP^{-4}, which induce the opening of non-selective pores on macrophages membranes. Subsequently, receptors sensitive to both ATP and uridine 5'-triphosphate (UTP) have been described (P2U receptors) as well as receptors sensitive to both ATP and diadenosine polyphosphates, named P2D receptors (Pintor, Diaz-Rey and Miras-Portugal, 1993).

A substantial change occurred when the cloning and expression of P2 receptors was performed. Two facts were extremely important for the cloned receptors. On the one hand the pharmacological profiles of some of the receptors cloned did not fit with the previous works developed in cells or tissues. On the other and most of the P2 receptor subtypes described above could be fitted into two main P2 receptors families: the P2X and P2Y receptor families.

P2X receptors

P2X receptors are ionotropic ATP receptors can be included together with the nicotinic, glycine, serotonin, N-methyl-D-aspartate (NMDA) and $GABA_A$ receptors as receptor operated ion channels. P2X receptors are directly opened by ATP and are selective to small ions such as Na^+, Ca^{2+} and K^+. These receptors are involved in fast synaptic transmission between neurons and also between autonomic nerves and smooth muscle, where ATP acts as main transmitter molecule. Expression experiments of the cloned P2X receptors in *Xenopus* oocytes showed that the ion pore is opened a few milliseconds after ATP application and that it does not require any difussible second messenger for this to occur (Kennedy, Hickman and Silverstein, 1998).

Molecular biology and expression studies have allowed identifying to date up to 7 P2X receptors named $P2X_1$–$P2X_7$. These receptors are proteins that share 35–48% of homology in the amino acid sequence and possess two hydrophobic transmembrane domains. Most of P2X receptors have 10 cysteine residues in areas with low homology that could be important for the tertiary structure of the receptor and ATP binding site. In analogy to other ionotropic receptors it is supposed that the assembly of several subunits moulding a multimeric receptor with allosteric properties forms functional P2X receptors. Some authors, studying the molecular weight of the receptor, suggest that 5 subunits are necessary to form a functional receptor. The main features of the different P2X receptors are summarised in Table 6.2. It is noteworthy to see that the P2Z receptor is now included in the superfamily of the P2X receptors as the $P2X_7$ (Evans, Surprenant and North, 1998).

P2Y receptors

P2Y receptors are a group of metabotropic receptors characterised by seven transmembrane domains that are coupled to seconds messengers systems by means of G proteins. Several positive charged residues in three transmembrane domains (TM3, TM6, and TM7) are probably involved in the binding of the nucleotide phosphate chain. In fact, mutagenesis experiments demonstrated that those residues (His^{262}, Arg^{265} in TM6 and Arg^{292} in TM7) have a crucial role in the nucleotide binding and receptor activation. Most of the studies on the P2Y family demonstrate that phospholipase Cβ (PLCβ) is the enzyme implicated in the signal transduction after P2Y receptor activation (Boeynaems *et al.*, 1988). Nevertheless, some authors have shown that P2Y receptors can be associated to changes in the cytosolic cyclic adenosine monophosphate (cAMP) levels.

TABLE 6.2

Purinergic ionotropic receptors for nucleotides

Subtypes	Agonist potency	Antagonists	Signal transduction	Tissues
$P2X_1$	2MeSATP>ATP>α,β-MeATP (EC_{50} ATP ≈ 1 μM) (Rapid desensitisation)	Suramin PPADS	$I_{Na/K/Ca}$	Vas deferens (rat) Urinary bladder (human, mouse)
$P2X_2$	2MeSATP>ATP (EC_{50} ATP ≈ 10 μM) (α,β-MeATP inactive)	Suramin PPADS	$I_{Na/K}$ (potentiated by Zn^{2+})	PC12 cells, cochlea (rat)
$P2X_3$	2MeSATP>ATP>α,β-MeATP (EC_{50} ATP ≈ 1 μM) (Rapid desensitisation)	Suramin PPADS	$I_{Na/K}$ (potentiated by Zn^{2+})	Dorsal root ganglia (rat)
$P2X_2/P2X_3$	α,β-MeATP $\approx$ ATP (EC_{50} ATP ≈ 1 μM)	Suramin PPADS	?	Sensory neurons (rat)
$P2X_4$	ATP>2MeSATP>α,β-MeATP (α,β-MeATP normally inactive) (EC_{50} ATP ≈ 10 μM)	No	$I_{Na/K/Ca}$ (potentiated by Zn^{2+})	Hippocampus, dorsal root ganglia, endocrine tissue, brain (human, rat)
$P2X_5$	ATP>2MeSATP>ADP (α,β-MeATP inactive) (EC_{50} ATP ≈ 15 μM)	Suramin PPADS	$I_{Na/K/Ca}$	Sensory ganlia, spinal cord (rat)
$P2X_6$	2MeSATP>ATP>ADP (α,β-MeATP inactive) (EC_{50} ATP ≈ 10 μM)	No	$I_{Na/K/Ca}$ (potentiated by Zn^{2+})	Brain, sensory ganglia (rat)
$P2X_7$	BzATP>ATP>UTP (α,β-MeATP inactive)	Suramin, PPADS (very weak antagonists)	Non-specific pore	Brain, macrophages (rat, mouse)

ADP, adenosine 5'-diphosphate; ATP, adenosine 5'-triphosphate; BzATP, 3'-O-(4-benzoyl)benzoyl ATP; α,β-MeATP, α,β-methyleneATP; 2MeSATP, 2-methylthioATP; PPADS, pyridoxalphosphate-6-azophenyl-2',4'-disulphonic acid; UTP, uridine 5'-triphosphate

The family of P2Y receptors is formed by seven members $P2Y_1$–$P2Y_7$, whose amino acid sequences are the smallest (329–379 amino acids) described within the superfamily of G protein-coupled receptors (Weisman *et al.*, 1998). Within P2Y receptors are now included those receptors sensitive to UTP (formerly P2U), to ADP (P2T) and diadenosine polyphosphates (P2D). An overview of the P2Y receptor subtypes is given in Table 6.3.

WHERE DO THE PURINES COME FROM?

There are different potential sources of extracellular ATP and adenosine. ATP is present and released from neurons as a transmitter or as a cotransmitter.

Sympathetic nerves are a model where noradrenaline and ATP act as cotransmitters and where they are released in variable proportions after stimulation (Burnstock, 1990; Evans and Cunnane, 1992). Evidence for purinergic cotransmission includes the block of prazosin-resistant component of the response to sympathetic nerve stimulation by the ATP antagonists suramin and 3'-O-(3-[N-(4-azido-2-nitrophenyl)amino]-propionyl)ATP ($ANAPP_3$); release of ATP during nerve stimulation, which is blocked by 6-hydroxydopamine but unaffected by selective depletion of noradrenaline by reserpine; selective block by ATP antagonists of excitatory junction potentials in response to sympathetic nerve stimulation and mimicry of the excitatory junction potentials by ATP but not by noradrenaline. Sympathetic cotransmission have been demonstrated in different isolated blood vessels (Schwartz and Malik, 1989; Starke *et al.*, 1991), in the circulation of skeletal muscle, intestine (Taylor and Parsons, 1989) and in rat iris (Fuder and Muth, 1993).

Although there is evidence that ATP is stored in the same vesicles with noradrenaline (Langercrantz, 1976), there may be different populations of sympathetic nerves some of which contain high levels of ATP compared to noradrenaline and vice versa (Trachte, 1988; Ellis and Burnstock, 1989).

Acetylcholine (ACh) and ATP are cotransmitters in parasympathetic neurons in the bladder (MacKenzie, Burnstock and Dolly, 1982; Hoyle and Burnstock, 1985). There is also indirect evidence for ATP being colocalised with ACh, neuropeptide Y or somatostatin in subpopulations of intrinsic neurons in the heart and airways that project to small blood vessels in the heart and lung (Inoue and Kannan, 1988; Saffrey *et al.*, 1992).

As commented before ATP can be also released from the so-called non-adrenergic non-cholinergic nerves presenting important roles in the gastrointestinal tract (Burnstock, 1995).

Blood vessel endothelial cells seems to be also a source for ATP after stimulation with various circulating agonists (Yang *et al.*, 1994). Surprisingly one of these compounds was ADP, which is released from platelets, producing an enhancement of ATP extracellular levels (Buxton and Cheek, 1995). ATP can be released from other cells such as platelets, together with ADP (as commented before) being another source of extracellular nucleotides to the blood stream.

It is extremely important to take into consideration the relation between the release of ATP and where the receptors are located. Indeed, local ATP can be transformed into adenosine by the actions of ecto-nucleotidases. These enzymes comprise a family of

TABLE 6.3

Purinergic metabotropic receptors for nucleotides

Subtypes	Agonist potencies	Signal transduction	Tissues
$P2Y_1$ (formerly P_{2Y})	2MeSATP>ATP>>UTP (α,β-MeATP inactive)	PLCβ/IP$_3$/Ca^{2+}	Brain (chick), placenta, brain, prostate, ovary, HELA cells (human), endothelium (bovine), insulinoma (mouse, rat)
$P2Y_2$ (formerly P_{2U})	ATP=UTP>>2MeSATP (α,β-MeATP inactive)	PLCβ/IP$_3$/Ca^{2+}	Bone (human), lung (cat, rat), pituitary (rat)
$P2Y_3$	ADP>UTP>ATP=UDP	PLCβ/IP$_3$/Ca^{2+}	Brain (chick)
$P2Y_4$ (pyrimidinoceptor)	UTP≥UDP>ATP≈ADP (2MeSATP inactive)	PLCβ/IP$_3$/Ca^{2+}	Placenta, chromosome X (human), brain (rat)
$P2Y_5$	ATP>ADP>2MeSATP>>UTP,α,β-MeATP (from binding studies)	(not determined)	Activated T cells (human), lymphocytes (chick)
$P2Y_6$	UDP>UTP>ADP>2MeSATP>ATP	PLCβ/IP$_3$/Ca^{2+}	Placenta, spleen (human), smooth muscle, aorta (rat)
$P2Y_7$	ATP>ADP=UTP	(not determined)	HELA cells (human)
$P2Y_8$	ATP=UTP=ITP=CTP=GTP	PLCβ/IP$_3$/Ca^{2+}	Neural plate (Xenopus)
$P2Y_{ADP}$	ADP	Not cloned	Platelets
$P2Y_{ApnA}$	Ap$_n$A	Not cloned	

ADP, adenosine 5′-diphosphate; ATP; adenosine 5′-triphosphate; Ap$_n$A, diadenosine polyphosphates; BzATP, 3′-O-(4-benzoyl)benzoyl ATP; CTP; cytidine 5′-triphosphate; GTP; guanosine 5′-triphosphate; IP$_3$, inositol 1 4,5-triphosphate; ITP, inosine 5′-triphosphate; α,β-MeATP; α,β-methyleneATP; 2MeSATP, 2-methylthioATP; PLCβ, phospholipase C β; UDP, uridine 5′-diphosphate; UTP, uridine 5′-triphosphate.

nucleotide hydrolases that have their active site facing outwards from the cell membrane. These enzymes degrade ATP sequentially, forming what is known as the ecto-nucleotidase cascade, which yields adenosine as the final product (for a review see Zimmermann, 1996a).

THE EYE: HOW MANY STRUCTURES ARE CROSSED BY LIGHT?

The eye is a highly specialised organ for photoreception, the process by which light from the environment produces electrical changes in specialised cells in the retina (Figure 6.2). The eyes are placed in the orbits which are bony pockets in the front of the skull. The eyes are moved by six extraocular muscles attached to the tough coat of the eye called sclera. These muscles cannot be seen because their attachments to the eyes are hidden by the conjunctiva.

The outer layer of most of the eye, the sclera, is opaque and thus do not permit entry of light. However, the cornea the outer layer at the front of the eye is transparent and admits light. The anterior chamber appears between the cornea and the pupil and contains the aqueous humour, which is produced by ciliary processes. The amount of light that enters the eye is regulated by the size of the pupil, which is an opening in the iris, the pigmented ring of muscle situated behind the cornea. The lens is situated behind the iris and consists of a series of transparent "onion-like" layers. The shape of the lens can be changed by contraction of ciliary muscles. These changes in shape permit the eye to focus images of near or distant objects on the retina in a process termed accommodation. After passing through the lens, the light traverses the vitreous humour.

Vitreous humour is a clear gelatinous substance that gives the eye its volume. After passing through the vitreous humour, light arrives at the retina which is the interior layer at the back of the eye. In the retina the receptor cells, cones and rods, are collectively known as photoreceptors. The human retina contains about 120 million rods and 6 million cones. Although they are greatly outnumbered by rods, cones provide most of the information about our environment since they are responsible for daytime vision. Cones provide information on the colour and also of small features of the environment and thus they are the source of vision of the highest sharpness, an aspect that is known as acuity. The central region of the retina, the fovea, is responsible for the most acute vision and contains only cones. Rods provide vision of poor acuity, and no information about colour, but they are extremely sensitive to light. In a poorly light environment rods are the photoreceptors which are stimulated.

Photoreceptors are anchored to the last cell layer of the eye: the retinal pigmented epithelium (RPE). This monolayer lies on Bruch's membrane and has many physical, optical, biochemical and transport functions which play a critical role in the normal visual process. These cells maintain the adhesion of photoreceptors to the back of the eye, and have the important mission of the photoreceptor discs turnover.

Environmental information seized by photoreceptors is transmitted to bipolar cells, sometimes directly, and sometimes through horizontal and amacrine cells. The information flows from bipolar cells to ganglion cells. Axons of ganglion cells gather the retina

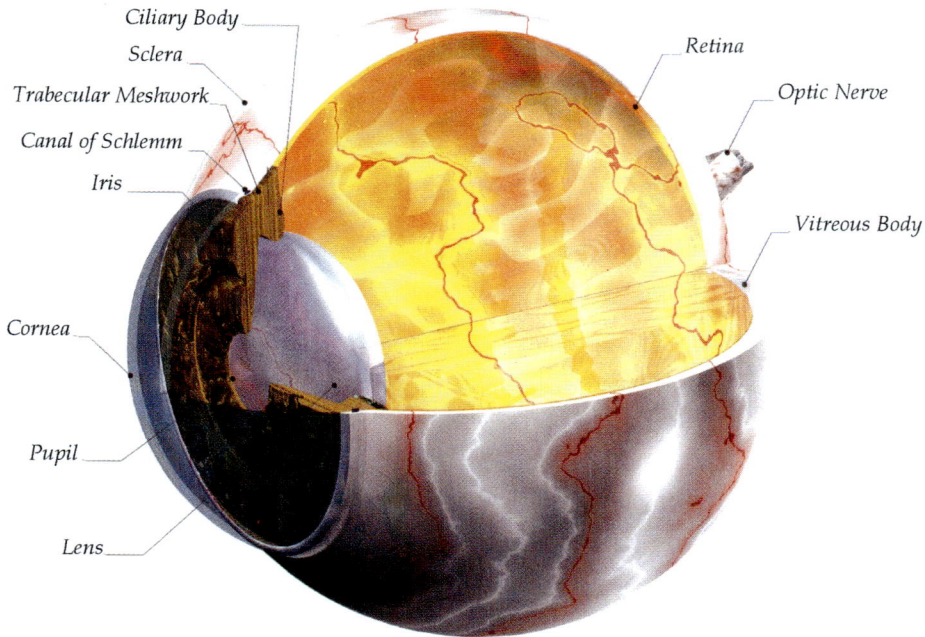

Figure 6.2   Three dimensional model of a human eye. All the structures here represented keep the anatomical sizes and distances of a human eye. Section has been performed in order to allow the vision of internal parts.

through the optic disk forming the optic nerve. The optic nerve ascends through the rest of the brain reaching the dorsal lateral geniculate nucleus (LGN) of the thalamus. The neurons of the LGN send their axons via the optic radiations to the primary visual cortex.

THE CORNEA

The cornea is an efficient combination of structure and function that gives the eye a clear refraction interface and protection from external factors. These functions have been reached by means of a very simple design. Five concentric layers form the cornea: the epithelium, Bowman's membrane, stroma, Descemet's membrane and endothelium (Figure 6.3). The cornea is an avascular structure. This property allows it to attain the right optical sharpness but forces it to capture oxygen from the tear film.

Transparency and tensile strength are obtained by the extremely small collagen fibres that are aligned in a special periodicity with one another. The anterior corneal surface represents the main refractive component of the eye, giving additional optical potency to converge images on the retina after passing through the lens.

Sensory nerves derived from the ophthalmic branch of trigeminal nerve innervate the cornea. Nerve fibres emerging from the deepest corneal layers penetrate Bowman's layer

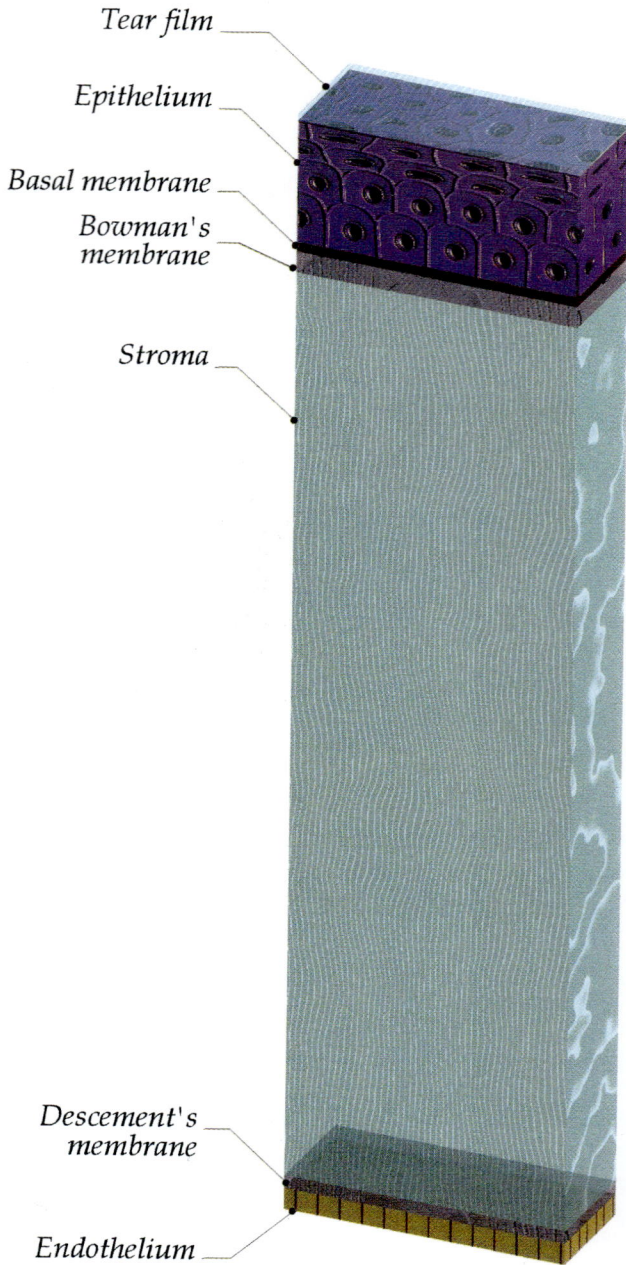

Figure 6.3   Three dimensional model of a corneal section, where the main parts are indicated. The distances, sizes and proportions of this structure are have been taken from real human corneas.

to form a subepithelial plexus. From here, axons project to the epithelium, the nerve endings being situated among the endothelial cells to avoid light disturbances. Sympathetic nerves, which are originated in the superior cervical ganglion, also innervate the cornea. The functions of these sympathetic nerves are not clear, but they could be involved in modulating ionic transport and mitotic activity. In this sense, stimulation of sympathetic nerves inhibits corneal epithelium renewal after injury.

Role of Purine Compounds in the Cornea

From a neurochemical point of view corneal nerves can be classified as sensory, adrenergic or cholinergic (for review see Tervo and Palkama, 1978). The rat cornea contains other terminals, characterised by the presence of agranular vesicles, apart from adrenergic and cholinergic terminals. These nerves are positive for quinacrine staining, suggesting they contain ATP as a neurotransmitter (Cavalotti *et al.*, 1982). This indicates that after nerve stimulation ATP could be present at the extracellular space. No studies on the possible existence of ATP receptors have been so far performed in the cornea. Nevertheless, the hydrolytic product of ATP, adenosine, has interesting actions on epithelial cells.

Epithelial cells transport Cl$^-$ actively from the aqueous side to the tear side, thus dehydrating the cornea (Zadunaisky, 1966). It has been found that compounds capable of increasing cytosolic cAMP, can stimulate chloride transport. Adenosine, applied to the epithelial surface, resulted in a prompt increase of 50% in the Cl$^-$ current. This effect was prolonged in time and it was slow when applied to the endothelial side, probably due to the time taken for it to penetrate the cornea to reach the receptors present in the epithelium. From a pharmacological point of view, this receptor is sensitive to adenosine, partially sensitive to AMP and insensitive to ATP. This receptor seems to be coupled positively to adenylate cyclase (Spinowitz and Zadunaisky, 1979). The pharmacology plus activation of adenylate cyclase indicates that this receptor is an adenosine A_2 receptors.

Nucleotides, particularly ATP, are present in corneal cells and also are important in corneal metabolism (Salla, Redbrake and Frantz, 1996). Apart from its well-known roles in cell metabolism, ATP has important roles in controlling ion transport systems. A Ca^{2+}-ATPase has been identified at the interface between axons, Schwann cells and the plasma membrane of epithelial cells. Further ultrastructural analysis has demonstrated that epithelial plasma membranes contain high levels of Ca^{2+}-ATPase activity. Bovine epithelial cells contain V-type H$^+$ pumps, which are ATP dependent and maintain the cytosolic pH. In this model ATP is the most potent nucleotide that stimulates H$^+$-pump activity guanosine 5′-triphosphate (GTP) and inosine 5′-triphosphate (ITP) can substitute for ATP, but UTP and cytidine 5′-triphosphate (CTP) are inactive (Torres-Zamorano *et al.*, 1992). Also, Na$^+$/K$^+$-ATPase activity has been observed in mammals. In a clear difference from the Ca-ATPase and V-ATPase, the Na$^+$/K$^+$-ATP is located on endothelial cell membranes (Tervo, Palva and Palkama, 1977).

The cornea is 75–80% water. It also contains glycosaminoglycans, which are negatively charged. It has been suggested that these glycosaminoglycans are responsible for the swelling of the cornea when fluid gains access to the stroma after injury. As the cornea becomes opaque when it swells, regulation of water content is essential for the transparency of this ocular medium. Endothelial cells are relatively permeable, so fluid enters to the stroma and needs to be removed to maintain normal corneal hydration. The normal

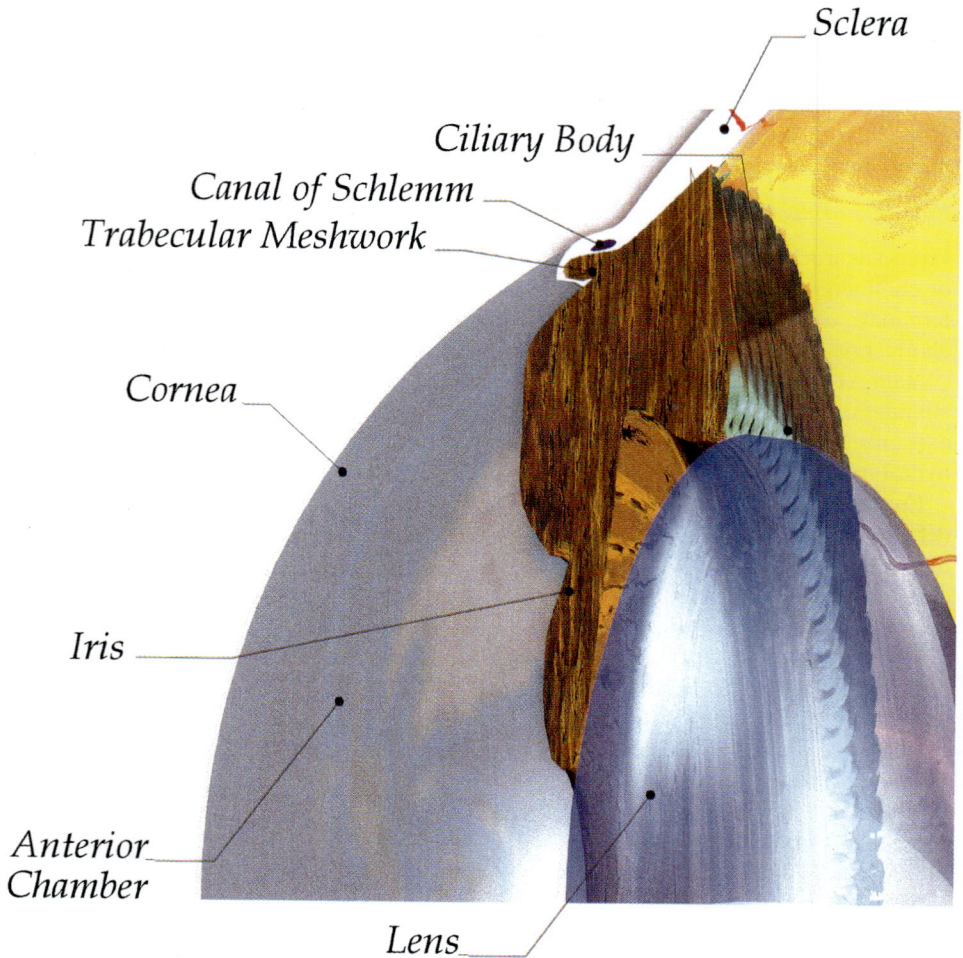

Figure 6.4   Three dimensional model of anterior segment where the main structures described in the text are indicated.

hydration level is maintained by the passive diffusion of fluid from the aqueous humour into the stroma (through the endothelial cells) and by ion pumps functioning to remove solute and thereby water from the stroma to the aqueous humour. The corneal endothelium and epithelium ion pumps are extremely important and directly dependent on the cytosolic ATP concentration to carry out their functions. Adenosine formed from the hydrolysis of the ATP released from NANC nerves or from adrenergic and cholinergic nerves (as a cotransmitter) is an important neural pacemaker that maintains the homeostasis of the cornea.

IRIS-CILIARY BODY

The iris-ciliary body constitutes part of what is called the anterior segment of the eye (Figure 6.4). The iris, ciliary muscle and choroid are tissues that contain significant amounts of melanin pigments, this being one of the most typical characteristic feature in this part of the eye. The iris is formed by sphincter and dilator muscles, and their function is to control the amount of light which enter in the eye.

Cholinergic parasympathetic nervous fibres, derived from the oculomotor nerve, predominantly innervate the iris sphincter and cause constriction. This part of the iris is also innervated by terminals derived from trigeminal nerve supplying a non-adrenergic, noncholinergic neuroactive component (Ueda *et al.*, 1981). Stimulation of the NANC nerves also induces constriction of the sphincter. The dilator muscle is innervated by the sympathetic (adrenergic) nervous system, derived from the superior cervical ganglion.

Cholinergic activation constricts the sphincter and relaxes the dilator muscle to reduce the entry of light (miosis). On the other hand, adrenergic stimulation of the dilator muscle induces contraction and sphincter dilatation, allowing more light into the eye. This phenomenon is known as mydriasis. The iris-ciliary body muscles are smooth muscles and have the same biochemical, physiological and pharmacological properties as other smooth muscles in other organs.

Beside the ciliary muscle is the ciliary epithelium. The ciliary epithelium is formed by pigmented and non-pigmented cells that are responsible for the production of aqueous humour into the posterior chamber. Aqueous humour passes to the anterior chamber through the pupil and there it leaves the eye by two mechanisms: the trabecular meshwork and the uveo-scleral pathway (Busch *et al.*, 1993). The aqueous humour drains into the trabecular meshwork and from there passes into the canal of Schlemm where it crosses the inner wall to pass into the collector channels which finally release it into the venous circulation. The Uveo-scleral pathway uses connective tissue between the muscle bundles of the iris root, then into the sclera and finally into the circulation. The first way (through the trabecular meshwork) is the main drainage system for the aqueous humour.

The drainage of the aqueous humour is extremely important since it is the main factor determining intraocular pressure (IOP), the resistance of the outflow pathway being what determines the IOP. Obstruction of the aqueous humour drainage is the basis for an increase in IOP and in glaucoma.

Role of Purine Compounds in the Iris-Ciliary Body

The prejunctional regulation of nerve function in the iris-ciliary body is of physiological importance since it regulates pupil size as well as lens accommodation (see below). Marumatsu, Kigoshi and Oda (1994) demonstrated that the component of the non-adrenergic fibres was ATP and that ATP, 2MeSATP and specially α,β-MeATP produced transient contractions of the dilator muscle. The lack of a proper pharmacological characterisation only allows the possible involvement of a P2X receptor subtype in the contraction of the dilator muscle to be suggested.

The cholinergic and NANC components of neurally mediated muscle contraction in rabbit iris is strongly affected by purinoceptor agonists. The nerve-induced cholinergic twitches are inhibited by adenosine analogues in the following order l-PIA (phenylisopropyl adenosine) > NECA (N-ethylcarboxamidoadenosine), indicating the participation of a prejunctional adenosine A_1 receptor. A cholinergic late phase was enhanced by postjunctional A_1 receptor activation. Moreover, the NANC component was enhanced by adenosine analogues with the order NECA > L-PIA suggesting the involvement of a prejunctional adenosine A_2 receptor (Gustafsson and Wiklund, 1986). Similar studies were performed in rat iris. In this model, the application of the A_1 antagonist 1,3-dipropyl-8-cyclopentyl xanthine (DPCPX), increased the stimulation-evoked noradrenaline release by 35%, indicating endogenous prejunctional inhibition mediated via this adenosine receptor (Fuder et al., 1992). Further studies by Crosson and Gray (1996) have demonstrated that prejunctional adenosine A_1 receptors are linked to a $G_{i/o}$-protein. Nevertheless, the expression of this response depended on the frequency of neural stimulation, concluding that A_1 receptors may act selectively to limit high frequency neurotransmission.

Experiments carried out in rat iris have demonstrated that ATP is equipotent to adenosine but acts through different receptors since the application of adenosine deaminase did not affect the ATP induced prejunctional inhibition but affected the adenosine one (Fuder et al., 1992). This clearly suggests the presence of receptors sensitive to ATP.

Fuder and Muth (1993) performed experiments in order to characterise the ATP receptor mediating the inhibition of field-stimulation evoked noradrenaline overflow. The action of ATP was not affected by suramin or 4,4'-diisothiocyanotostilbene-2,2'-disulphonate (DIDS) (at micromolar concentrations) but was blocked by the P2Y antagonist Cibacron blue. This antagonist was unable to inhibit adenosine effect on noradrenaline overflow. More detailed pharmacological characterisations is necessary to determine which of the $P2Y_n$ receptors is involved in this action.

The ciliary epithelium is another tissue where nucleotides and adenosine can exert their actions. Rabbit ciliary epithelium responds to extracellular ATP by increasing intracellular Ca^{2+} as revealed by confocal microscopy. Changes in cytosolic Ca^{2+} levels are measurable in both pigmented and non-pigmented ciliary epithelium but changes are slightly better in non-pigmented ciliary epithelium. Studies performed with ryanodine and theophylline, which interfere with the effect of ATP, indicate the involvement of innositol triphosphate (IP_3) generation (Suzuki, Nakano and Sears, 1996). In bovine ciliary epithelial cells ATP also mobilises intracellular calcium. ATP, UTP and ADP induce dose-dependent Ca^{2+} transients that decrease rapidly. Different natural agonists presented the following potency order: ATP = UTP > ADP > AMP. Adenosine, α,β-MeATP and 2-MeSATP were completely ineffective as other transmitters such as adrenaline, noradrenaline or ACh. The effect of ATP and UTP was present even when extracellular Ca^{2+} was removed, indicating that Ca^{2+} transients came from intracellular stores. Moreover, nucleotide effects were inhibited by pertussis toxin indicating the coupling of the receptor to a G protein. Taking into account the features of this receptor it is concluded that the purine receptor present in bovine ciliary epithelial cells (pigmented and not pigmented) belong to the $P2Y_2$ subclass (Shahidullah and Wilson, 1996).

Ciliary Processes and Intraocular Pressure (IOP)

The balance between aqueous humour secretion by ciliary epithelial cells and its drainage regulates intraocular pressure. IOP is potently modulated by α_2-adrenergic and D_2

dopaminergic agonists (Hasslinger, 1969; Mekki and Turner, 1985). Adenosine is a substance which also significantly modulates IOP (Crosson, 1992). The non-selective adenosine receptor agonist, NECA, applied topically to rabbit eyes produces a biphasic response, initially it causes hypertension (first 30–60 min after application) followed by an hypotensive phase from 2 to 5 h. To identify the receptor subtype experiments were performed with the A_1 agonists N^6-cyclopenthyl adenosine (CPA) and the A_2 agonist CV-1808 (2-phenylaminoadenosine). Application of CPA produced a decrease in the IOP while CV-1808 did not modify the intraocular pressure (Crosson and Gray, 1994). A more careful analysis with different adenosine receptor agonists showed the following potency order: R-PIA (R-N^6-phenylisopropyl adenosine) = CHA (N^6-cyclohexyl-adenosine) > CGS21680 (2-[p-(2-carbonylethyl)-phenylethylamino]-5′-N-ethylcarboxamidoadeno-sine) > s-PIA (s-N^6-phenylisopropyl adenosine) >> CV-1808. These data and the antagonism of adenosine by DPCPX are consistent with the involvement of an adenosine A_1 receptor in the reduction of IOP (Crosson, 1995). Additional evidence for this has been provided by studies performed in order to detect changes in the cytosolic levels of cAMP, which demonstrated that the A_1 receptor is coupled to adenylate cyclase inhibition (Crosson and Gray, 1992; Wax et al., 1993; Crosson and Gray, 1994). R-PIA and CHA, which have been used for pharmacological characterisation of this receptor, are also agonists of the A_3 adenosine receptor, nevertheless this receptor is insensitive to neutral methylxanthines and the involved in IOP regulation is blocked (Linden, 1994).

The role of sympathetic nervous system in the control of intraocular pressure seems to be very relevant (Potter, 1981). As described before, sympathetic nerve terminals possesses prejunctional adenosine receptors (Fuder et al., 1992). The possible relation of these nerves with IOP was studied by denervating the iris and ciliary body. The hypotensive effect mediated by A_1 receptors was maintained even without sympathetic innervation. These experiments do not exclude the possibility of adenosine receptors present prejunctionaly but indicate that the changes in the IOP are due to a postjunctional action of adenosine (Crosson, 1995).

One of the ways to reduce IOP can be the reduction in the rate of flow of aqueous humour. CPA and R-PIA act in this way: both lower IOP by reducing aqueous humour flow, but an activation of the outflow facility by adenosine cannot be discarded (Camras et al., 1994). Neufeld and Sears (1975), demonstrated that the outflow facility is stimulated by analogues of cAMP. To fit these results to the adenosine receptor model it would be necessary to associate an adenosine A_2 receptor to the trabecular meshwork or the Schlemm's wall, but this has not been yet described.

THE LENS

The lens is situated behind the iris and between the aqueous humour and the vitreous. It has a passive role in accommodation process by which the light that passed through the cornea and aqueous humour is focused in the retina (Figure 6.4). This adjustable focus occurs by tension or relaxation of the zonular fibres by the ciliary muscle (ciliary body).

The lens is contained within a collagenous capsule. Inside the anterior part of the capsule is a single layer of epithelial cells that divide, move to equator and there elongate into the fibre cells that occupy most of the lens. This growth put new cells outside and the old inside the lens. The tips of fibre cells meet in a region called the lens suture. The fibroblast growth factor and insulin-like growth factor (IGF-1) induce the differentiation of epithelial cells into fibre cells.

Transparency is the main physical feature and essential for an accurate vision. The lens is neither innervated nor vascularised so nutrition comes from both humours the aqueous and the vitreous. The molecular structure of the lens is very specific since it consists of 2/3 water and 1/3 proteins, other constituents representing about 1% of the total weight. The high protein content is necessary for the high refraction index of the lens and also to focus the light beams at the retina. The main proteins in the lens are the crystallines, subdivided in α, β and γ, and which fulfil a structural role.

The lens metabolism is crucial to maintain and adequate transparency. One of these characteristics is the low-resistant gap-junction among cells. This fact transforms the lens in a syncitium and thus a more efficient metabolic interchange between the inner part of the lens and periphery. Since the lens does not loose any cell during its life, the oldest are located in the centre of this structure. This fact produces a loss of transparency, which in some cases reduce visual acuity. When this circumstance occur it is said there is a cataract. Cataracts can be also associated to other pathological states such as diabetes or physical stress such as radiations.

Purine compounds and the lens

The lens contains high amounts of nucleotides. The concentration of nucleotides in different species varies in part due to the anatomical characteristics for the accommodation process. Birds possesses the higher nucleotide contents followed by mammals and fish (Klethi and Mandel, 1963; Hockwin *et al.*, 1968). Among all the nucleotides studied in several species, ATP is always the most representative nucleotide followed by ADP and other nucleotide triphosphates. This particular distribution has been observed in sheep and goldfish lenses by high performance liquid chromatography (HPLC). The mammal lens contain 80 times more adenine nucleotides than the fish (Hernández *et al.*, 1994). HPLC as well as nuclear magnetic resonance techniques have been used to measure the nucleotide content of rabbit and human lenses (Greiner *et al.*, 1981, 1982; Deussen and Pau, 1991).

The crystalline lens is not a tissue with high metabolic activity, such as muscle or liver but it requires ATP for active transport processes. Electrolyte balance is vital for lens transparency, this being mainly performed by Na^+/K^+-ATPases keeping the right electrolyte balance in the organ. This enzymatic activity is located in the epithelium as well as in lens fibres and demand part of the ATP present in this structure (Delamere and Dean, 1981). ATP is used also in other active transport systems such as the Ca^{2+}-ATPases present in lens epithelial cells which control the Ca^{2+} levels and thus prevent a series of pathological processes (Paterson *et al.*, 1996).

Nevertheless other functions for this nucleotide have been described in the lens. ATP interacts with α-crystallines establishing the protein quaternary structure. The moderate strong binding of ATP to α-crystallines is related to differentiation and maturation of the lens. The decreasing metabolic activity and consequent metabolite gradient as the fibre cells are displaced inward is concomitant with supramolecular organisation.

The turnover of lens proteins is also important to maintain the structure of this eye medium. Proteolysis is important to guarantee the renewal of all those proteins related to the molecular organisation in the lens. Ubiquitin is one of the polypeptides necessary, together with other proteases, for α-crystalline turnover. The ubiquitin degradative system requires ATP for its activation and this being other biochemical duty for ATP in this system (Jahngen-Hodge et al., 1991). The role of ATP/ubiquitin proteolytic system is not only important in the lens but also in the retina following light stress.

Another role for ATP and adenine nucleotides is their possible involvement in protection of the retina from UV light. Although cornea retains 45% of UV wavelengths below 280 nm (Lerman, 1980), 55% can arrive to the retina producing potential damages in this part of the eye. The lens, due to its high content in adenine nucleotides can exert a protective action of the retina by absorbing wavelengths near 260 nm (which correspond to absorbance maximum of adenine ring). In this way, animals living on earth surface posses 80 times more nucleotides per mg of protein than a fish. An explanation for this is that the sea water absorbs 99% of UV light in the first 1.8 m. During evolution animals living on earth developed a system to further protect their retinas form UV light, whereas fish had natural medium to partially protect them against those radiations (Hernández et al., 1994). To support this hypothesis, Klethi and Mandel (1963) described that birds have the highest levels in nucleotides when compared to mammals and fish. This could be the first time in which ATP is described as biological molecule with important physical actions.

Purine receptors in the lens

Since the most external part of the lens is an epithelial tissue, receptors for adenosine and ATP similar to those described in other epithelial tissues would be expected.

Concerning adenosine receptors, there are no evidences for the actions of this nucleoside in the lens. Moreover, Kvanta et al. (1997) studying the possible receptors for adenosine in eye structures, concluded that rat lenses did not express any mRNA of the known adenosine receptors (A_1, A_{2A}, A_{2B} and A_3). Apart from these in situ hybridisation studies, nothing has been reported in the literature on adenosine receptors in the lens.

The volume of the lens is a factor, which needs to be controlled to avoid opacification processes. Epithelial cells present in the anterior side of the lens have a high chloride permeability which significantly contributes to the lens-volume regulatory system (Jacob and Zhang, 1993). This channel is also present in fibre cells and its open probability is increased by membrane depolarisation. This Cl^- channel presents a similar pharmacological profile to the whole cell Cl^- currents generated by the activation of P-glycoprotein in multidrug resistant transfected cells (Valverde et al., 1992). It has been found that these volume-activated currents require ATP binding but not hydrolysis for activation (Gill et al., 1992). ADP was unable to produce the rapid activation ATP does but it did a slow long-term increase in the channel activity. The use of an antibody of the chloride conductance activated by ATP (Zhang and Jacob, 1994). The chloride channel in the lens is closely related to P-glycoprotein and needs ATP for its activation. This channel plays an important role in lens physiology functioning as a volume regulatory mechanism to prevent changes in the refractive index that would cause the scatter of light and result in cataracts.

Human lens epithelial cells present receptors sensitive to ATP and ADP (but not for adenosine) which induce intracellular calcium mobilisation (Crawford, 1992). Pharmacological studies demonstrated the following profile: ATP = UTP > ATPγS > ADP = GTP >> AMP = Adenosine = α,β-MeATP. Suramin is an antagonist, which corresponds with a P2Y$_2$ receptor (formerly P2U), activation of which leads to an oscillatory release of Ca^{2+} from lens cell intracellular stores, and IP$_3$ generation, which was thapsigargin sensitive (Riach et al., 1995).

In sheep lens, a P2Y$_2$ receptor coupled to phospholipase C (PLCβ) and increasing intracellular Ca^{2+} was demonstrated. Studies carried out on single cell demonstrated that 92% of lens cells did respond to extracellular ATP and responses to other neurotransmitters such as adrenaline and histamine were also observed. ATP was partially sensitive to suramin and desensitised the receptor present in these cells after prolonged exposure. This fact can be due either to the intracellular stores depletion or by down regulation of an intermediate of in the signal transduction pathway (Churchill and Louis, 1997).

Many different agonists have been tested for their ability to mobilise intracellular Ca^{2+} in the lens of several species. Histamine, noradrenaline, ACh and ATP mobilised Ca^{2+} from intracellular stores in human, rabbit and sheep, nevertheless not all of them responded in all the systems under study (Williams et al., 1993; Riach et al., 1995; Duncan et al., 1996). The only transmitter molecule which was active in the three models was ATP, this indicating a conserved response regardless of species.

Concerning the possible role of P2Y$_2$ receptors in the lens, some authors suggest their involvement in regulating cell division by inhibiting growth factor-stimulated cell division (Duncan, Williams and Riach, 1994). On the other hand, ATP acting on lens P2Y$_2$ receptors seems to be involved in differentiation from lens epithelial cells to fibre cells (Churchill and Louis, 1997). It should be remembered that Ca^{2+} mobilisation has been reported to regulate lens fluid transport (Berridge, 1993; Jacob and Zhang, 1993). This is very important for maintenance of the volume of crystalline lens, because apart from the ATP-activated Cl$^-$ channels to regulate cell volume there are Ca^{2+}-activated Cl$^-$ channels (Valverde et al., 1993).

Where does ATP come from?

Since the lens is not vascularised or innervated, the only sources for ATP are the aqueous and vitreous humours apart from mitochondrial ATP synthesis. ATP has been detected in both media but the studies demonstrate that it is present at higher concentration in the vitreous compared with the aqueous (Greiner and Meidman, 1991). The iris and ciliary body could be the source for ATP in the aqueous humour but this has not yet been demonstrated. ATP can appear extracellularly from injured cells (Neary et al., 1996), probably indicating a pathological event in the surrounding cells. The release of cytosol after cellular rupture has been demonstrated in single lens cells. In this case an increase in intracellular Ca^{2+} was observed, probably due to the ATP released after cellular disruption (Churchill, Atkinson and Louis, 1996). In pathological conditions such as uveitis and general inflammation, ATP and other substances like histamine are increased, thus being available at the lens surrounding media (Unger and Butler, 1988).

Cataracts

The lens has been designed for the transmission of the light. It responds to any insult that disturbs normal development or metabolism by opacification. Many types of cataracts have been described in the literature and all of them are characterised by an increasing of light scatter by lens proteins. Since crystallines are so highly organised within the lens, any distribution of their supramolecular organisation leads to opacification and thus light scatter (for a review see Harding and Crabbe, 1984). Certain metabolic conditions are associated with cataracts, the best known being diabetes and galactosaemia.

In diabetes and galactosaemia, the elevated concentrations of glucose and galactose in the aqueous humour lead to an increase in the glucose levels in the lens cells. Glycolysis becomes saturated and accordingly the polyol pathway is up regulated (Hockwin, Fink and Beyer, 1973). The increase in polyols increases the osmotic drag of water into the cells. One of the main problems is the reduction of intracellular ATP and glutathione levels (Iwata and Takehana, 1982). As has been described before, ATP is connected with important intracellular processes in the crystalline lens. A reduction in the ATP cytosolic levels would affect the stabilisation of the quaternary structure, this probably favouring a less organised crystalline organisation and a higher light scattering (Palmisano *et al.*, 1995). Moreover, intracellular ATP is closely related to active transport systems, particularly the Na^+/K^+-ATPase that is responsible for electrolyte balance, which partially regulates cell volume (Delamere and Dean, 1993). Also, Ca^{2+}-ATPase could be involved in cataract formation. This transporter is responsible for the maintenance of Ca^{2+} homeostasis since the extracellular Ca^{2+} concentration is about 100 times higher than inside the lens cells. Ca^{2+}-ATPase is present in intracellular stores although it is also present in cell membranes (Hightower, Duncan and Harrison, 1985). In cataractous lenses the activity of Ca^{2+}-ATPase is reduced to 50% compared with normal lenses and Ca^{2+} homeostasis is severely affected (Duncan, Williams and Riach, 1994). This could be partially due to the lower intracellular ATP concentration and the consequent reduction in the enzyme activity (Paterson *et al.*, 1996).

ATP is reported to be associated with proteolytic systems as described above. The general decrease in the ATP cytosolic content would affect the crystalline turnover, subsequently affecting supramolecular structure of the lens (Jahngen-Hodge *et al.*, 1991).

Extracellular adenine nucleotides could be related to cataract formation at least to some extent. For example, ATP activates the Cl^- conductance in lens fibres (Jacob and Zhang, 1993). Since this channel contributes to lens volume regulation, an over activation of this channel would lead to a change in the refractive index that would cause light scatter and result in cataract (Zhang and Jacob, 1994).

It has been demonstrated that cataracts appear after inflammation, which produces the release of ATP and other substances (Duke-Elder, 1969; Secchi, 1982). Since an increase in lens Ca^{2+} seems to play an important role in the development of cataract, any interference with systems regulating lens calcium is likely to be cataractogenic in nature (Marcantonio, Duncan and Rink, 1986). For example, the application of acetylcholinesterase reduces the risk of cataract due to ACh breakdown, since this substance has receptors in epithelial lens cells which mobilises intracellular Ca^{2+} (Schaffer and Hethrington, 1966). In this way ATP released in large amounts after cell damage or inflammation would behave as a potential cataract inducer by acting on $P2Y_2$ receptors in the lens.

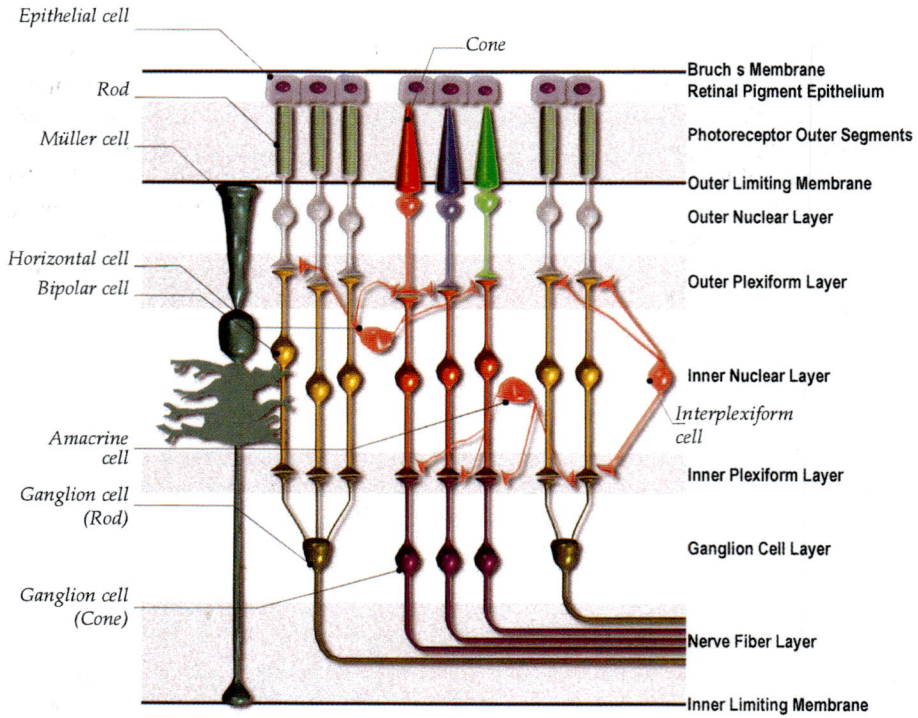

Figure 6.5   Three dimensional model of the retina. The most representative retinal cells (neural and non-neural) and the main layers are indicated.

THE RETINA

The retina is the photosensitive part of the eye containing photoreceptor cells. There are two types of photoreceptors, cones that respond to moderate light levels, responsible for daylight vision and colour, and rods, which respond to low light levels and are mainly used for night vision. When photoreceptors cells are stimulated a series of biochemical changes occur in the disks containing rhodopsin, which lead to an inhibition of the photoreceptor transmitter release (glutamate). This affects the neural excitability of neighbouring cells which transmit this information to the ganglion cells. Axons of ganglion cells project to the thalamus (mainly lateral geniculate nucleus and superior colliculus) and from there to visual cortex.

The organisation of the retina is rather complicated. Cells are geometrically oriented with the retina presenting two cell planes, a vertical one with photoreceptors, bipolar cells and ganglion cells, and an horizontal one with horizontal cells and amacrine cells (Figure 6.5). Further cell types are important constituents of the retina, epithelial cells where rods and cones are anchored and glial cells (Müller cells). Two main synaptic regions occur in the retina, the first among photoreceptors, bipolar and horizontal cells, and the second among horizontal, amacrine and ganglion cells. These two regions are termed external plexiform layer and internal plexiform layer respectively and are easily identified in histological preparations.

When the light stimulates a photoreceptor, this cell, by inhibition of the release of glutamate, modifies the activity of a bipolar cell. The bipolar cell transmits this information to ganglion cell and from there to the thalamus. Sometimes photoreceptors do not stimulate bipolar cells but activate horizontal cells and these act on photoreceptors which establish synapses with bipolar cells. In other cases, photoreceptors establish synaptic connections with bipolar cells but these connect to a ganglion cell by means of an amacrine cell.

The pigmented epithelium is a monolayer of 10–20 μm thickness, which lies on Bruch's membrane through which nutrition and oxygenation are derived. These cells are tightly stacked against each other and they maintain close relations by gap junctions. This epithelium does not perform any neural role but is crucial for appropriate turnover of photoreceptor. Retinal pigmented epithelium phagocytoses apical portions of photoreceptors including the oldest disks because they embrace one third of the external segment of rods and a little less of cones. The term pigmented indicates the high abundance of granules containing melanolipofuchsin. This pigment prevents reflection of light, and thus diffuse illumination of the retina is avoided. It helps the retina to get a better acuity and contrast between clear and dark images.

Glial cells are very important in the retinal physiology. The most important representative glial cells are Müller cells. These cells are large and cross the retina vertically, permitting the retina to have the right cell distribution. Their role is to help neuronal cells of the retina to eliminate extracellular K^+ in the plexiform layers. Astrocytes are also present and they appear mostly in the central part of the retina. Their role is to supply nerve cells with a glycogen supply for metabolic purposes.

Purinergic transmission in the retina

Adenosine receptors   Several neurotransmitters candidates have been described for photoreceptors, horizontal, bipolar and amacrine cells (Brecha, 1983). Biogenic amines, peptides and amino acids have been associated with retinal neurons, and actions mediated by adenosine and ATP have also been demonstrated.

Identification of adenosine receptors has been performed in retinal membranes by means of adenosine radioactive agonists. Specific binding sites for [³H]PIA have been detected in bovine retinal membranes with the following pharmacology: R-PIA = CPA > CHA > CADO (2-chloroadenosine) > S-PIA > NECA which correlates well with an A_1 receptor. In this membrane preparation the adenosine A_1 receptor antagonist, DPCPX, was effective preventing the binding of radiolabelled PIA (Woods and Blazynski, 1991). Membranes pre-incubated with the stable GTP analogue guanyl-imidodiphosphate (GMP-PNP) decreased the specific binding, this result being in agreement with the idea of guanine nucleotides shift the equilibrium of high and low affinity states of the receptors to favour the low affinity state (Yeung and Green, 1983; Stiles, 1988; Woods and Blazynski, 1991). The adenosine A_1 receptor present in retinal membranes was sensitive to Mg^{2+}; this ion increases the affinity of agonist for this receptor (Goodman and Snyder, 1982).

It is reasonable to think that there should be a connection between the adenosine receptor and cAMP. Some authors reported that adenosine did not modify cAMP levels (Campau, Kinsherf and Ferrendelli, 1983). Nevertheless later studies offer different conclusions. The rabbit retina possesses adenosine A_1 receptors which inhibit adenylate cyclase (Blazynski et al., 1986; Blazynski and Perez, 1991). Inhibition was only observed when the basal activity of adenylate cyclase was elevated with forskolin. Under these conditions, adenosine or an A_1 agonist reduced the cytosolic levels of cAMP. In other work, Blazynski et al. (1986) demonstrated that rabbit retina accumulates cAMP whereas mouse, guinea pig and rat retina did not. This increase in the cAMP levels suggests the involvement of an adenosine A_2 receptor. The presence of adenosine A_2 receptors has been described in bovine retina (Blazynski and McIntosh, 1993). Radioligand binding studies with [^{3}H]NECA, [^{3}H]CGS21680 and non-labelled adenosine agonists, indicated the presence of both A_{2A} and A_{2B} receptors.

Isolated bovine retina pre-loaded with [^{3}H]dopamine, and the K^+-induced release of this substance was studied in relation to adenosine receptor agonists (Crosson, DeBenedetto and Gidday, 1994). CPA inhibited dopamine release and this effect was reversed after application of cyclopentyl theophylline (CPT), indicating the involvement of an adenosine A_1 receptor.

Although there is not a clear correlation between adenosine receptors and retinal cell types, autoradiographic analysis gives some information. Adenosine inmunoreactivity has been observed in ganglion cells and in the optic nerve fibre layer (Zarbin et al., 1976). The inner plexiform layer and some cells in the inner nuclear layer were stained by adenosine A_1 labelling with PIA in rat, guinea-pig, monkey and human retinas (Braas, Zarbin and Snyder, 1987). The colocalisation of adenosine inmunoreactivity and adenosine binding sites is noteworthy, since this suggests these sites are physiologically relevant. Ganglion cells are divided into X, Y and W cells according to their receptive fields. Adenosine A_1 receptors seems to be present in W cells, which have small bodies and large dendritic arbours with projecting axons mainly to the superior colliculus (Stone and Hoffman, 1972; Hoffman, 1973). This is in accordance with immunoreactivity for endogenous adenosine and adenosine deaminase in this superior colliculus (Nagy, La Bella and Buss, 1984; Braas et al., 1986). In addition, stimulation of the optic tract terminals in the superior colliculus promotes adenosine release, and adenosine analogues inhibit neuronal transmission in these species (Heller and McIlwain, 1973; Kostopoulos and Phillis, 1977).

Specific functions of adenosine in retinal cells have not been completely studied. Adenosine has been involved in retinal transmission modulation. The concentration-dependent inhibition of depolarisation-induced ACh release by PIA has been demonstrated (Perez and Ehinger, 1990). Cholinergic cell processes reside in the inner plexiform layer, where a subset of cholinergic cells contain adenosine (Blazinsky, 1989). In this retinal area, A_1 receptors have been described (Blazynsky, 1990). A similar role to one found in other locations for adenosine acting by means of A_1 receptors can be supposed in this part of the retina.

Adenosine A_2 receptors have been identified in both outer and inner segments of photoreceptors (Blazynski, 1990). The role of A_2 receptors in photoreceptors is not clear. Since these cells contain an active Na^+/Ca^{2+} exchanger, adenosine might regulate this, as occurs in cardiac sarcolema (Brechler et al., 1990).

We cannot forget about the non-neural components of the retina: retinal pigment epithelial cells (RPE) and Müller cells. Concerning the first, human RPE cells have adenosine A_1 receptors that are positively coupled to adenylate cyclase (Friedman et al., 1989). Adenosine agonist-stimulated accumulation of cAMP was inhibited by alkylxanthines indicating that it was a receptor-mediated event. The possible presence of A_1 receptors in this tissue was discarded since activation of adenylate cyclase by isoprenaline was not inhibited by the A_1 agonists PIA and CPA. Radioligand binding studies confirmed the present of A_2 receptors in RPE cell membranes. These experiments helped to identify this receptor as an A_{2B} by using different adenosine agonists (Blazynski, 1993).

The role of adenosine A_2 receptors in RPE cells has not been fully investigated. Due to the receptor-effector coupling these receptors might regulate the transmembrane electrical current across the RPE, as suggested by Frambach et al. (1990). Another possibility is that A_2 receptors could be involved in the maintenance of the photoreceptor length since, for example, NECA inhibits phagocytosis of photoreceptors in culture (Gregory, Hall and Abrams, 1991). This is in clear correlation with those studies which indicate that elevation of cAMP levels inhibits phagocytosis (Edwards et al., 1986). In summary, adenosine in RPE cells acting via an A_{2B} receptor would be an important mechanism for controlling photoreceptor length.

Little is know on the presence of adenosine receptors in Müller cells. Due to its location in the retina is difficult to investigate the presence of such receptors. Only autoradiographic studies may indicate the possible labelling of Müller cells but until now no studies have been performed to investigate the nature of the putative adenosine receptors in these cells (Blazynski, 1987).

Adenosine sources   The origin of extracellular adenosine has been investigated by pre-loading isolated retina with [^{3}H]adenosine. Release was evoked by stimulating with K^+, and it was a Ca^{2+}-dependent process. Light was also able to induce labelled purine release but not as great as by K^+ depolarisation. Adenosine metabolites accounted for more than 90% of purines but these adenine compounds can be rapidly transformed by ecto-nucleotidases (Henderson, 1979; Perez et al., 1986) or deaminases (Senba, Daddona and Nagy, 1986). Retinal cells posses a system for adenosine transport which is sensitive to nitrobenzylthioinosine (Perez and Bruun, 1987; Blazynski, 1989; Blazynski, 1991; Paes de Carvalho et al., 1992). Several cell bodies appear labelled in the innermost and middle cell rows of the inner nuclear layer as well as in ganglion cells. The selective accumulation of nucleosides in certain retinal neurons is what would be expected if nucleosides are neurotransmitters or neuromodulators in the retina (Ehinger and Perez, 1984).

Cytochemical techniques have helped to localise 5'-nucleotidase, which converts AMP into adenosine (Kreutzberg, Barron and Schubert, 1978). It is noteworthy that this enzyme is not only associated with the cytosol but also with the plasma membrane, the catalytic site being in an ecto-position. This ecto-5'-nucleotidase is present in rod cells, Müller cells, and pigmented epithelial cells (Kreutzberg and Hussain, 1984). These findings suggest that adenosine might play a role in synaptic function, especially in the synapses close to the enzyme localisation.

P2 receptors   Apart from its metabolic tasks, ATP interacts with a protein present in photoreceptor cells called arrestin (Glitscher and Rüppel, 1989). This protein is responsible for light-dependent transducin and cGMP phosphodiesterase activity in rods. Some studies have been performed on the extracellular functions of ATP in the retina.

The early embryonic chick neural retina possesses receptors sensitive to ATP that are also activated by UTP (Laasberg, 1990; Lohman et al., 1991). The receptors are slightly more sensitive to UTP than to ATP, while suramin and reactive blue 2 completely block their responses. A developmental study of this receptor has shown that it is most active during the period between the third and eighth days of development, and that its activity decreases until the eleventh day (Sugioka, Fukuda and Yamashita, 1996). Regarding the second messengers coupled to this $P2Y_2$ (P2U) receptor, it stimulates PLCβ by acting through a G protein, which is pertussis toxin-sensitive. This second messenger system mobilises Ca^{2+} from intracellular stores via IP_3 generation (Sakaki, Fukuda and Yamashita, 1996).

Neal, Cunningham and Paterson (1995) reported the binding of α,β-MeATP to rabbit eye-cup preparations, suggesting the existence of P2X receptors in the retina. The expression of $P2X_2$ receptors in the retina has been recently reported (Greenwood, Yao and Housley, 1997). These authors used in situ hybridisation studies to demonstrate that photoreceptors cell bodies, neurons in the inner nuclear layer and retinal ganglion cells were labelled whereas the outer segments of photoreceptors and the inner plexiform layer were unlabelled. Also, experiments performed with $P2X_2$ antiserum demonstrated the same pattern of expression as that determined with an antisense probe (Kanjhan et al., 1996). Concerning the site of the receptor protein, retinal ganglion cells were stained as well as the dendritic processes which penetrate in the inner plexiform layer.

This colocalisation of the mRNA hybridisation and the inmunoreactivity is strong evidence that extracellular ATP participates actively in the retinal network, especially at the ganglion cells, due to the high degree of labelling reported (Greenwood, Yao and Housley, 1997). The presence of inmunoreactivity in the outer segment of the photoreceptors also suggests a non-neural site of action of extracellular ATP in the retina.

ATP can be synthesised by neural retina since adenosine is rapidly phosphorylated to ATP (Perez et al., 1986). Cholinergic neurons in the retina, which include amacrine cells, are sensitive to light, and release ACh. In eye-cup preparations pre-incubation of the P2 antagonist PPADS (pyridoxalphosphate-6-azophenyl-2',4'-disulphonic acid) potentiates the release of ACh by 40% (Neal and Cunningham, 1994). In contrast, ATP is able to inhibit ACh release by 20%. The A_1 antagonist, DPCPX, did not block the action of ATP. The possible corelease of ATP with ACh demonstrated by Neal and Cunningham by studying the release of glycine from inhibitory neurons suggests that P_2 receptors can be present in glycinergic amacrine cells. Light would stimulate the release of ACh and ATP, and both would stimulate glycinergic neurons, which in turn provide inhibitory feedback onto the cholinergic amacrine cells (Neal and Cunningham, 1994).

Photoreceptor cells and RPE cells maintain a close physiological relationship. RPE cells are stimulated by extracellular application of ATP, ADP or UTP (Nash, Carvalho and Osborne, 1996), the stimulation being coupled to inositol phosphate turnover and intra-cellular Ca^{2+} mobilisation. The receptor is coupled to a pertussis-sensitive G protein, which, together with the pharmacology, defines the receptor as a $P2Y_2$ (Sullivan et al., 1997). Although in these cells an A_2 receptor has been described, it is not coupled to an increase in the intracellular Ca^{2+} levels (Friedman et al., 1989). The response induced by ATP or UTP is biphasic, with an initial peak and followed by a sustained plateau. This could be due to the activation of two different signal transduction pathways, but needs further study (Sullivan *et al.*, 1997). The physiological role of these receptors is not clear. It has been suggested that they could be involved in the turnover of photoreceptor outer segment (as described for A_2). The ingestion of photoreceptor is inhibited by increased intracellular Ca^{2+} levels or by the activation of protein kinase C (PKC), presumably by diacylglycerol (Hall, Abrams and Mittag, 1991).

Another possible role of $P2Y_2$ receptors in the maintenance of the appropriate volume and composition of the subretinal space. Peterson *et al.* (1997) have demonstrated that the extracellular application of ATP or UTP increases basolateral membrane Cl^- conductance and decreases apical K^+ conductance, which produce apical to basolateral fluid absorption. These effects were blocked by suramin and partially by DIDS. The ability of ATP or UTP to affect Cl^- conductance is related to the intracellular increase of Ca^{2+}. In other epithelia, Ca^{2+}-activated Cl^- channels have been proposed to play an important role in fluid secretion (Jiang *et al.*, 1993; Smith and Welsh, 1993). In toad RPE cells intracellular Ca^{2+} increases significantly inactivate inward rectifying K^+ channels (Segawa and Hughes, 1994). Regulation of ionic conductances may play an important role in maintaining tight apposition between the retina and RPE (Negi and Marmor, 1986).

Retinal oligodendrocytes also possess receptors sensitive to ATP and UTP. It seems that the level of expression can change depending on the stage of development but activation always produces an increase in the cytosolic Ca^{2+} levels (Kirishuk *et al.*, 1995). The rank order of agonist potency at this receptor is: UTP $\geq$ ATP > ADP >> AMP, which is in agreement with it being a $P2Y_2$ receptor. The role of this receptor in oligodendrocyte physiology needs to be elucidated.

What is the source of nucleotides in the retina?

It is generally accepted that extracellular nucleotides can appear as a consequence of cellular lysis. This is can occur under pathological wound-healing situations and when inflammatory processes occur. Nevertheless there is a common process that is continuous in the retina: photoreceptor phagocytosis. This is a potential source of ATP, since these cells contain this nucleotide to carry out structural changes in the rod outer segment (Thatcher, 1983; Borys *et al.*, 1986). This extracellular source of ATP would permit the activation of those receptors present in the RPE cell membranes with their corresponding physiological roles.

On the other hand, when considering a normal retina, nucleotides are released from synaptic vesicles or granules. In this case the nucleotide release is produced in response to a stimulus. Work described by Perez and coworkers has demonstrated that apart form adenosine, cells were able to release measurable amounts of ATP (Perez *et al.*, 1986; Perez, Arner and Ehinger, 1988; Perez and Ehinger, 1990). It is not clear whether the authors took into consideration the presence of ecto-nucleotidases and the possible involvement in the final adenosine/ATP ratio. The experiments performed in cholinergic neurons by Neal and Cunningham (1994), described above, demonstrated that after exposure to light some cholinergic cells are able to release ATP as a cotransmitter. It cannot be excluded that other neural cells in the retina may release adenine nucleotides together with other neurotransmitters.

Retinitis pigmentosa and RPE detachment

$P2Y_2$ receptors seem to be related to photoreceptor phagocytosis. Disturbances in the functioning of the purinergic system at this level, such as an increase in the nucleotide supply, may lead to an accumulation of retinal pigments, although it is important to note that the relationship between an elevation of the extracellular nucleotide levels and retinitis pigmentosa has not been yet demonstrated.

The fact that ATP and UTP increase the levels of fluid absorption may have important implications. Accumulation of fluid in the subretinal space often produces retinal detachment from the epithelium with a corresponding loss of vision (Anand and Tasman, 1994). Analyses performed by Peterson *et al.* (1997) have demonstrated that apical application of ATP or UTP would remove approximately 0.75 ml of fluid per day in the human eye. This is strongly suggests that these nucleotides could be used therapeutically to reduce pathological accumulation of fluid in the subretinal space.

VASCULATURE AND PURINE COMPOUNDS

Vasodilatation in response to hypoxia in blood vessels is mainly due to the release of ATP from endothelial cells. This ATP interacts with $P2Y_1$ receptors, present in the same cells, which trigger synthesis of nitric oxide (NO) and consequent vasodilatation (see Burnstock, 1993). Adenosine is known to be a vasoactive substance in vascular smooth muscle (Berne, 1963), but it seems likely that adenosine is generated after ATP breakdown by ecto-nucleotidases. This would explain the participation of adenosine in hypoxic vasodilatation. Nevertheless, the vasodilator potency of ATP is from 10 to 100 times greater than that of adenosine (Forrester, Harper and MacKenzie, 1979; Alborch and Martin, 1980). It is clear that both adenosine and ATP can exert important actions on vascularised eye structures and, moreover, these actions can be mediated by different purine receptors.

VASOACTIVE PROPERTIES OF ADENOSINE IN THE RETINA

The physiological mechanisms underlying the control of retinal blood flow have yet to be elucidated. Although metabolic regulation of flow is expected, it is not completely understood how this comes about. Studies performed in retinal arterioles have demonstrated that vasodilator responses of these vessels to hypoxia was attenuated to 55% by the adenosine receptor antagonist, 8-sulphophenyltheophylline (8-SPT). Similarly, arteriolar vasodilatation was inhibited 76% by the same antagonist. This indicates that potentiation and inhibition of endogenous adenosine action affected retinal arteriolar dilator responses to hypoxia and hypotension suggesting that adenosine mediates autoregulatory adjustments in the eye (Gidday and Park, 1993a; Portellos *et al.*, 1995). Application of adenosine analogues by intravitreal topical microinfusion produces vasodilatation in retinal arterioles and venules with a potency order of NECA > CADO > adenosine = CHA. This suggests the presence of an adenosine A_2 receptor mediating the vasodilator actions of adenosine. Potentiation of adenosine was obtained when the adenosine transport inhibitor, nitrobenzylthioinosine, or the adenosine kinase inhibitor, iodotubercidin, were applied to the system. Adenosine, by acting on A_2 receptors mediates the retinal arteriolar dilatation (Gidday and Park, 1993b; Takagi *et al.*, 1996). Thus, adenosine may be responsible for the hyperaemic response to systemic hypoxia in retinal arterioles (Gidday *et al.*, 1992). More recently Zhu and Gidday (1996), have demonstrated that acute hypoglycaemia causes a compensatory increase in retinal flow, effect which is partially sensitive to 8-SPT. In this case adenosine would be involved in eliciting an increase in the retinal flow that accompanies hypoglycaemia.

Antagonism of the adenosine receptor attenuates hypoxic vasodilatation in cat retina (Roth and Pietrzyk, 1993). It is important to point out that although adenosine is involved in the vasodilatation that occurs when blood flow is restored after retinal ischaemia, adenosine receptor blockade does not abolish the hyperaemia, indicating that other vasoactive substances also affect post ischaemic hyperaemia in the retina (Roth, 1995).

Experiments performed on cats subjected to retinal ischaemia by ligation of the central retinal artery have shown that adenosine, presumably originating from ATP hydrolysis, appears in retinal vessels along with a gradual increase in the adenosine metabolites, inosine and hypoxanthine. Prolonged ischaemia results in the production of xanthine, which serves as a precursor for oxygen radical free formation with the consequent pathological implication (Roth *et al.*, 1997).

VASOACTIVE PROPERTIES OF ATP IN THE RETINA

Apart from the effects elicited by adenosine, ATP is an active nucleotide with important vasoactive properties. Studies performed on retinal capillary endothelial cells demonstrated that ATP was an active substance producing an increase of cytosolic IP_3 levels. Concomitantly with this effect an elevation of the intracellular Ca^{2+} concentration was also measured in these cells after the extracellular application of ATP. The pharmacological profile for these effects is ATP > ADP > β,γ-MeATP (β,γ-methylene ATP). This profile together with the second messenger system, which is activated, suggest that action of ATP is mediated by a P2Y receptor (Tao *et al.*, 1992). Studies performed in the retinal microvascular endothelium have demonstrated that ATP stimulates the synthesis of IP_3 and 6-keto-prostaglandin $F_{1\alpha}$ in a dose-dependent way (Robertson, Ar and Goldstein, 1990). AMP and adenosine have also been tested but were found to be inactive. Other nucleotide triphosphates such as CTP, UTP, GTP are very weak agonists. These results indicate that after receptor activation by ATP an elevation of intracellular Ca^{2+} occur. This transient increase stimulates phospholipase A_2, which releases arachidonic acid from membrane phospholipids (Lapetina, Billah and Cuatrecasas, 1981), and initiates prostaglandin formation. Prostaglandins are known to be potent vasodilators (Wiskler *et al.*, 1978), and their release by the retinal endothelium could cause vasodilatation of retinal blood vessels. On the other hand, stimulation of IP_3 synthesis may be linked to NO formation (formerly endothelium derived relaxation factor). NO is synthesised by vascular endothelium and possesses very important vasodilator properties (Pirotton *et al.*, 1987). The enzyme for NO synthesis, nitric oxide synthase, is present in retinal endothelial cells and it is highly modulated by various vasoactive substances (Ostwald *et al.*, 1997).

Under normal conditions, ATP is present at very low concentrations in the blood. Marked increases of this nucleotide occur when vascular beds are under hypoxic conditions. Vascular cells release intracellular stores of ATP after hypoxic insults and the concentration that can be reached extracellularly is more than sufficient to stimulate P2Y receptors present on endothelial cells (Pearson and Gordon, 1979). Endothelial ecto-nucleotidases transform ATP into adenosine in a relatively short period of time (Zimmermann, 1996b). Indeed, after stimulating P2Y receptors, ATP can be transformed to ADP by an ecto-ATPase, and the ADP formed to adenosine by an adenosine diphosphate phosphohydrolase present in the eye (Shoukrey and Tabbara, 1987). This fact might indicate that the vasoactive effects mediated by adenosine are produced, at least in some cases, as a secondary process depending on the ATP hydrolysis in blood vessels.

VASCULATURE, PURINES AND OTHER OCULAR STRUCTURES

There is a lack of information about the effect of purines in other vascularised ocular structures. Adenosine applied intravitreally affected the ocular blood flow by increasing it in iris and iris root ciliary body, in addition to the retina as previously indicated. In these two structures (iris and iris root ciliary body), adenosine produces an increase in the blood flow which is potentiated by adenosine transport inhibitors, such as dipyridamole, and is blocked by the adenosine receptor antagonist, 8-phenyltheophylline (Braunagel et al., 1988). Although there is no exhaustive study on the adenosine and P2 receptors present on vascular endothelial cells in other areas of the eye, it seems probable that other structures have a receptor system closely related to the one described in retinal vascular endothelial cells.

FUTURE DEVELOPMENTS

Little is known about the action of purines in the eye structures in normal and pathological conditions. Despite the lack of experimental evidence to demonstrating the possible roles of purine compounds in some ocular structures, the following paragraphs outline future trends of purinergic systems involved in the functioning of the eye.

TROPHIC ACTIONS OF PURINE COMPOUNDS

Most of the studies of the actions of purine compounds are based on receptors and actions from a short-term point of view. The long-term actions of purines have been studied for a relatively short period of time. There are some studies on embryonic development (Knudsen and Elmer, 1987; Laasberg 1990) and cell growth and proliferation (Gonzales et al., 1990; Erlinge et al., 1995). The trophic role of purines on neural or glial cell activation has been described, as has a relationship between nucleotides and growth factors (Neary and Noremberg, 1992; Abbracchio et al., 1996). This is a point of great interest in the eye, since high-density neural areas such as the retina can be modulated not only by the short-term but also by the long-term action of purines.

PURINE COMPOUNDS AND EXTRAOCULAR MUSCLES

Six extraocular muscles are responsible for the eye movement. They are innevated by three of the cranial nerves, namely, the oculomotor, abducens and trochlear. Lesions of these nerves produce several problems such as the inability to look upward or downward, external strabismus and diplopia. Lesions in the autonomic sympathetic nerves supplying to the eye and related structures may produce ptosis (in this case sympathetic nerves), dilated non-reactive pupils, and lack of accommodation. Extraocular muscles are stimulated by means of ACh released from the nerve endings. Nothing is known about the possible modulation of ACh release by adenosine as occurs at other neuromuscular junctions. If adenosine receptors are present in the neuromuscular junction (presumably A_1 receptors) some of the problems described before could be modified by application of adenosine agonists or antagonists. It is necessary to know whether these muscles posses adenosine or nucleotide receptors prior to studying the possible therapeutic applications of purines.

IOP AND GLAUCOMA

The control of IOP is the main target in glaucoma formation and development. Glaucoma is a disorder that can be divided into two main groups, primary and secondary glaucoma both subdivided into open angle and closed angle. In primary open angle glaucoma a slow rise in the IOP is accompanied by occlusion of the posterior ciliary artery, so that ischaemic optic atrophy occurs. Damage to optic nerve fibres may be due to pressure induced vascular disease of the optic nerve or to the effect of direct mechanical pressure on axoplasmic flow. Primary closed angle glaucoma is characterised by a decrease in size of the ocular globe and an increase the size of the lens. The lens displaces the pupil and the anterior chamber becomes smaller, and since the angles become narrower the pressure builds up behind the iris and pushes it towards the trabecular meshwork. This produces an enhancement of the IOP and again axoplasmic flow is altered.

Open angle secondary glaucoma occurs after inflammation, lens capsule degeneration or haemorrhage. Closed-angle secondary glaucoma is characterised by a closure of the anterior angle either by anterior displacement of the lens or during uveitis in which fibrin initiates formation of adhesions between the peripheral iris and the trabecular meshwork.

In all cases a general increase in the IOP applies pressure to axons and blood vessels which gradually causes retinal degeneration. Adenosine and its analogues seem to be affective in regulating IOP by activating adenosine receptors (see earlier). It is possible that in the same way as β-blockers, pilocarpine or adrenaline are used to reduce IOP, adenosine and adenine nucleotides could exert similar actions and thus contribute together with other pharmacological tools to help reduce intraocular pressure.

VASCULAR PATHOLOGIES

Endothelial dysfunction occurs under pathological conditions. An imbalance between NO and contracting factors would be important for the development of vascular ophthalmic complications arising from such conditions as hypertension, diabetes, arteriolosclerosis and retinal ischaemia. Endothelial malfunctions could contribute to vasospastic events in retinal migraine and some forms of low-tension glaucoma. Since a relationship exists between P2Y receptors present on endothelial cell membranes and a vasodilator effect of extracellular ATP, this nucleotide could be an important key in the regulation of all these pathologies. Presumably, slowly hydrolysable ATP analogues could be used as therapeutic tools in those cases in which a vasoconstrictive process is pathologically enhanced.

FINAL REMARKS

There is a clear therapeutic potential for adenosine and ATP acting through P1 and P2 receptors in the different eye structures. Although there is still a long way to go until the receptor identification and effects are completely elucidated, nucleosides and nucleotides are interesting new pharmacological tools which might be suitable for the treatment of some ocular pathologies. The future points to finding specific ways to allow the purines access to the receptors that have been identified *in vitro*. Only when the presence of these substances is confirmed in areas with difficult access, such as the retina, can a proper therapeutic role of the purines will be developed.

AKNOWLEDGEMENTS

I would like to thank Professor G. Burnstock for his continuous support and encouragement during the preparation of this chapter and for his critical reading of the manuscript. I am indebted to Dr. C.H.V. Hoyle for his valuable discussion and comments on all the topics treated in this chapter. I would like to thank Sunti Peral for her valuable comments in the field of the optometry, also to "Los Chiquis" for their kind comprehension during the development of this chapter. EYE REALITY (Jose Joaquín Infantes and Jesús Pintor) designed the 3D images prepared for this chapter.

REFERENCES

Abbracchio, M.P., Ceruti, S., Bolego, C., Puglisi, L., Burnstock, G. and Cattebeni, F. (1996). Trophic roles of P2-purinoceptors in central nervous system astroglial cells. In *P2 Purinoceptors: Localization, Function and Transduction Mechanisms. Ciba Foundation Symposium 198*, pp. 142–148. Chichester: John Wiley and Sons.

Alborch, E. and Martin, G. (1980) Adenosine and adenine nucleotides: action on the cerebral blood and cerebral vascular resistance. *Blood Vessels*, **17**, 143.

Anand, R. and Tasman, W.S. (1994). Non-rhegmatogenous retinal detachment. In *Retina, Volume 3*, edited by B.M. Glaser, pp 2463–2488. St. Louis: Mosby.

Bayliss, W.M. and Starling, E.H. (1900). The movements and innervation of the small intestine. *Journal of Physiology*, **26**, 125–138.

Bennett, D.W. and Drury, A.N. (1931). Further observations relating to the physiological activity of adenine compounds. *Journal of Physiology*, **72**, 288–320.

Berne, R.M. (1963). Cardiac nucleotides in hypoxia: possible role in regulation of coronary blood flow. *American Journal of Physiology*, **204**, 317–322.

Berridge, M.J. (1993). Inositol triphosphate and calcium signalling. *Nature*, **361**, 3_5–325.

Blazynski, C. (1987). Adenosine A$_1$ receptor mediated inhibition of adenylate cyclase in rabbit retina. *Journal of Neuroscience*, **7**, 2522–2528.

Blazynski, C. (1989). Displaced cholinergic, GABAergic amacrine cells in the rabbit retina also contain adenosine. *Visual Neuroscience*, **3**, 425–431.

Blazynski, C. (1990). Discrete distributions of adenosine receptors in mammalian. *Journal of Neurochemistry*, **54**, 648–655.

Blazynski, C. (1991). The accumulation of [^{3}H(phenylisopropyl adenosine ([^{3}H]PIA) and [^{3}H]adenosine into rabbit retinal neurons is inhibited by nitrobenzylthioinosine (NBI). *Neuroscience Letters*, **121**, 1–4.

Blazynski, C. (1993). Characterization of adenosine A$_2$ receptors in bovine retinal pigment epithelial membrane. *Experimental Eye Research*, **56**, 594–599.

Blazynski, C. and McIntosh, H. (1993). Characterization of adenosine A$_2$ receptors in bovine retinal membranes. *Experimental Eye Research*, **56**, 585–593.

Blazynski, C. and Perez, M.T.R. (1991). Neuroregulatory functions of adenosine in the retina. *Progress in Retina Research*, **11**, 293–332.

Blazynski, C., Kinsherf, D.A., Geary, K.M. and Ferrendelli, J.A. (1986). Adenosine-mediated regulation of cyclic AMP levels in isolated incubated retinas. *Brain Research*, **366**, 224–229

Boeynaems, J.M., Communi, D., Janssens, R., Motte, S., Robaye, B. and Pirotton, S. (1988). Nucleotide receptor coupling to the phospholipase C signaling pathway. In *The P2 Nucleotide Receptors*, edited by J.T. Turner, G.A. Weisman and G.S. Fedan, pp 169–184. Totowa, NJ: Humana Press.

Borys T.J., Gupta, B.D., Deshpande, S. and Abrahamson, E.W. (1986). The structural changes in bovine rod outer segments in the presence of ATP. *Photochemistry and Photobiology*, **43**, 183–187.

Braas, K.M., Newby, A.C., Wilson, V.S. and Snyder, S.H. (1986). Adenosine-containing neurones in the brain localized by inmunocytochemistry. *Journal of Neuroscience*, **6**, 1952–1961.

Braas, K.M., Zarbin, M.A. and Snyder S.H. (1987). Endogenous adenosine and adenosine receptors localized to ganglion cells of the retina. *Proceedings of the National Academy of Sciences USA*, **84**, 3906–3910.

Braunagel, S.C., Xiao, J.G. and Chiou, C.G. (1988). The potential role of adenosine in regulating blood flow in the eye. *Journal of Ocular Pharmacology*, **4**, 61–73.

Brecha, N. (1983). *Chemical Neuroanatomy*, edited by P.C. Emson. New York: Raven.

Brechler, V., Pavoine, C., Lotersztajn, S., Garbarz, E. and Pecker, F. (1990). Activation of Na$^+$/Ca^{2+} exchange by adenosine in ewe heart sarcolemma is mediated by a pertussis toxin-sensitive G protein. *Journal of Biological Chemistry*, **265**, 16851–16855.

Bruns R.F. (1981). Adenosine antagonism by purines, pteridines and benzopteridines in human fibroblasts. *Biochemical Pharmacology*, **30**, 325–333.

Burnstock, G. (1969). Evolution of the autonomic innervation of visceral and cardovascular systems in vertebrates. *Pharmacological Reviews*, **21**, 247–324.

Burnstock, G. (1972). Purinergic nerves. *Pharmacological Reviews*, **24**, 509–560.

Burnstock, G. (1978). A basis for distinguishing two types of purinergic receptor. In *Cell Membrane Receptors for Drugs and Hormones: A Multidisciplinary Approach*, edited by R.W. Straub and L. Bollis, pp. 107–118. New York: Raven Press.

Burnstock, G. (1990). Cotransmission. The Fifth Heymans Lecture-Ghent, February 17, 1990. *Archives of International Pharmacodynamics and Therapeutics*, **304**, 7–33.

Burnstock, G. (1993). Hypoxia, endothelium and purines. *Drug Development Research*, **28**, 301–306.

Burnstock, G. (1995). Noradrenaline and ATP: Cotransmitters and Neuromodulators. *Journal of Physiological Pharmacology*, **46**, 365–384.

Burnstock, G. and Kennedy, C. (1985). Is there a basis for distinguishing more than one type of P$_2$-receptor? *General Pharmacology*, **16**, 433–440.

Burnstock, G., Campbell, G., Bennett, M. and Holman, M.E. (1963). The effects of drugs of the transmission of inhibition from autonomic nerves to the smooth muscle of the guinea-pig taenia coli. *Biochemical Pharmacology*, **12** (Suppl.), 134–135.

Busch, M.J.W.M., Kobayashi, K., Hoyng, P.F.J. and Mittag, T.W. (1993). Adenylyl cyclase in human and bovine trabecular meshwork. *Investigative Ophthalmology and Visual Science*, **34**, 3028–3034.

Buxton, I.L.O. and Cheek, D. (1995). On the origin of extracellular ATP in cardiac blood vessels: a dual role for endothelium. In *Adenosine and Adenine Nucleotides: From Molecular Biology to Integrative Physiology*, edited by L. Belardinelly and A. Pelleg pp. 193–197. Boston; Kluwer Academic Publishers.

Campau, K.M., Kinsherf, D.A. and Ferrendelli, J.A. (1983). Different effect of adenosine on cAMP regulation in retina and brain. *Neuroscience Abstracts*, **9**, 89.

Camras, C.B., Zahn, G.L., Wang Y.L., Toris, C.B. and Yablonski, M.E. (1994). Effect of (R)-phenylisopropyladenosine, an adenosine A$_1$ agonist, on aqueous humor dynamics in rabbits. *Investigative Ophthalmology and Visual Science*, **35** (Suppl), 2052.

Cavalotti, C., Caeccarelli, E., Evangelisti, F. and Amenta, F. (1982). Nerve firbers with a selective affinity for quinacrine in the cornea. *Acta Anatomica*, **112**,14–17.

Churchill, G.C. and Louis, C.F. (1997). Stimulation of P2U purinergic or α_{1A} adrenergic receptors mobilises calcium in the lens cells. *Investigative Ophthalmology and Visual Science*, **38**, 855–865.

Churchill, G.C., Atkinson, M.M. and Louis, C.F. (1996). Mechanical stimulation initiates cell-to-cell calcium signalling in ovine cells epithelial cells. *Journal of Cell Science*, **109**, 355–365.

Courtache, D. and Guyon, J.F. (1897). Influence motrice du grand sympathique et du nerf érecteur sur le gros intestin. *Archives of Physiology*, **9**, 880–890

Crawford, K.S., Kaufman, P.L. and Bito, L.Z. (1990). The role of the iris in accommodation of rhesus monkeys. *Investigative Ophthalmology and Visual Science*, **31**, 2185–2190.

Crosson, C.E. (1992). Ocular hypotensive activity of the adenosine agonist (R)-phenylisopropyladenosine in rabbits. *Current Eye Research*, **11**, 453–458.

Crosson, C.E. (1995). Adenosine receptor activation modulates intraocular pressure in rabbits. *Journal of Pharmacology and Experimental Therapeutics*, **273**, 320–326.

Crosson, C.E. and Gray, T. (1992). Evidence for adenosine A1 receptor involvement in the modulation of intraocular pressure and cAMP formation in the iris/ciliary body of the rabbit. *Investigative Ophthalmology and Visual Science*, **33**, 903S.

Crosson, C.E. and Gray, T. (1994). Modulation of intraocular pressure by adenosine agonists. *Journal of Ocular Pharmacology*, **10**, 379–383.

Crosson, C.E. and Gray, T. (1996). Response to prejunctional adenosine receptors is dependent on stimulus frequency. *Current Eye Research*, **16**, 359–364.

Crosson, C.E., DeBenedetto, R. and Gidday, J.M. (1994). Functional evidence for retinal adenosine receptors. *Journal of Ocular Pharmacology*, **10**, 499–507.

Daly, J.W. (1982). Adenosine receptors: target for future drugs. *Journal of Medical Chemistry*, **25**, 197–207.

De Mey, J.G. and Vanhoutte, P.M. (1981). Role of the intima in cholinergic and purinergic relaxation of isolated canine femoral arteries. *Journal of Physiology*, **316**, 347–355.

Delamere, N.A. and Dean, W.L. (1993). Distribution of lens sodium-potassium-adenosine triphosphatase. *Investigative Ophthalmology and Visual Science*, **34**, 2159–2163.

Deussen, A. and Pau, H. (1991). Nucleotide levels in human lens: regional distribution in different forms of senile cataract. *Experimental Eye Research*, **48**, 37–47.

Dolphin, A.C., Forda, S.R. and Scott, R.H. (1986). Calcium-dependent currents in cultured rat dorsal root ganglion neurons are inhibited by an adenosine analogue. *Journal of Physiology*, **373**, 47–61.

Drury, A.N. and Szent-Györgi, A. (1929). The physiological activity of adenine compounds with special reference to their action in mammalian heart. *Journal of Physiology*, **68**, 213–237.

Duke-Elder, S. (1969). diseases in the lens and vitreous. In *System of Ophthalmology, volume II*, edited by S. Duke-Elder, pp. 63–165. London: Henry Kimpton.

Duncan, G., Williams, M.R. and Riach, R.A. (1994). Calcium, cell signalling and cataract. *Progress in Retinal Eye Research*, **13**, 623–652.

Duncan, G., Riach, R.A., Williams, M.R., Webb, S.F., Dawson, A.P. and Reddan J.R. (1996). Calcium mobilisation modulates growth of lens cells. *Cell Calcium*, **19**, 83–89.

Edwards, R.B. and Bakshiam, S. (1986). Phagocytosis of outer segments by cultured rat pigmented epithelial cells. Reduction by cyclic AMP and phosphosdiesterase inhibitors. *Investigative Ophthalmology and Visual Science*, **19**, 1184–1188.

Ehinger, B. and Perez, M.T.R. (1984). Autoradiography of nucleoside uptake into the retina. *Neurochemistry International*, **6**, 369–381.

Ellis, J.L. and Burnstock, G. (1989). Angiotensin neuromodulation of adrenergic and purinergic co-transmission on guinea-pig vas deferens. *British Journal of Pharmacology*, **97**, 1157–1164.

Erlinge, D., You, J. Wahlestedt, C. and Edvinsson, L. (1995). Characterisation of an ATP receptor mediating mitogenesis in vascular smooth muscle cells. *European Journal of Pharmacology Molecular Pharmacology*, **289**, 135–149.

Evans, R.J. and Cunnane T.C. (1992). Relative contributions of ATP and noradrenaline to the nerve evoked contraction of the rabbit jejunal artery. *Naunyn-Schmiedeberg's Archives of Pharmacology*, **345**, 424–430.

Evans, R.J., Suprenant, A. and North, R.A. (1998). P2X receptors, cloned and expressed. In *The P2 Nucleotide Receptors*, edited by J.T. Turner, G.A. Weisman and G.S. Fedan, pp 43–62. Totowa, NJ: Humana Press.

Forrester, T., Harper, A.M. and MacKenzie, E.T. (1979). Effect of adenosine triphosphate and some derivatives on cerebral blood flow and metabolism. *Journal of Physiology*, **296**, 343–355.

Frambach, D.A., Fain, G.L., Farber, D.B. and Bok, D. (1990). Beta adrenergic receptors on cultured human retinal pigment epithelium. *Investigative Ophthalmology and Visual Science*, **31**, 1767–1772.

Fredholm, B.B., Johansson, B., van der Ploeg, Y., Hu, P.S. and Jin, S. (1993). Neuromodulatory roles of purines. *Drug Development Research*, **28**, 349–353.

Fredholm B.B., Abbracchio, M.P., Burnstock, G., Daly, J.W., Harden, T.K., Jacobson, K.A., Leff, P. and Williams, M. (1994). Nomenclature and classification of purinoceptors. *Pharmacological Reviews*, **46**, 143–156.

Friedman, Z., Hackett, S.F., Linden, J. and Campochiaro, P.A. (1989). Human retinal pigment epithelial cells in culture possess A_2-adenosine receptors. *Brain Research*, **492**, 25–35.

Fuder, H. and Muth, U. (1993). ATP and endogenous agonists inhibit evoked [^{3}H(-noradrenaline release in rat iris via A_1 and P_{2Y}-like purinoceptors. *Naunyn-Schmiedeberg's Archives of Pharmacology*, **348**, 352–357.

Fuder, H., Brink, A., Meinke, M. and Tauber, U. (1992). Purinoceptor-mediated modulation by endogenous and exogenous agonists of stimulation-evoked [^{3}H(noradrenaline release on rat iris. *Naunyn-Schmiedeberg_s Archives of Pharmacology*, **345**, 417–423.

Gaddum, J.H. and Holz, P. (1933). The localization of the action of drugs on the pulmonary vessels of dogs and cats. *Journal of Physiology*, **77**, 139–158.

Gerwins, P., and Fredholm, B.B. (1992). Stimulation of adenosine A_1 receptors and bradykinin receptors, which act via different G-proteins, synergistically raises inositol 1,4,5–trisphosphate and intracellular free calcium in DTT_1 MF-2 smooth muscles cells. *Proceedings of the National Academy of Sciences USA*, **89**, 7330–7334.

Gidday, J.M. and Park, T.S. (1993a). Microcirculatory responses to adenosine in the newborn pig retina. *Pediatric Research*, **33**, 620–627.

Gidday, J.M. and Park, T.S. (1993b). Adenosine mediated autoregulation of retinal arteriolar tone in the piglet. *Investigative Ophthalmology and Visual Science*, **34**, 2713–2719.

Gidday, J.M., Lanius, T.M., Shah, A.R. and Park, T.S. (1992). Endogenous adenosine contributes to the mediation of hypoxic dilation of retinal arterioles in the newborn pig. *Investigative Ophthalmology and Visual Science* (Suppl.), **33**, 1048.

Gill, D.R., Hyde, S.C., Higgins, C.F., Valverde, M.A., Minteng, G.M. and Sepulveda, E.V. (1992). Separation of drug transport and chloride channel functions of the human multidrug resistant P-glycoprotein. *Cell*, **71**, 23–32.

Glitscher, W. and Rüppel, H. (1989). Arrestin of bovine photoreceptors reveals strong ATP binding. *FEBS Letters*, **1**, 101–105.

Goodman, R.R. and Snyder, S.H. (1982). Autoradiographic localization of adenosine receptors in rat brain using [^{3}H]-cyclohexyladenosine. *Journal of Neuroscience*, **2**, 1230–1241.

Gordon, J.L. (1986). Extracellular ATP: effect, sources and fate. *Biochemical Journal*, **233**, 309–319.

Green, H.N. and Stoner, H.B. (1950). The effect of purine derivatives on the cardiovascular system. In *Biological Actions of Adenine Nucleotides*, pp. 65–103. London: H.K. Lewis.

Greenwood, D., Yao, W.P. and Housley, G.D. (1997). Expression of the $P2X_2$ receptor subunit of the ATP-gated channel in the retina. *Neuroreport*, **8**, 1083–1088.

Gregory, C.Y., Hall, M.O. and Abrams, T.A. (1991). Adenosine and adenosine analogs inhibit ROS phagocytosis suggesting there are adenosine receptors on the surface of the RPE. *Investigative Ophthalmology and Visual Science*, **32**, 839–849.

Greiner, J.V. and Weidman T.A. (1991). Comparative histogenesis of Bruch's membrane (complexus basalis). *Experimental Eye Research*, **53**, 47–54.

Greiner, J.V., Kopp, S.J., Sanders, D.R. and Glonek, T. (1981). Organophosphates of crystalline lens: a nuclear magnetic resonance spectroscopy study. *Investigative Ophthalmology and Visual Science*, **31**, 700–713.

Greiner, J.V., Kopp, S.J., Mercola, J.M. and Glonek, T. (1982). Organophosphate metabolites of the human and rabbit crystalline lens: a phosphorus-31 nuclear magnetic resonance spectroscopic analysis. *Experimental Eye Research*, **34**, 545–552.

Gustafsson, L.E. and Wiklund, N.P. (1986). Adenosine-modulation of cholinergic and non-adrenergic non-cholinergic neurotransmission in the rabbit iris sphincter. *British Journal of Pharmacology*, **88**, 197–204.

Hall, M.O., Abrams, T.A. and Mittag, T.W. (1991). ROS ingestion by RPE cells is turned off by increased protein kinase C activity and by increased calcium. *Experimental Eye Research*, **52**, 591–598.

Harding J.J. and Crabbe, M.J. (1984). The lens: development, proteins, metabolism and cataract. In *The Eye*, edited by H. Davson, pp 207–492. London: Academic Press.

Hasslinger, C. (1969). catapres (2–[2,6–dichlorophenylamino]-2–imidazoline hydrochloride) — a new drug lowering intraocular pressure. *Klinishe Monatsblatter fur Augenheilkunde*, **154**, 95–105.

Heller, I.H. and McIlwain, H. (1973). Release of [^{14}C]adenine derivatives from isolated subsystems of the guinea pig brain: actions of electrical stimulation and of papaverine. *Brain Research*, **53**, 105–116.

Henderson, J.F. (1979). Regulation of adenosine metabolism. In *Physiological and Regulatory Functions of Adenosine and Adenine Nucleotides*, edited by H.P. Baer and G.I. Drummond. New York: Raven.

Hernández, F., Laiz, R., Infantes, J.J. and Pintor, J. (1994). Nucleotides in the lens: Do they protect from UV light? *Gaceta Optica*, **278**, 10–15.

Hightower, K.R., Duncan, G. and Harrison, S.E. (1985). Intracellular calcium concentration and calcium transport in the rabbit lens. *Investigative Ophthalmology and Visual Science*, **26**, 1032–1034.

Hockwin O., Khabbazadeh, H., Noll, E. and Licht, W. (1968). A study of the content of free nucleotides and other acid soluble phosphate compounds in lenses of various mammals in relation to their age. *Albrecht Von Graefes Archives Klinische Experimental Ophthalmology*, **174**, 245–253.

Hockwin, O., Fink, H. and Beyer, D. (1973). Carbohydrate metabolism of the lens depending on age. Factor analysis of changes in the content of adenine nucleotides inorganic phosphate and lactate in bovine lenses. *Mechanisms of Ageing and Development*, **2**, 409–419.

Hoffman, K.P. (1973). Conduction velocity in pathways from retina to superior colliculus in the cat: A correlation with receptive field properties. *Journal of Neurophysiology*, **36**, 409–424.

Holton, P. (1959). The liberation of adenosine triphosphate on antidromic stimulation of sensory nerves. *Journal of Physiology*, **145**, 494–504.

Holton, F.A. and Holton, P. (1953). The possibility that ATP is a transmitter at sensory nerve endings. *Journal of Physiology*, **119**, 50–51P.

Holton, F.A and Holton, P. (1954). The capillary dilator substance in dry powders of spinal roots: a possible role of adenosine triphosphate in chemical transmission from nerve endings. *Journal of Physiology*, **126**, 124–140.

Hoyle, C.H.V. and Burnstock, G. (1985). Atropine resistant excitatory junction potential in rabbit bladder are blocked by α,β-methyleneATP. *European Journal of Pharmacology*, **114**, 239–240.

Inoue, T. and Kannan, M.S. (1988). Nonadrenergic and noncholinergic excitatory neurotransmission in rat intrapulmonary artery. *American Journal of Physiology*, **254**, H1142–H1148.

Iwata, S. and Takehana, M. (1982). Biochemical studies on human cataract lens. Opacity-related changes of cations, ATP and GSH in various types of human senile cataracts. *Yakugaku Zasshi*, **102**, 940–945.

Jacob, T.J.C. and Zhang, J.J. (1993). Electrophysiological properties of isolated anterior and posterior surfaces of the lens. *Investigative Ophthalmology and Visual Science*, **34**, 1255.

Jacobson, M.A. (1995). Molecular biology of adenosine receptors. In *Adenosine and Adenine Nucleotides: from Molecular Biology to Integrative Physiology*, edited by L. Bellardinelli and A. Pelleg, pp. 5–13. Boston: Kluwer Academic Publishers.

Jahngen-Hodge, J., Laxman, E., Zuliani, A. and Taylor, A. (1991). Evidence for ATP and ubiquitin dependent degradation of proteins in cultured bovine lens epithelial cells. *Experimental Eye Research*, **52**, 341–347.

Jiang, C., Finkbeiner, W., Widdicombe, J., McCray, P. Jr. and Miller, S.S. (1993). Altered fluid transport across airway epithelium in cystic fibrosis. *Science*, **262**, 424–427.

Kanjhan, R, Housley, G.D., Thorne, P.R., Christie, D.L., Palmer, D.J., Luo, L. and Ryan, A.F. (1996). Localization of ATP-gated ion channels in cerebellum using P2x2R subunit-specific antisera. *Neuroreport*, **7**, 2665–2669.

Kennedy, C., Hickman, S.E. and Silverstein S.C. (1998). Characteristics of ligand gated ion channel P2 nucleotide receptors. In *The P2 Nucleotide Receptors*, edited by J.T. Turner, G.A. Weisman and G.S. Fedan, pp 211–230. Totowa, NJ: Humana Press.

Kirishuk, S., Scherer, J., Kettenmann, H. and Verkhransky A. (1995). Activation of P_2-purinoceptors triggered Ca^{2+} release from InsP₃-sensitive internal stores in mammalian oligodendrocytes. *Journal of Physiology*, **483**, 41–57.

Klethi, J. and Mandel, P. (1963). Eye lens nucleotides of different species of vertebrates. *Nature*, **4976**, 1114–1115.

Knudsen, T.B. and Elmer, W.A. (1987). Evidence for negative control of growth by adenosine in the mammalian embryo: induction of $Hm^{x/+}$ mutant limb outgrowth by adenosine deaminase. *Differentiation*, **33**, 270–279.

Kostopoulos, G.K. and Phillis, J.W. (1977). Purinergic depression of neurons in different areas of the rat brain. *Experimental Neurology*, **55**, 719–724.

Kreutzberg, G.W. and Hussain, S.T. (1984). Cytochemical localization of 5′-nucleotidase activity in retinal photoreceptor cells. *Neuroscience*, **11**, 857–866.

Kreutzberg, G.W., Barron, K.D. and Schubert, P. (1978). Cytochemical localisation of 5'-nucleotidase in glial plasma membranes. *Brain Research*, **158**, 247–257.

Kvanta, A., Seregard, S., Sejersen, S., Kull, B. and Fredholm B.B. (1997). Localisation of adenosine receptor messenger RNAs in the rat eye. *Experimental Eye Research*, **65**, 569–602.

Laasberg, T. (1990). Ca^{2+} mobilising receptors of gastrulating chick embryo. *Comparative Biochemistry and Physiology*, **97C**, 9–12.

Langercrantz, H. (1976). On the composition and function of large dense-core vesicles in sympathetic nerves. *Neuroscience*, **1**, 81–92.

Langley, J.N. (1898). On the inhibitory fibres in the vagus for the end of the oesophagus and stomach. *Journal of Physiology*, **23**, 407–414.

Langley, J.N. (1900). The sympathetic and other related systems of nerves. In *Textbook of Physiology, Volume 2*, edited by E.A. Shafer, pp. 616–696. Edinburgh and London: I.J. Pentland.

Lapetina, E.G., Billah, M.M. and Cuatrecasas, P. (1981). The phosphatidylinositol cycle and the regulation of arachidonic acid production. *Nature*, **292**, 367–369.

Lerman, S. (1980). *Radiant energy and the eye*. New York: Macmillan.

Liang, B.T. (1992). Adenosine receptors and cardiovascular function. *Trends in Cardiovascular Medicine*, **2**, 100–108.

Linden, J. (1994). Coned adenosine A_3 receptors: pharmacological properties, species differences and receptor functions. *Trends in Pharmacological Sciences*, **15**, 298–306.

Lohman, F., Drews, U., Donie, F. and Reiser, G. (1991). Chick embryo muscarinic and purinergic receptors activate cytosolic Ca^{2+} via phosphatidylinositol metabolism. *Experimental Cell Research*, **197**, 326–329.

MacKenzie, I., Burnstock, G., and Dolly, J.O. (1982). The effects of purified botulinum toxin Type A on cholinergic, adrenergic and non-adrenergic atropine-resistant autonomic neuromuscular transmission. *Neuroscience*, **1**, 997–1006.

Marcantonio, J.M., Duncan G. and Rink, H. (1986). Calcium-induced opacification and loss of protein in the organ-cultured bovine lens. *Experimental Eye Research*, **42**, 617–630.

Martison, J. and Muren, A. (1963). Excitatory and inhibitory effects of vagus stimulation on gastric motility in the cat. *Acta Physiologica Scandinavica*, **57**, 308–316.

Marumatsu, Y., Kigoshi, S. and Oda, Y. (1994). Evidence for sympathetic, purinergic transmission in the iris dilator muscle of the rabbit. *Japanese Journal of Pharmacology*, **66**, 191–193.

Mekki, Q.A. and Turner, P. (1985). Stimulation of dopamine receptors (type 2) lowers human intraocular pressure. *British Journal of Ophthalmology*, **69**, 909–910.

Nagy, J.I., La Bella, L.A. and Buss, M. (1984). Inmunocytochemistry of adenosine deaminase: implications for adenosine neurotransmission. *Science*, **224**, 166–168.

Nash, M.S., Carvalho A.L. and Osborne, N.N. (1996). Cultured retinal pigment epithelium (RPE) possesses P_{2U} purinergic receptors. *Vision Research* (Suppl.), **36**, S68.

Neal, M.J. and Cunningham, J.R. (1994). Modulation by endogenous ATP of the light evoked release of ACh from retinal cholinergic neurones. *British Journal of Pharmacology*, **113**, 1085–1087.

Neal, M.J., Cunningham, J.R. and Paterson, S.J. (1995). Binding sites for α,β-methylene ATP are present in the vertebrate retina. *British Journal of Pharmacology*, **114** (Suppl.), 101P.

Neary, J.T. and Norenberg, M.D. (1992). Signaling by extracellular ATP: physiological and pathological considerations in neuronal-astrocytic interactions. *Progress in Brain Research*, **94**, 145–151.

Neary, J.T., Rathbone, M.P. Cattabeni, F., Abbracchio, M.P. and Burnstock, G. (1996). Trophic actions of extracellular nucleotides and nucleosides on glial and neuronal cells. *Trends in the Neurosciences*, **19**, 13–18.

Negi, A. and Marmor, M.F. (1986). Mechanism of subretinal fluid resorption in the cat eye. *Investigative Ophthalmology and Visual Science*, **27**, 1560–1563.

Neufeld, A.H. and Sears, M.L. (1975). Adenosine 3',5'-monophosphate analogue increases the outflow facility of the primate eye. *Investigative Ophthalmology*, **14**, 688–689.

Olson, R.A. and Pearson, J.P. (1990). Cardiovascular purinoceptors. *Physiological Reviews*, **70**, 761–845.

Ostwald, P., Park, S.S., Toledano, A.Y. and Roth, S. (1997). Adenosine receptor blockade and nitric oxide synthase inhibition in the retina: impact upon post-ischaemic hyperemia and the electroretinogram. *Vision Research*, **37**, 3453–3461.

Paes de Carvalho, R., Braas, K.M., Adler, R. and Snyder S.H. (1992). Developmental regulation of adenosine A_1 receptors, uptake sites and endogenous adenosine in the chick retina. *Developmental Brain Research*, **70**, 87–95.

Palmer, T.M. and Stiles, G.L. (1995). Adenosine receptors. *Neuropharmacology*, **34**, 683–694.

Palmisano, D.V., Groth-Vasselli, B., Farnsworth, P.N. and Reddy, M.R. (1995). Interaction of ATP and lens α-crystallin characterized by equilibrium binding studies and intrinsic tryptophan fluorescence spectroscopy. *Biochimica et Biophysicl Acta*, **1246**, 91–97.

Paterson, C.A., Zeng, J., Husseini, Z., Borchman, D., Delamere, N.A., Garland, D. and Jimenez-Asensio, J. (1996). Calcium ATPase activity and membrane structure in clear and cataractous human lenses. *Current Eye Research*, **16**, 333–338.

Pearson, J.D. and Gordon, J.L. (1979). Vascular endothelial and smooth muscle cells in culture selectively release adenine nucleotides. *Nature*, **281**, 384–386.

Perez, M.T.R. and Bruun, A. (1987). Colocalisation of [^{3}H]-adenosine accumulation and GABA inmunoreactivity in the chicken and rabbit retinas. *Histochemistry*, **87**, 413–417.

Perez, M.T.R. and Ehinger, B. (1990). Inhibition of acetylcholine release from rabbit retina by adenosine analogues. *Investigative Ophthalmology and Visual Science*, **31** (Suppl.), 534.

Perez, M.T.R., Ehinger, B.E., Lindström, K. and Fredholm, B.B. (1986). Release of endogenous and radio-active purines from the rabbit retina. *Brain Research*, **398**, 106–112.

Perez, M.T.R., Arner, K. and Ehinger, B. (1988). Stimulation evoked release of purines from the rabbit retina. *Neurochemisty International*, **13**, 307–318.

Peterson, W.M., Meggyesy, C., Yu, K. and Miller, S.S. (1997). Extracellular ATP activates calcium signaling, ion, and fluid transport in retinal pigment epithelium. *Journal of Neuroscience*, **17**, 2324–2337.

Pintor, J., Díaz-Rey, M.A. and Miras-Portugal, M.T. (1993). Ap$_4$A and ADP-β-S binding to P2 receptors present on rat brain synaptic terminals. *British Journal of Pharmacology*, **108**, 1094–1099.

Pirotton, S., Raspe, E., Demolle, D., Erneux, C. and Boeynaems, J. (1987). Involvement of inositol 1,4,5–triphosphate and calcium in the action of adenine nucleotides on aortic endothelial cells. *Journal of Biological Chemistry*, **262**, 17461–17466.

Portellos, M., Riva, C.E., Cranstoun, S.D., Petrig, B.L. and Brucker, A.J. (1995). Effects of adenosine on ocular blood flow. *Investigative Ophthalmology and Visual Science*, **36**, 1904–1909.

Potter, D.E. (1981). Adrenergic Pharmacology of aqueous humor dynamics. *Pharmacological Reviews*, **33**, 133–151.

Ramkumar, V., Olah, M.E., Jacobson, K.A. and Stiles, G.L. (1991). Distinct pathways of desensitisation of A$_1$- and A$_2$-adenosine receptors in DDT1 MF-2 cells. *Molecular Pharmacology*, **40**, 639–647.

Ramkumar, V., Stiles, G.L, Beaven, M.A. and Ali, H. (1993). The A$_3$ adenosine receptor is the unique adenosine receptor which facilitates release of allergic mediators in mast cells. *Journal of Biological Chemistry*, **268**, 16887–16890.

Riach, R.A., Duncan, G., Williams, M.R. and Webb, S.F. (1995). Histamine and ATP mobilize calcium by activation of H$_1$ and P$_{2U}$ receptors in human lens epithelial cells. *Journal of Physiology*, **486**, 273–282.

Robertson, P.L., Ar, D. and Goldstein, G.W (1990). Phosphoinositide metabolism and prostacyclin formation in retinal microvascular endothelium: stimulation by adenine nucleotides. *Experimental Eye Research*, **50**, 37–40.

Roth, S. (1995). Post-ischemic hyperemia in the cat retina: the effects of adenosine receptor blockade. *Current Eye Research*, **14**, 323–328.

Roth, S. and Pietrzyk, Z. (1993). Adenosine antagonism attenuates hypoxic vasodilation in the retina in cats. *Anesthesia and Analgesia (Suppl.)*, **76**, S358.

Roth, S., Rosenbaum, P.S., Osinski, J., Park, S.S., Toledano, A.Y., Li, B. and Moshfeghi, A.A. (1997). Ischemia induces significant changes in purine nucleoside concentration in the retina-choroid in rats. *Experimental Eye Research*, **65**, 771–779.

Saffrey, M.J., Hassall, C.J.S., Hoyle, C.H.V., Belai A., Moss, J., Schmidt, H.H.H.W., Förstermann, U., Murad, F. and Burnstock, G. (1992). Nitric oxide synthase and NADPH diaphorase activity in cultured myenteric neurones. *Neuroreport*, **3**, 333–336.

Sakaki, Y., Fukuda, Y. and Yamashita, M. (1996). Muscarinic and purinergic Ca^{2+} mobilizations in the neural retina of early embryonic chick. International *Journal of Developmental Neuroscience*, **14**, 691–699.

Salla, S., Redbrake, C. and Frantz, A. (1996). Employment of bioluminiscence for the quantification of adenosine phosphates in the human cornea. *Greafe's Archives of Clinical and Experimental Ophthalmology*, **234**, 521–526.

Schaffer, R.N. and Hethrington, J. (1966). Anticholinesterase drugs and cataracts. *American Journal of Ophthalmology*, **62**, 613–618.

Schwartz, D.D. and Malik, K.U. (1989). Renal periarterial nerve stimulation-induced vasoconstriction at low frequencies is primarily due to the release of a purinergic transmitter in the rat. *Journal of Pharmacology and Experimental Therapeutics*, **250**, 764–771.

Secchi, A.G. (1982). Cataracts in uveitis. *Transactions of Ophthalmological Societies of the UK*, **102**, 390–394.

Segawa, Y. and Hughes, B.A. (1994). Properties of the inwardly rectifying K$^+$ conductance in the toad retinal pigment epithelium. *Journal of Physiology*, **476**, 41–53.

Senba, E., Daddona, P.E. and Nagy, J.I. (1986). Inmunohistochemical localization of adenosine deaminase in the retina of the rat. *Brain Research Bulletin*, **17**, 209–217.

Shahidullah, M. and Wilson, W.S. (1996). Mobilisation of intracellular calcium by P2Y$_2$ receptors in cultured, non-transformed bovine ciliary epithelial cells. *Current Eye Research*, **16**, 1006–1016.

Shoukrey, N.M. and Tabbara, K.F. (1987). An adenosine diphosphate phosphohydrolase in limbal vasculature. *Experimental Eye Research*, **44**, 149–153.

Smith, J.J. and Welsh M.J. (1993). Fluid and electrolyte transport by cultured airway human epithelia. *Journal of Clinical Investigation*, **91**, 1590–1597.

Spielman, W.S., Arend, L.J. and Forrest, J.N. (1987). The renal and epithelial actions of adenosine. In *Topics and Perspectives in Adenosine Research*, edited by B.F. Becker, pp 249–261. Germany: Gerlach E.

Spinowitz, B.S. and Zadunaisky, J.A. (1979). Action of adenosine on chloride active transport of isolated frog cornea. *American Journal of Physiology*, **237**, F121–F127.

Starke, K., von Kügelgen, Y., Bullock, J.M. and Illes, P. (1991). Nucleotides as co-transmitters in vascular sympathetic neuroeffector transmission. *Blood Vessels*, **28**, 19–26.

Stiles, G.L. (1988). A_1 adenosine receptor-G protein coupling in bovine brain membranes: effects of guanine nucleotides, salt and solubilization. *Journal of Neurochemistry*, **51**, 1592–1598.

Stone, J. and Hoffman, K.P. (1972). Very slow conducting ganglion cells: a major new functional group. *Brain Research*, **43**, 610–616.

Sugioka, M. Fukuda, Y. and Yamashita, M. (1996). Ca^{2+} responses to ATP via purinoceptors in the early embryonic chick retina. *Journal of Physiology*, **493**, 855–863.

Sullivan, D.M., Erb, L., Anglade, E., Weisman, G., Turner, J.T. and Csaky, K.G. (1997). Identification and characterization of $P2Y_2$ nucleotide receptors in human retinal pigment epithelial cells. *Journal of Neuroscience Research*, **49**, 43–52.

Suzuki, Y., Nakano, T. and Sears, M. (1997). Calcium signals from intact rabbit ciliary epithelium observed with confocal microscopy. *Current Eye Research*, **16**, 166–175.

Takagi, H. King G.L., Ferrara, N. and Aiello, L.P. (1996). Hypoxia regulates vascular endothelial growth factor receptor KDR/Flk gene expression through adenosine A_2 receptors in retinal capillary endothelial cells. *Investigative Ophthalmology and Visual Science*, **37**, 1311–3211.

Tao, L., Zhan, Y., Yanoff, M., Cohen, S. and Li, W. (1992). Active P_{2Y} purinoceptor on retinal capillary endothelial cells (RCEs) *in vitro*. *Investigative Ophthalmology and Visual Science* (Suppl.), **33**, 1062.

Taylor, E.M. and Parsons, M.E. (1989). Adrenergic and purinergic neurotransmission in arterial resistance vessels of the cat intestinal circulation. *European Journal of Pharmacology*, **164**, 23–33.

Tervo, T. and Palkama A. (1978). Ultrastructure of the corneal nerves after fixation with potassium permanganate. *Anatomical Record*, **190**, 851–859.

Tervo, T., Palva, M. and Palkama A. (1977). Transport adenosine triphosphatase activity in the rat cornea. *Cell and Tissue Research*, **176**, 431–443.

Thatcher, S.M. (1983). ATP causes a structural change in retinal rod outer segments: disk swelling is not involved. *Journal of Membrane Biology*, **74**, 95–102.

Torres-Zamorano, V., Ganapathy, V., Sharawy, M. and Reinach, P. (1992). Evidence for an ATP-driven H+-pump in the plasma membrane of the bovine corneal epithelium. *Experimental Eye Research*, **55**, 269–277.

Trachte, G.T. (1988). Angiotensin effects on vas deferens adrenergic and purinergic neurotransmission. *European Journal of Pharmacology*, **146**, 261–269.

Trussel, L.O. and Jackson, M.B. (1985). Adenosine-activated potassium conductance in cultured striatal neurones. *Proceedings of the National Academy of Sciences USA*, **82**, 4857–4861.

Ueda, N., Marumatsu, I., Sakakibara, Y. and Fujiwara, M. (1981). Noncholinergic, nonadrenergic contraction and substance P in rabbit iris sphincter muscle. *Japanese Journal of Pharmacology*, **31**, 1071–1079.

Unger, W.G. and Butler, J.M. (1988). Neuropeptides in the uveal tract. *Eye*, **2**, 202–212.

Valverde, M.A., Diaz, M., Sepulveda, F.V., Gill, D.R., Hyde, S.C. and Higgins, C.F. (1993). Volume-regulated chloride channels associated with the human multidrug-resistant P-glycoprotein. *Nature*, **355**, 830–833.

Wax, M., Sanghavi, D.M., Lee, C. and Kapadia, M. (1993). Purinergic receptors in ocular ciliary epithelial cells. *Experimental Eye Research*, **57**, 89–95.

Wedd, A.M., (1931). The action of adenosine and certain related compounds on the coronary flow of the perfused heart of the rabbit. *Journal of Pharmacology and Experimental Therapeutics*, **41**, 355–366.

Weisman, G.A., Gonzalez, F., Erb, L., Garrad, R. and Turner J.T. (1998). The cloning and expression of G protein-coupled P2Y nucleotide receptors. In *The P2 Nucleotide Receptors*, edited by J.T. Turner, G.A. Weisman and G.S. Fedan, pp 63–81. Totowa, NJ: Humana Press.

Williams, M. (1987). Purine receptors in mammalian tissues: Pharmacology and functional significant. *Annual Reviews in Pharmacology and Toxicology*, **27**, 315–345.

Williams, M. (1989). Adenosine antagonists. *Medical Research Reviews*, **9**, 219–243.

Williams, M.R., Duncan, G., Riach, R.A. and Webb, S.F. (1993). Acetylcholine receptors are coupled to mobilisation of intracellular calcium in cultured human lens cells. *Experimental Eye Research*, **57**, 381–384.

Wiskler, B.B., Ley, C.W. and Jaffe, E.A. (1978). Stimulation of endothelial cell prostacyclin production by thrombin, trypsin, and ionophore A23187. *Journal of Clinical Investigation*, **62**, 923–930.

Woods, C.L. and Blazynski, C. (1991). Characterisation of adenosine A_1-receptor binding sites in bovine retinal membranes. *Experimental Eye Research*, **53**, 325–331.

Yang, S., Cheek, D.J., Westfall, D.P. and Buston, I.L.O. (1994). Purinergic axis in cardiac blood vessels. *Circulation Research*, **74**, 401–407.

Yeung, S.M. and Green, R.D. (1983). Agonists and antagonist affinities for inhibitory adenosine receptors are reciprocally affected by 5′-guanidylimidodiphosphate or *N*-ethylmaleimide. *Journal of Biological Chemistry*, **258**, 2334–2339.

Zadunaisky, J.A. (1966). Active transport of chloride in the frog cornea. *American Journal of Physiology*, **211**, 506–512.

Zarbin, M.A., Wamsley, J.K., Palacios, J.M. and Kuhar, M.J. (1976). Autoradiographic localization of high affinity GABA, benzodiazepine, dopaminergic, adrenergic and muscarinic cholinergic receptors in the rat, monkey and human retina. *Brain Research*, **374**, 75–92.

Zhang, J.J. and Jacob, J.C. (1994). ATP-activated chloride channel inhibited by an antibody to P-glycoprotein. *American Journal of Physiology*, **267**, C1095–C1102.

Zhu, Y. and Gidday, J.M. (1996). Hypoglycaemic hyperemia in retina of newborn pigs. Involvement of adenosine. *Investigative Ophthalmology and Visual Science*, **37**, 86–92.

Zimmermann, H. (1996a). Biochemistry, localisation and functional roles of ecto-nucleotidases in the nervous system. *Progress in Neurobiology*, **49**, 589–618.

Zimmermann, H. (1996b). Extracellular purine metabolism. *Drug Development Research*, **39**, 337–352.

7 Retinal Transplantation and the Pupillary Light Reflex

Raymond D. Lund[1], Henry Klassen[1,2] and
Michael J. Young[3]

[1]*Institute of Ophthalmology, Neural Transplant Program, 11-43 Bath Street, London EC1V 9EL, UK*
[2]*Children's Hospital of Orange County, Orange, CA 92868, USA*
[3]*The Schepens Eye Research Institute, Harvard Medical School, 20 Stamford Street, Boston, MA 02114, USA*

It is clear that cells transplanted into the central nervous system can differentiate and integrate with host brain structures. To assess whether these transplants can also encode input signals, relay them to the correct brain regions, and effect the appropriate responses, it is desirable to make use of a simple relay system that is relatively isolated from other brain functions. The pupillary light reflex provides one such system. The dynamics of pupilloconstriction can provide information as to graft efficacy following a variety of transplantation paradigms involving the primary optic system. In this review, attention is directed at three such approaches: (i) implantation of embryonic retina into the brainstem of host rats; (ii) implantation of cells into the eye of rats with inherited retinal degeneration; and (iii) replacement of the optic nerve with a sciatic nerve graft which promotes regeneration of severed axons. In each case the pupillary light reflex serves as a good indicator of functional efficacy. By quantifying output dynamics it is possible to compare transplant-related responses to normal ones, and to evaluate factors compromising graft performance. To understand these response patterns it has become clear that more needs to be known about the normal reflex, and conversely, that transplantation can provide insights into the normal reflex that may not have otherwise been considered.

KEY WORDS: retinal transplantation; pupillary light reflex; olivary pretectal nucleus.

Correspondence: Raymond D. Lund, Institute of Ophthalmology, Neural Transplant Program, Bath Street, London EC1V 9EL, UK; Tel: +44 171 608 6893; Fax: +44 171 608 6881; E-mail: r.lund@ucl.ac.uk

INTRODUCTION

The foundations of developmental neurobiology owe much to the application of cell and tissue transplantation technology to nonmammalian vertebrate systems. Transplantation of tissue to the central nervous system (CNS) of mammals was first attempted in the latter part of the nineteenth century (Thompson, 1890) and several isolated studies were subsequently reported. It was not until the 1970s, however, that a concerted effort was made to use this approach to address a variety of questions regarding the mammalian brain and associated structures (Das and Altman, 1971; Lund and Hauschka, 1976; Björklund and Stenevi, 1977, 1979). Major emphasis has been given to the problems of development and repair, but many other issues such as molecular interactions among nerve cells, immune responses in the CNS, information processing, and the role of trophic factors have proven amenable to transplantation techniques.

One important issue is whether transplanted cells can positively impact on host neural functions. This could occur either by modulating the activity of a relay system in the host CNS or by functioning as part of a relay system, i.e., receiving input signals, encoding them and relaying them to appropriate brain regions. In either situation it is essential to be able to measure the level of functional efficacy specifically attributable to the presence of the transplant. To achieve this, a preparation is required in which the relay function to be assessed can be clearly separated from other functions and where simple quantifiable tests can be devised. These criteria are particularly well met by the primary optic pathway. Tests of visual function serve to evaluate a range of responses, from simple reflexes to complex behavioural discriminations. Among the simple reflexes is that of pupilloconstriction following retinal illumination. This response is readily quantifiable and can be used to assess efficacy in a variety of retinal transplantation paradigms, including evaluations of how function can be enhanced by various manipulations and how host and transplant inputs interact to produce an output signal. Besides the evident value in exploring the use of transplants to repair damaged retinal and central neural circuitry, retinal transplantation provides insights into how minimally functioning systems can contribute to neural processing and how integrative events occur in the CNS.

RETINAL TRANSPLANTATION

The primary optic system comprises the two retinae and their connections with as many as 10 different regions in the brainstem by way of the optic nerves, optic chiasm, and optic tracts. It is a system which, because of its simplicity of intrinsic organisation, lends itself particularly well to transplantation experiments. There are three different transplantation paradigms that have emerged: (i) whole eyes or retinae, implanted heterotopically to the CNS where central connections can be formed, (ii) orthotopic implantation of cells into the subretinal space to either prevent degeneration of photoreceptors or replace those already lost, and (iii) replacing the optic nerve with a sciatic nerve graft that provides a permissive substrate for the regeneration of host retinal ganglion cell axons. We now describe these three transplantation paradigms in more detail.

TRANSPLANTATION OF WHOLE RETINAE

Whole-eye transplantation was the first of the approaches to be attempted and, even in the late nineteenth century, there are reports of attempts to implant a rabbit eye into the orbit of a human recipient (May, 1887). While this was notably unsuccessful, a fruitful approach to transplantation was subsequently developed, in amphibians capitalising on early observations which revealed an impressive innate capacity for regeneration of the primary optic projection in these species (Stone and Zaur, 1940; Sperry, 1945). Ocular transplantation in amphibians provided the foundation for a substantial investigation into the biological bases of the specificity of neural connectivity, not only in the context of regeneration but also during development (Sperry, 1963; Gaze, 1970). The success of this work led us to explore the possibility of pursuing similar studies in the developing mammalian visual system (Lund and Hauschka, 1976), culminating in a series of experiments in which embryonic retinae were implanted into the brain of neonatal rats, as reviewed in a recent article (Lund et al., 1992).

Despite the wealth of information that had come from the amphibian studies, and an intriguing report that retinae placed in the brain of newborn rats would survive or even regenerate (Tansley, 1946), it was not until the studies of McLoon and Lund (1980a,b) that it was evident that embryonic retinae also survived transplantation to the rat brain, developed a remarkably normal cytoarchitecture with clearly identifiable histological layers, and even extended projections to visual centres of the host brainstem. These graft-host projections were substantially heavier after visual deafferentation of the host at the time of transplantation. Further studies showed that either electrical or photic stimulation of the exposed graft resulted in correlated electrophysiological activity in the superior colliculus (Simons and Lund, 1985), indicating the presence of connections capable of transmitting information to host brain structures. Together with some preliminary exploration of grafting to adult hosts (McLoon and Lund, 1983; Lund et al., 1987b) this body of work suggested that intracranial retinal transplants presented an effective system in which to examine the potential of grafts to function as a relay elements and to investigate factors that would enhance, or compromise, such functional capabilities.

INTRARETINAL TRANSPLANTATION

The first intraretinal grafting studies concentrated on introducing immature retinal tissue into the vitreous cavity to abut the remaining retina (Turner and Blair, 1986). While there were some suggestions of limited connectivity with the host retina (Aramant and Seiler, 1995), in general there was little indication of neural integration between graft and host. A more fruitful approach was to introduce cells into the subretinal space as a means of combating the loss of photoreceptors that occurs in a variety of conditions. This strategy provides a model therapy for a series of clinically important diseases, such as macular degeneration, retinitis pigmentosa, and other retinal dystrophies (Bird, 1995). The field of subretinal transplantation encompasses two quite different approaches. One is to protect the photoreceptors at an early stage from incipient degeneration; the other is to replace lost photoreceptors at a later stage with healthy ones that secondarily establish connections with the remaining retina.

The first approach has been developed with considerable success in the Royal College of Surgeons (RCS) rat, an animal in which most photoreceptors are lost during the first 2–3 months of life as a result of a genetic defect that appears to target the retinal pigment epithelium (RPE). The addition of healthy RPE cells to the subretinal space has been shown to result in substantial, long-term preservation of photoreceptors (Li and Turner, 1988; Lopez *et al.*, 1989). While attention has been largely focused on anatomical rescue, much less effort has been devoted to evaluating the functional impact of the transplants. The work that has been done in this regard has centred primarily on electroretinographic (ERG) activity and, until recently, the recovery of specific centrally-mediated visual responses have received less attention (Yamamoto *et al.*, 1993; Jiang and Hamasaki, 1994).

A second approach has been studied in mice in which an intrinsic mutation specifically affecting the photoreceptors leads to degeneration of these cells (Silverman *et al.*, 1992; Gouras *et al.*, 1994). Newly introduced photoreceptors appear to re-establish connections in the inner nuclear layer; here the impact such a procedure has on centrally-mediated functions has been explored very little.

OPTIC NERVE REPLACEMENT

For many years it was thought that neurons of the mammalian central nervous system, including those of the primary visual pathway, were intrinsically incapable of regeneration following injury. A series of transplantation studies have subsequently shown that if severed CNS axons are presented with a suitable environment, considerable regeneration is possible (David and Aguayo, 1981; Aguayo *et al.*, 1987). For example it is possible to cut the optic nerve immediately behind the eye and replace it with a length of sciatic nerve, the other end of which is implanted into a suitable brain region (Vidal-Sanz *et al.*, 1987). Axons regenerate along this nerve and form connections in a variety of visual centres from which physiological responses can be elicited (Kierstead *et al.*, 1989; Sauve, Sawai and Rasminsky, 1995).

THE TRANSPLANT-MEDIATED LIGHT REFLEX

In the early exploration of each of the above transplant paradigms, the lack of a simple behavioural assay limited any evaluation of transplant function through reconstructed, or preserved, circuitry. This problem was first addressed in a study in which it was found that illumination of an intracerebral retinal transplant could effect a pupilloconstrictor response in the host animal's eye (Klassen and Lund, 1987). This approach took advantage of the close relationship between the visual and autonomic nervous systems and the elegant simplicity of the pupillary light reflex (PLR). The initial discovery raised a number of additional issues which are reviewed here. These include:

(i)   The development of a practical system with which to measure various parameters of the PLR including latency, amplitude, and rate of constriction, as well as the degree of post-stimulus dilatation.

(ii) Clarifying the substrates of the normal pupilloconstriction response.

(iii) Comparison of graft-mediated versus normal response properties.

(iv) Defining the interactions that occur between transplant and host inputs and how these compare with interactions normally occurring between inputs from the two eyes.

(v) Identifying conditions that enhance or diminish PLR responses.

MEASUREMENT OF THE PUPILLARY LIGHT REFLEX

When measuring the PLR in experimental animals, three sets of problems must be addressed: anaesthesia, stimulus presentation and data recording. Addressing first the issue of anaesthesia, in humans it is usual to record the PLR in unanaesthetised subjects (Loewenfeld, 1993). In animals such as rats or mice, while possible (Trejo and Cicerone, 1982), this is much more difficult. Animals can be maintained in a sling while recording changes in pupillary diameter. Unfortunately the animals often go to sleep while dark adapting between trials and, when stimulated, show alerting responses accompanied by pupillary dilatation which can obfuscate the details of the light-induced constriction. In addition, head and eye movements can seriously disrupt pupillary recordings. As a result it is usually necessary to average a number of recordings to obtain a satisfactory record and this inevitably results in some artifactual damping of the responses obtained. Anaesthesia, on the other hand, is not without its drawbacks: most importantly, many anaesthetics have a significant impact on pupillary responsivity. In a study aimed at finding an anaesthetic suitable for long-term data collection, it was found that halothane, delivered in a mixture of nitrous oxide and oxygen, was the most effective (Radel et al., 1995). This combination had the advantage of maintaining a stable, well-dilated baseline pupillary diameter in the dark with a light-induced constriction of comparable amplitude to that achieved in unanaesthetised animals. Furthermore, repeated stimulation with light at a given luminance level produced the same amplitude of constriction over sequential trials. The main difficulty with many anaesthetics is that repeated use over a short period of time results in a diminished baseline pupillary diameter, even in the dark, accompanied by decreased responsiveness to light.

Photic stimuli are ideally delivered to the eye through a diffuser (Trejo and Cicerone, 1984). However, this method imposes significant limitations on the amount of space available for camera placement, which is necessary for recording data, as well as presenting problems of light leakage to the contralateral eye. Direct stimulation using a fibre optic light source, on the other hand, has difficulties associated with even and repeatable stimulation of the eye but, as long as absolute measures are not required, control studies have shown that this is an effective way of delivering light.

Data are best collected via an infrared-sensitive video (CCD) camera coupled with infrared illumination of the recorded eye. This allows pupillary recordings to be made in the dark. The simplest method of measuring pupillary diameter, directly from a magnified image of the eye in real time (Klassen and Lund, 1987; Thanos, 1992). This is sufficient for measuring the amplitude of constriction to a continuous light but is not sufficient for more detailed analyses. Video taping pupillary records, in conjunction with stop-frame capability, allows more complex data to be measured, albeit in a laborious

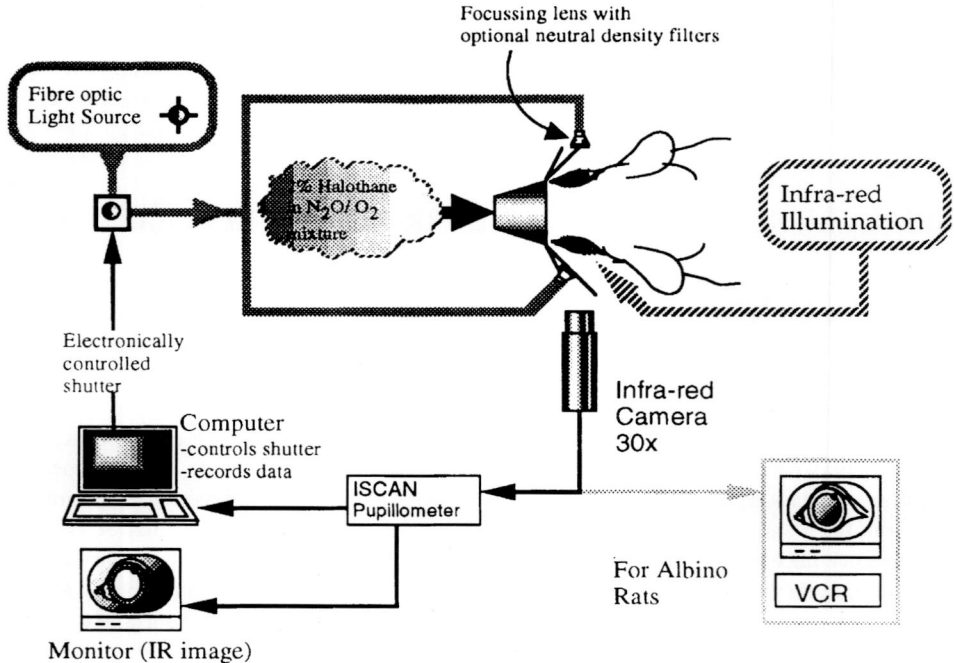

Figure 7.1   Experimental system for automated pupillometry (Reprinted from Chan, Young and Lund, 1995, with approval of the European Neuroscience Association).

manner (Klassen and Lund, 1990a). Ideally the pupillary diameter should be extracted from the video image and digitised by a pupillometer (Klassen and Lund, 1990b; Radel, Kustra and Lund, 1995). Automated pupillometry allows for the rapid collection of many data points and, once digitised, derived information such as percent amplitude and rate of constriction can be calculated. These data can then be transferred to a personal computer and further analysed using appropriate statistical software packages. A schematic diagram of a recording system incorporating the features described in this section is shown in Figure 7.1.

SUBSTRATES OF THE PUPILLARY LIGHT REFLEX

As a background to transplantation studies, it is important to have a good understanding of the basic biology of the PLR. This is now reviewed, with particular reference to the rat.

The pupillary light reflex in normal animals is subserved by a relatively simple polysynaptic pathway that adjusts the pupillary aperture in response to changes in ambient

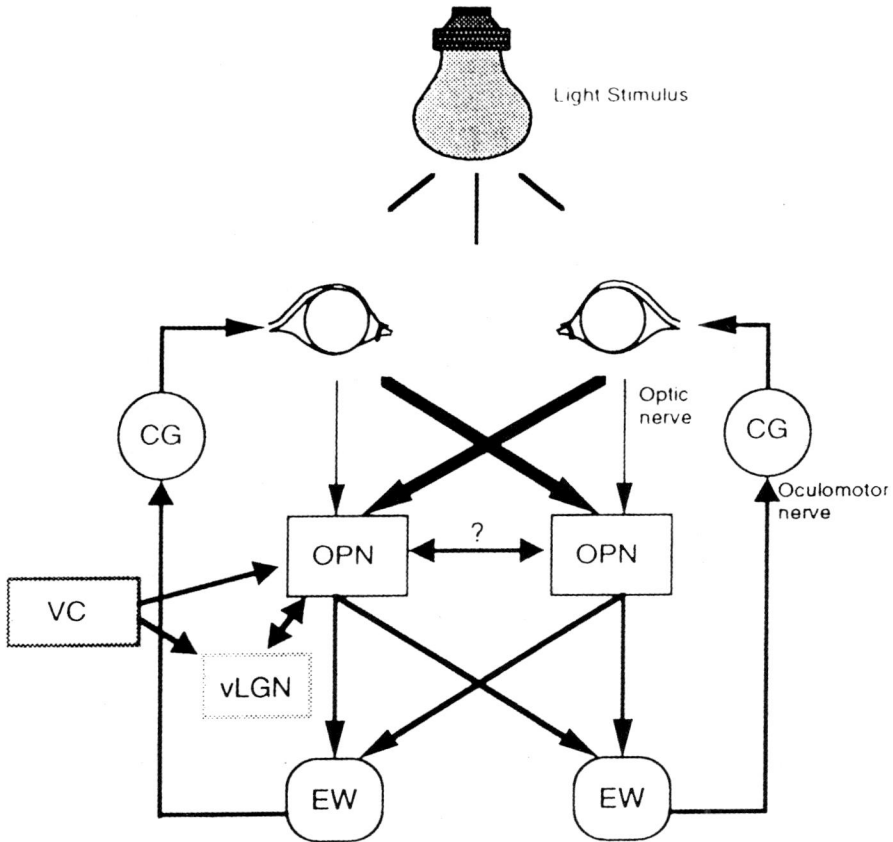

Figure 7.2   Neural circuit diagram of the pupillary light reflex. The retinae project bilaterally to the olivary pretectal nucleus (OPN), each of which in turn project bilaterally to the Edinger-Westphal nucleus (EW). Each pathway leaving EW is unilateral, synapsing in the ipsilateral ciliary ganglion (CG). In addition to the direct retinal projection, the OPN also receives inputs from the ventral lateral geniculate nucleus (vLGN) and visual cortex (VC). (Reprinted from Chan, Young and Lund, 1995, with approval of the European Neuroscience Association).

luminance (Warwick, 1954). This pathway, summarised in Figure 7.2, begins with the photoreceptors and crosses the layers of the retina to reach the retinal ganglion cells (RGCs). A subset of ganglion cells sends axons to the olivary pretectal nucleus (OPN), a small cluster of cells at the rostral edge of the dorsal mesencephalon. Axons from the OPN in turn project to the Edinger-Westphal nucleus (EW), terminating on preganglionic parasympathetic neurons. Autonomic efferents from the EW then join the oculomotor nerve and enter the orbit before branching off to innervate the ciliary ganglion (CG). The CG contains the parasympathetic postganglionic neurons that provide the final link in the pathway, entering the eye and terminating on the sphincter muscle of the iris.

PHOTORECEPTORS

There is strong evidence that both rods and cones are capable of initiating a pupillary response in a number of species ranging from moths, lizards and birds to humans and non-human primates (Denton, 1956; Alpern and Campbell, 1962; Gamlin et al., 1984; Nordtug and Melφ, 1992; Schaeffel and Wagner, 1992). The respective contributions of rods and cones to the PLR can be studied relatively selectively by comparing scotopic to photopic PLRs or by testing patients with complete colour-blindness or night blindness (Alpern and Campbell, 1962; Young, Han and Wu, 1993). Under normal conditions the rod and cone contributions to the PLR appear to be thoroughly integrated; there is no evidence to suggest that these two components subserve distinct functions. This is probably also true with respect to the different subtypes of cones (red, green, blue) since the spectral sensitivity curves for both the PLR and colour vision correspond closely in humans (Lowenstein and Loewenfeld, 1969). In rodents the relative contributions to the PLR from the different classes of photoreceptors have yet to be determined. Interestingly, progressive rod loss precedes the loss of cones in rodent models of retinal degeneration (rd), such as the RCS rat and the rd mouse. This sequential pattern of cell loss affords an opportunity to investigate the relative contribution of the two photoreceptor types to the PLR in rodents. Work in our lab has shown that in the RCS rat, this progressive loss of rods can be detected as a discontinuity in the constriction phase of the PLR (Whiteley et al., 1993), suggesting that the two inputs differentially drive the PLR.

RETINAL GANGLION CELLS

A small number of studies provide insight into retinal ganglion cell participation in the PLR. For instance, there is now sufficient evidence to say that only a subset of RGCs is directly involved in the generation of this reflex. For example, one study in rats reported that less than 10% of the total RGC population projects to the OPN (Young et al., 1994). Physiological studies of response latency in the pretectum following optic nerve stimulation have shown that, as a population, the pupillary afferents are among the slowest primary visual fibres (Trejo and Cicerone, 1984). Furthermore, the majority of RGCs back-labelled by OPN injections in the rat are small, corresponding in size to the type III variety (analogous to the W class in the cat) with a much lower number of large RGCs, tentatively identified as type I, also labelled (Young et al., 1994).

The retinal distribution of back-labelled RGCs following OPN injections has also been examined (Young et al., 1994). Although labelled RGCs could be found throughout the retina, the vast majority were found to lie in the ventral hemiretina, a finding that appears to be consistent across a variety of species (Gamlin et al., 1984; Distler and Hoffmann, 1989).

OLIVARY PRETECTAL NUCLEUS

Although there is one report to suggest otherwise (Legg, 1975), it is generally believed that the PLR is mediated through an obligatory synapse in the OPN: focal lesions of this nucleus in different animals result in loss of the reflex (Magoun and Ranson, 1935;

Klassen and Lund, 1988; Young and Lund, 1994). In mammals the pretectum is located in the dorsal part of the rostral brainstem, at the junction of the mesencephalon and diencephalon, and comprises several cytoarchitecturally distinct cell groups. One of these, the OPN, is a small nucleus situated close to the dorsal surface of the pretectum, midway between the habenular commissure and the rostral border of the superior colliculus. The OPN may in fact consist of two different nuclei, the largest part of which is known as the pars oralis, or rostral head. The pars oralis is approximately 250 μm in diameter and its shape gave rise to the nucleus olivary designation. This part of the nucleus receives the largest amount of retinorecipient innervation, while the caudally extending "tails" of the nucleus receive less (Campbell and Lieberman, 1985). There appear to be two distinct types of neurons in the OPN. The perimeter of the nucleus is formed by densely packed large, sparsely branched cells (Class I, somal diameter 15–30 μm) and smaller, usually bipolar, cells with complex dendritic appendages (Class II, somal diameter $7–10 \times 15–20$ μm). The centre of the OPN shows an extreme dearth of cell bodies and myelinated fibres (Campbell and Lieberman, 1985).

In the rat, as well as several other species, the OPN has been shown to receive a bilateral retinal projection, although the contralateral projection predominates (Scalia, 1972). Furthermore, this projection is topographically ordered: the caudal OPN receives fibres from dorsal retina, the lateral OPN from temporal retina, with rostral OPN receiving projections from ventral retina and medial OPN from nasal retina (Scalia and Arango, 1979).

The importance of the OPN as a central nucleus in the pupilloconstrictor reflex was demonstrated unequivocally by two studies, both using electrophysiological recordings in the OPN following light stimulation of the retina (Trejo and Cicerone, 1984; Clarke and Ikeda, 1985). These experiments demonstrated that the OPN contained "luminance detector cells", which responded to graded increases in luminance levels at the retina with graded increases in their firing rates. Significantly, it was also established that stimulation of these OPN cells resulted in pupilloconstriction. Finally, bilateral removal of the OPN results in complete abolition of the PLR (Chan, Young and Lund, 1995).

SECONDARY PROJECTION

The large output neurons of the OPN project to a number of different regions within the rostral brainstem (Itaya and Van Hoesen, 1982; Campbell and Lieberman, 1985; Klooster et al., 1993; Young and Lund, 1994). Those that project to EW, both by ipsilateral and contralateral routes, provide the direct pathway of the reflex but there is also a potential indirect relay through the nucleus of the posterior commissure (NPC) (Young and Lund, 1994). A reciprocal connection also exists between the OPN and the ventral lateral geniculate nucleus, although its role is unclear. One suggested role for the pathway from the OPN to the ventral lateral geniculate body offered was that it might play a role in the light driven behaviours of the hypothalamus (Young and Lund, 1994). Since these depend on actual light intensity levels, and pupillary activity will tend towards maintaining uniform light levels on the retina, it would be important for the hypothalamus to have information as to the degree of pupilloconstriction. This pathway could provide that information via the ventral lateral geniculate body.

OUTPUT PATHWAY

The EW nucleus, a small collection of cells within the oculomotor nuclear complex, is the preganglionic parasympathetic nucleus responsible for pupilloconstriction. This is based on anatomical evidence showing that EW neurons project directly to the intrinsic eye musculature (Warwick, 1954; Jaeger and Benevento, 1980), and electrophysiological studies indicating that cells in this region are indeed the preganglionic parasympathetic pupilloconstrictor cells (Sillito and Zbrozyna, 1970a,b). The neurons within EW drive the pupillary responses of the ipsilateral eye exclusively; they do not connect reciprocally across the midline of the brainstem (Klooster et al., 1993) and all project to the ipsilateral ciliary ganglion (Warwick, 1954; Loewy, Saper and Yamodis, 1978; Jaeger and Benevento, 1980; Martin and Dolivo, 1983). Presynaptic fibres leave the brainstem with the oculomotor nerve; postsynaptic fibres travel to the eye in the short ciliary nerves before terminating on the sphincter muscle of the iris.

SITE OF GENERATION OF THE CONSENSUAL RESPONSE

One important feature of the PLR is that illumination of one eye results in constriction of not only the eye being illuminated but also the contralateral eye. This latter component is termed the consensual response, in contrast to the direct response seen in the stimulated eye. There are two sites at which the input from each eye diverges bilaterally to provide the substrates of the consensual response: the partial decussation of OPN afferents at the chiasm and the bilateral secondary projection from the OPN and NPC to EW.

Studies on pigmented rats indicate that only about 3% of the optic axons projecting to the OPN derive from the ipsilateral retina (Young et al., 1994), with this number further reduced in albino rats (Chan, Young and Lund, 1995). Despite this markedly reduced uncrossed pathway, the amplitude of the consensual response is the same as the direct in pigmented rats, and only slightly reduced in albinos.

Lesion studies made it possible, for the first time, to estimate the relative functional weightings of the crossed and uncrossed connections between various nuclei in the PLR pathway. It was found that the contralateral connections were dominant and carried 71% of the function from retina to OPN and 52% of that from OPN to EW. While the ipsilateral retinal — OPN connection contributes only 3% of the total projection to the OPNs, it carries 29% of the function (Young et al., 1994; Chan, Young and Lund, 1995). It must be noted that these figures are only predictions based on the behavioural responses under a variety of experimental conditions. However, the results do suggest that the functional weighting found for the contralateral projection is much lower than would be expected from the anatomical data alone.

By showing that pupillary responses in rats compare closely with those from humans, these experiments support the validity of using this animal as a model system. Furthermore, the site of integration for luminance information was examined by way of experimental lesions, together with variations in the size of the uncrossed pathway. Results indicate that the primary inputs to the OPN, together with the outputs from OPN to the EW, are responsible for this integration. Of course the relative importance of each integration site may be dynamic, varying with experimental conditions.

INTRACEREBRAL TRANSPLANTATION STUDIES

In the initial studies of intracerebral retinal transplantation, donor retinae taken from embryonic rats were grafted to a location overlying the midbrain of newborn rat hosts. To promote innervation of the host brain by optic axons from the transplant, one host eye was removed at birth thereby reducing competition for central synaptic sites (McLoon and Lund, 1980a). It was found that transplants initially placed over the midbrain tended to be progressively translocated to the dorsal surface of the cerebellum during development. This was fortuitous in that this location is readily accessible in mature animals, thereby facilitating studies of graft function: after performing a craniotomy the exposed grafts could be identified and directly illuminated. Because of the importance of controlling for unintentional stimulation of the remaining host eye by light scatter, the remaining optic nerve was transected prior to testing, taking care to leave the autonomic efferents undisturbed. This model therefore allowed the full impact of transplant inputs on pupillary responses to be assessed in the absence of contaminating host visual inputs. It was found that illumination of the retinal transplant did in fact elicit a characteristic pupilloconstriction response in the host eye (Klassen and Lund, 1987). This response was brisk and reproducible, and was eliminated by destruction of the graft. The transplant in this situation therefore provided an input similar to that underlying the consensual response of the normal animal.

A typical record of a graft-mediated PLR response is shown in Figure 7.3. At best the maximum constriction is about 60% of baseline diameter and the latency of response is in the order of 500 ms, the latter being very close to that seen normally (Klassen and Lund, 1990b; Radel, Kustra and Lund, 1991; Young, Klassen and Lund, 1991). In general, the luminance level required to produce a given degree of pupillary constriction via the transplant was 1 to 2 log units greater than that needed to produce the same response through the host eye. This diminished response is perhaps not surprising given the number of factors which might compromise transplant function in this situations. These include the disruption of graft optics resulting from the harvesting process, as well as the removal of the RPE. Variability in graft position, degree of retinal tissue differentiation, and connectivity to the host were all found to influence graft performance, as will be discussed in the next section. Of particular importance was the finding that, for an individual transplant, the amplitude of constriction was correlated directly with luminance (Figure 7.4). Furthermore, under standardised conditions both the threshold and maximum responses remained remarkably consistent upon repeated testing, either during a single session or on different occasions. Despite the consistency of within-animal tests there was considerable variation in responsiveness among animals, as well as when the preparation was modified experimentally. This combination of consistent baseline conditions in a given animal together with systematic variance among animals make it possible to assess the factors that determine graft-mediated PLR performance. These factors will now be discussed in more detail.

VARIABILITY IN GRAFT-MEDIATED PLR AMONG ANIMALS

The functional efficacy of intracerebral retinal grafts can be modified in a variety of ways.

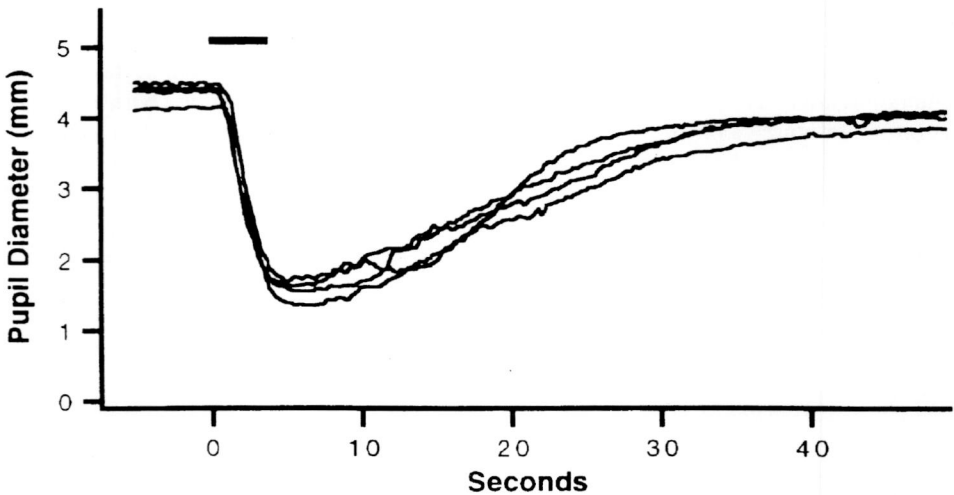

Figure 7.3 Transplant-mediated pupillary light reflex. Multiple responses to stimuli of the same intensity (2.5×10^4 cd/m^2; solid bar), presented over a 2-h period, are superimposed so as to demonstrate the robust character of the phenomenon. (Reprinted from Radel *et al.*, 1995, with kind permission from Elsevier Science Ltd.).

The first variable is the position of the graft: clearly grafts located under a venous sinus, or lying within the substance of the brain, are less likely to function optimally simply because of inadequate exposure to the light source. There is also evidence that, with time, the grafted retina shows a progressive loss of photoreceptors and, as a result, will become less responsive to light over time (Radel *et al.*, 1995). A potent factor determining the amplitude and latency of graft-mediated responses is the amount of innervation of the host OPN by the transplant. This was assessed in a light microscopic study in which the volume and density of projection were both measured and a significant direct correlation with function established (Klassen and Lund, 1990b). No effect on latency of the PLR was evident.

DEVELOPMENT

In normal animals the first indication of a PLR to illumination of the eye comes at postnatal day 7 (Radel, Das and Lund, 1992) corresponding to the appearance of photoreceptor outer segments. The nascent response is quite sluggish and of low amplitude. Over the next week the response increases in briskness until by postnatal day 9, when the eyes have just opened and synaptic ribbons are first seen in the retina, the PLR exhibits the typical mature wave form. Direct visual stimulation of intracerebral retinal transplants, taken from embryos aged 12–13 days (E12–13), fails to elicit a PLR in the host until they reach a chronological age similar to that at which the normal eye is capable of driving one. Specifically, since the graft is a week younger than the host at the time of trans-

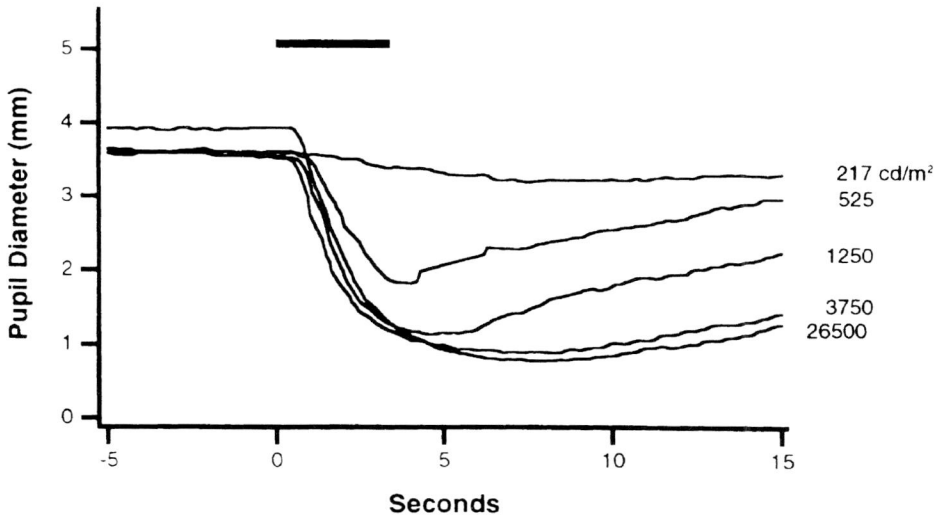

Figure 7.4   Graded transplant-mediated responses correspond to changes in stimulus intensity. The solid bar indicates presence of stimulus. Transplant response patterns resemble those of normal retinae, although sensitivity is generally reduced by 2–3 log units. (Reprinted from Radel *et al.*, 1995, with kind permission from Elsevier Science Ltd.).

plantation, the graft-mediated PLR first appears on postnatal day 14, one week delayed relative to the normal response. This indicates that the limiting factor in PLR development is maturation of the retina, not the retinofugal circuitry.

EFFECTS OF GRAFT REJECTION ON THE GRAFT-MEDIATED PLR

Healthy retinal grafts can continue to function for prolonged periods, but spontaneous rejection can occur if donor and host are sufficiently disparate in major and minor histocompatibility antigens, as between rat and mouse. It is also possible to induce graft rejection experimentally in a controlled manner when there is sufficient immunological disparity between donor and host. Retinae transplanted from one strain of rat to the neonatal brain of a different strain normally survive without evidence of rejection (Lund *et al.*, 1988; Rao *et al.*, 1989). In contrast, when a similar graft is introduced to most sites outside the CNS, immunological rejection occurs within 2 weeks. Intracerebral grafts can, however, be induced to reject by placing a skin graft from the donor strain on the flank of the host animal (Rao *et al.*, 1989; Young, Rao and Lund, 1989). Within days of placement the skin graft is rejected. A few days later the blood-brain barrier begins to leak in the vicinity of the intracerebral graft and lymphocytes are seen infiltrating the graft. By about 2 weeks, the graft has been largely destroyed (Young, Rao and Lund, 1989; Banerjee, Lund and Radel, 1993). The graft-mediated PLR provides a simple method for monitoring the progress of the rejection process *in vivo*. In a study investigating this, a surprising observation was made: even before the invasion of lymphocytes, there was an upregulation of microglia and Müller glia within the graft and at this stage there was an associated increase in amplitude and rate of pupilloconstriction (Banerjee, Lund and Radel, 1993). After about 10 days (the detailed timing varies considerably among animals), the constriction rate and amplitude drop and the graft becomes nonresponsive. While the enhanced performance at the early stage of the rejection process is difficult to explain, a plausible suggestion is that changes in the extracellular environment within the graft, associated with the upregulation of the Müller cells, might improve the efficiency of the response. This provides an example of intraretinal events modulating pupillary function.

EFFECT OF PRIOR STIMULATION CONDITIONS

Under normal circumstances the PLR is consistent from test to test. In transplanted animals, however, it was found that a repetitive stimulus regimen can, under certain circumstances, result in improved performance on subsequent trials (Radel *et al.*, 1995). This effect could be seen when transplants were stimulated at near-threshold levels for either 3 or 30 s. Transplants were then presented with a regimen of 3 s with the light on followed by 20 s of darkness, repeated over 20 trials, for a total of 460 s. On subsequent testing with a single 30 s stimulus of the same intensity, response amplitude was substantially enhanced. This effect was generally seen when stimulus intensity was just above threshold, but was most prominent in animals in which the transplant was relatively ineffective in driving a pupilloconstrictor response, notably in aged rats. This potentiation phenomenon seems to be particular to transplant inputs and has not been demonstrated in normal animals.

INTERACTION WITH HOST OPTIC INPUT

Of particular interest is the question of whether the interaction that normally occurs between inputs from the two eyes to produce an integrated pupilloconstrictor response also occurs between inputs from transplant and host. This question has been examined in a number of ways – both anatomically and functionally — including simultaneous, or sequential, stimulation of transplant and host eyes, host optic nerve section and unilateral OPN lesions. Each will be described separately.

Anatomical modifications

The inputs from the two eyes to the OPN are largely segregated from one another with the crossed projection occupying the head and tail of the nucleus and the uncrossed projection situated dorsolaterally to the main focus of the crossed projection. If one eye is removed at birth, the contralateral OPN shrinks and the uncrossed pathway from the remaining eye distributes over a larger area of the nucleus than normal. In animals with transplants, the host eye innervating the OPN to which the transplant projects by way of the large, crossed projection was typically removed at birth. Again this nucleus appeared smaller and less organised at later ages and the uncrossed input from the remaining eye expanded throughout the OPN (Radel, Kustra and Lund, 1991), as has been shown in other nuclei of the primary optic pathway after neonatal eye removal (Lund, Cunningham and Lund, 1973). The input from the transplant was found to overlap substantially with the expanded uncrossed projection. At the electron microscopic level, terminals from both transplant and host eye were found in close proximity on OPN neurons, frequently on the same dendrites (Lawson and Lund, 1995).

Stimulation of host eye and transplant

In normal animals, as has already been indicated, the diameter of the pupil is determined, not only by the amount of light entering that same eye, but also by the amount of light presented to the other eye. The resultant diameter is a logarithmic addition of the effect of stimulating each eye separately (Chan, Young and Lund, 1995). After neonatal transplantation and unilateral host eye removal, stimulation of either the transplant or the remaining host eye results in pupilloconstriction. Stimulation of both results in an added constriction. The resultant diameter was the logarithmic, rather than arithmetic, sum of the individual amplitudes and in this respect is similar to the situation found for stimulation of both eyes in normal animals.

Effect of host optic nerve section

Since pupillary diameter in this model is determined by integration of host and transplant inputs, the effect of host optic nerve section on transplant responsiveness was then examined. With the optic nerve intact, and the host eye occluded, it was found that some transplants drove a pupilloconstrictor response well, while others functioned poorly or not at all. If the host optic nerve was then sectioned, the transplant response frequently improved (Radel *et al.*, 1995). The greatest and most consistent improvements were from transplants that functioned poorly prior to nerve section. Anatomical studies showed quite heavy projections from transplant to pretectum in some of these animals, suggesting that the barrier to function was not simply an inadequate graft-host projection but rather an effect related to the presence of intact host input to the OPN, albeit from an unstimulated eye. Why such an improvement was seen is not readily apparent.

A number of theoretical possibilities can be suggested:

(i) The final pupil diameter is determined by averaging the inputs from the two retinas, host and transplant. If the host eye remains in the dark while the retinal transplant is illuminated, removal of the negative input from the host retina would result in improved transplant performance. This possibility was not, however, supported by examination of the analogous situation in normal animals: if one optic nerve was cut the response to stimulation of the other eye was unaffected. It should be noted that the unilateral enucleation performed at the time of transplantation substantially alters the interrelation of retinal terminals from host and transplant, and this may be important.

(ii) There may be a dark discharge from the host eye which effectively reduces the signal/ noise ratio of transplant inputs. Removal of the host eye would therefore increase transplant efficacy by removing the noise.

(iii) Removal of the host input may induce receptor upregulation on postsynaptic cells with the result that the transplant input is now much more effective in driving them.

(iv)   A similar effect could also be achieved by gating effects due to synapses lying close
to one another on the same dendrite and effectively reducing the impact of each
other's inputs. Removal of proximal inputs on a dendrite would make distally located
inputs more effective.

EFFECT OF UNILATERAL OPN LESIONS

In the previous cases in which host and transplant inputs were studied in relation to one
another, the transplant input was largely confined to one OPN, while the input from the
remaining host eye predominantly innervated the opposite OPN with only a small projection to the same OPN as the transplant. Interactions between the two retinal inputs
would most likely occur at the convergence of the second order pathways from the two
OPNs upon EW. To address this further, a series of studies was conducted in which one
OPN was lesioned to confine the transplant and host inputs to a single OPN (Young and
Lund, 1994). For comparison purposes similar lesions were made in untransplanted
control rats.

In normal animals, the PLR pathway from each eye to the ipsilateral efferent nerve
converges on EW by way of both OPNs and crossed and uncrossed projections from these
nuclei. If the uncrossed projection from the stimulated eye and the crossed projection from
the unstimulated eye are eliminated by a unilateral OPN lesion, then the stimulated eye
becomes more effective (Figure 7.5). One explanation for this is that the unstimulated
eye exerts a negative influence on the consensual response and is not simply a neutral
bystander. The reduced response seen when the driving input is delivered by the uncrossed
pathway might reflect a stronger inhibitory role from the unstimulated crossed pathway
than the stimulated uncrossed input. This concept is further supported by the situation
in which both eyes are stimulated and one OPN removed. Here the response is always
maximal. Conceptually this suggests that rather than regarding stimulation of one eye as
normal and binocular summation of the response resulting from stimulation of both eyes
as an additive event, it may be more appropriate to regard binocular stimulation as the
norm and reduced stimulation of one eye as causing binocular inhibition.

The impact of unilateral OPN lesions on transplant efficacy parallels the situation in
normal animals; host eye stimulation is much less effective in driving the PLR, while
transplant stimulation is much more effective, in fact transplant-mediated
pupilloconstriction is increased much more than in the comparable paradigm in normal
animals. This is illustrated in Figure 7.6. This differential effect may relate to unilateral
eye removal performed at the time of transplantation, which results in a more substantial
overlap of host and transplant terminals in the OPN. As before, the lesion-related depression of the normal eye stimulation could be explained by the negative effect presented
by the unstimulated transplant. The magnitude of the enhancement of transplant stimulation is the more surprising. This suggests an imbalance between transplant and host
inputs such that the transplant substantially dominants the host, more so than would be
the case between crossed and uncrossed inputs in normal animals. It should be noted that
while the changes described here relate to changes in amplitude of the PLR, OPN lesions
had no impact on latency measures for transplant stimulation, although they did result
in a slight reduction of latency to host eye stimulation.

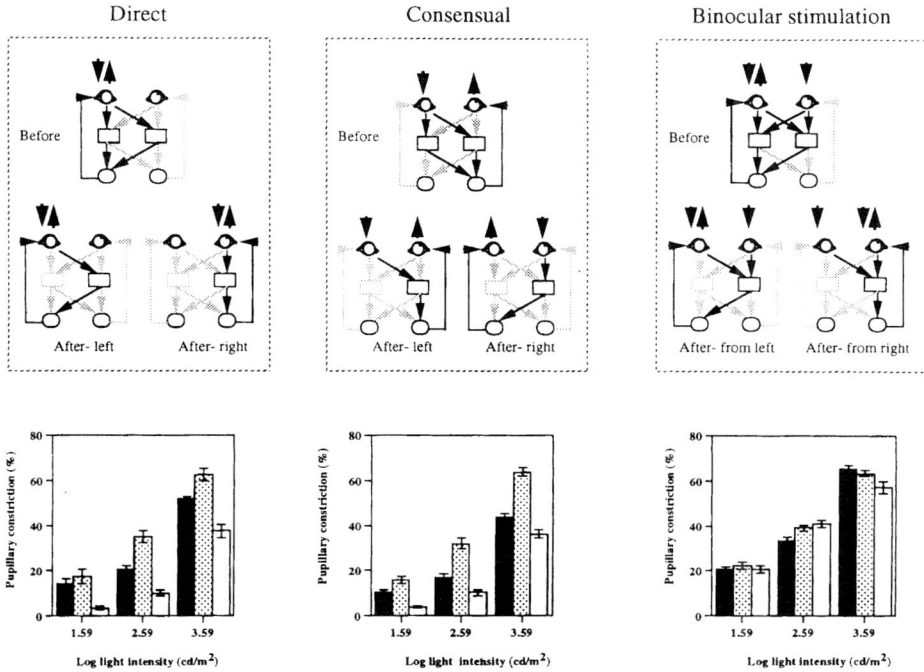

Figure 7.5   The effect of olivary pretectal nucleus (OPN) lesions on the various pupillary light reflexes (PLRs) in terms of anatomical circuitry and behaviour. In the circuit diagrams: rectangle = OPN; oval = Edinger-Westphal nucleus; down-arrow above eye = photic stimulation; up-arrow above eye = recording. For bar charts: solid bars = response before lesion; stippled bar = left response after lesion; open bars = right response after lesion. (Reprinted from Chan, Young and Lund, 1995, with approval of the European Neuroscience Association).

EFFECT OF HOST AGE

The introduction of grafts into neonatal rats results in a 'best case scenario' in which the immunological and pathological consequences of transplantation are minimal and the environment for outgrowth of transplant axons is relatively permissive. Older hosts present a very different situation. There is the potential for immune rejection of the graft, if care is not taken to use syngeneic strains (Lund et al., 1987a) or to provide an appropriate immunosuppressive therapy, and there is a substantial glial reactive response, involving both astrocytes and microglia, which may both compromise graft differentiation and limit access by outgrowing axons to the host neuropil. Furthermore, the environment of the mature brain is much less permissive for axonal growth. Despite these challenges transplants can survive, differentiate and send axons into adjacent host brain structures. They generally differentiate less well than when introduced into neonatal host brains, the projections are less robust and extend for only short distances in the host brain and the connections they form are much less substantial. Nevertheless it is still possible to elicit

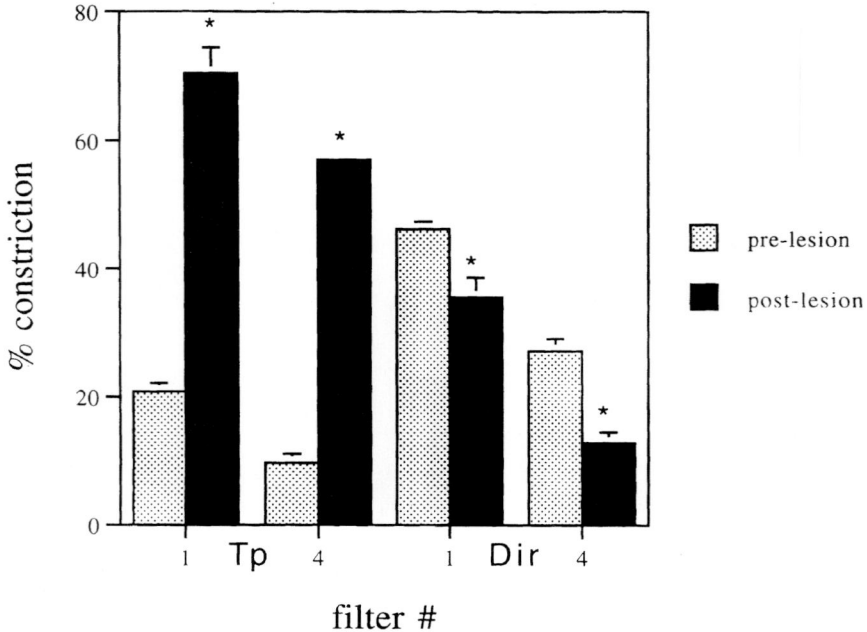

Figure 7.6   An example of the amplitude of pupilloconstriction seen in one animal that received a right olivary pretectal nucleus (OPN) lesion. The graft-mediated pupilloconstrictor response (Tp) is markedly upregulated at both luminance levels (filter #1 = brighter, filter #4 = dimmer). The host-mediated pupillary light reflex (Dir) shows a typical pattern of decreased response post-lesion. (Michael J. Young, doctoral thesis, University of Cambridge, previously unpublished).

a PLR to stimulation as long as the transplant is placed in close proximity to the OPN (Klassen and Lund, 1990a). The best responses compare favourably with those seen after transplantation to neonatal hosts, but there are many more animals with suboptimal responses.

SUMMARY

The intracerebral retinal transplants described here are of interest in that they provide a situation in which to address how well functional circuits can be restored by ectopic neural arrays. Furthermore by placing grafts close to the nuclei they would normally innervate, attention can be focused on how well they reconstruct synaptic circuitry to provide a substrate for function without also being concerned about the difficulties of navigating axons to their target. The pupillary light reflex provides a simple assay for testing whether a function can be restored, how good the response is and how efficacy can be compromised or improved. Clearly such studies can only be effective if suitable quantification of the response is established, permitting measures not only of amplitude but also of rate of constriction, latency of onset of responses, instability of response and components of the dilatation phase.

The observations summarised here show first that such grafts can mediate a relatively normal response. For example, the pupil of the host eye constricts rather than dilates on graft stimulation. Furthermore, response parameters such as amplitude and latency, although usually less good than normal, are nevertheless remarkably similar in the best cases. The grafts provide a useful feature not normally present in the regular CNS in that there is a wide range of functional efficacy. This provides opportunity to examine the substrates of performance. Intragraft factors are important as is the density of innervation of the host OPN by the graft. The studies of graft rejection show that the PLR can serve as a sensitive functional monitor for experimentally induced tissue damage, in this case an immune-mediated event. In addition, examination of the interactions between graft and host inputs, together with the variability in graft efficacy, reveals how inputs from different sources are integrated to produce an output function. This system can be further exploited by the use of selective lesions, allowing an examination of the sites of integration. Finally, by varying the host age it is possible to use this system to examine the age-related factors that compromise neural circuit reconstruction.

INTRARETINAL TRANSPLANTATION

The purpose of intraretinal transplantation is primarily to investigate the potential of such an approach to prevent degeneration or to repair retinae in which photoreceptor loss occurs as a result of a genetic disease or acquired event. Studies on experimental animals are largely to serve as a model for possible application to humans. However, there is still tremendous value in using these animals to study the information processing in retinae undergoing a selective and protracted loss of photoreceptors. The RCS rat is the model to which most attention has been given. In this animal extensive loss of photoreceptors occurs over the first 3 months of life. This is accompanied by a deterioration of visual function, including loss of acuity-related behavioural responses, visual reflexes, and a recordable ERG (Dowling and Sidman, 1962; Trejo and Cicerone, 1982; Wecker and Ison, 1986; Coffey et al., 1995). It was shown some years ago that the PLR is persistent even at 2 years of age and that this correlates with the persistence of a small number of residual photoreceptors, mostly located in the far peripheral retina (Trejo and Cicerone, 1982). With time the amplitude of the response diminishes and both threshold luminance and response latency increase. One interesting phenomenon is the interruption in the constriction phase of the response (see above section on photoreceptors) which can last as long as 1500 ms, especially in older animals (Whiteley et al., 1994). The reason for this effect is not known although it might reflect the dual driving input to OPN mentioned earlier and the fact that the defect in these animals primarily affects rods over cones. It should be noted that although a diminished PLR response has been recorded in two laboratories, a third was unable to show any reduction in amplitude of response with age (Kovalevsky et al., 1995). It is possible that the luminance levels used were too high, since as in humans with comparable diseases, the major deficits are always seen at lower luminance levels.

If RPE cells are introduced into the subretinal space during the first month of age the degenerative process is slowed in the region adjacent to the RPE cells (Li and Turner, 1988; Lopez et al., 1989). To examine whether this also affects the PLR, a series of animals was tested with such transplants. The results show that there are improvements, but these are not as obvious as might be expected. First, sham injections of just the carrier medium, a balanced salt solution, to the subretinal space result in some photoreceptor rescue (Silverman and Hughes, 1990; Li and Turner, 1991). Along with this physiological rescue there is concomitant sparing of the PLR although this may not be permanent (Whiteley et al., 1996). A prominent feature of the PLR in dystrophic RCS rats is the break in the constriction phase of the response. This is not eliminated by transplantation, but that may not be surprising because the response is an average of the whole retina and the transplant may rescue as little as 10% of the photoreceptors.

In summary, the PLR in this preparation provides a sensitive test of residual function, which can detect activity even when a traditional ERG recording reads negative. The finding of a break in the constriction phase may provide insight into the input drive of the reflex and the possibility that components of this can be dissociated. While improvement can be seen after transplantation, the fact that this is a minimum function test means that it may be less useful than a more selective test such as acuity function for assessing the impact of a transplant.

OPTIC NERVE REPLACEMENT

Studies in which the optic nerve of mature rodents has been replaced by a segment of peripheral nerve show that a small percentage of the severed axons are capable of regeneration along the peripheral nerve tube, presumably in response to the presence of Schwann cells (So and Aguayo, 1985). If the distal end of this nerve is implanted adjacent to visual centres in the brainstem, some of the regenerating axons enter the brainstem and form synapses in these centres (Vidal-Sanz et al., 1987). In contrast to the intracerebral grafts, this preparation requires that the axons follow the surrogate pathway from the in situ eye to the brain in addition to simply forming synapses in the target region. Electrophysiological studies show that these connections can relay discrete information about changes in stimulus intensity (Kierstead et al., 1989; Sauve, Sawai and Rasminsky, 1995). It has been shown that if the axons project to the vicinity of the OPN, pupilloconstrictor responses can be elicited by photic stimulation of the grafted eye (Thanos, 1992). The degree to which the responses via the regenerated pathway compare with the normal dynamics of the PLR has yet to be explored, but the study does show that in this, as in the previous sections, the reflex is an effective indicator of restored connectivity and that the connections are sufficiently specific to produce an appropriate response.

In summary, much less has been done using the PLR as a monitor of the efficacy of regeneration. As attempts are made to find ways of improving the number of regenerating axons (Thanos, Mey and Wild, 1993; Lawrence et al., 1996), it will be valuable to use readily quantifiable responses, such as the PLR, to assess the efficacy of reconstruction strategies.

GENERAL OVERVIEW

This review has shown how the pupillary light reflex serves as a simple test of function that is being applied to several transplant paradigms involving the retina and its output connections. This system has certain advantages that recommend it over other assays of transplant function:

(i)   The PLR depends purely on a one-way relay system and, as such, serves as a good model system in which to examine how effective transplants may be in relaying sensory information.
(ii) Because of the relatively circumscribed nature of the pathways involved in this reflex it is possible to dissect experimentally how the transplant input interacts with the host visual system at various levels.
(iii) There are a number of parameters of the pupillary response that reflect different aspects of transplant efficacy.
(iv) Interactions that occur between graft and host optic inputs to determine the final pupillary response provide an opportunity to examine the integration of host and transplant information in considerable detail.
(v)   The PLR is a robust reflex that can survive considerable disruption of the visual pathways, and can therefore serve as a sensitive test of minimal function.

ACKNOWLEDGEMENTS

Much of this work has been conducted in our laboratory over a 10 year period, and would not have been possible without the contributions of many colleagues, particularly Jeffrey Radel, Ranjita Betarbet Banerjee, Simon Whiteley, and Carmen Chan.

REFERENCES

Aguayo, A.J., Vidal-Sanz, M., Villegas-Perez, M.P. and Bray, G.M. (1987). Growth and conductivity of axotomized retinal neurons in adult rats with optic nerves substituted by peripheral nervous system grafts-linking the eye and the midbrain. *Annals of the New York Academy of Sciences*, **495**, 1–9.

Alpern, M. and Campbell, F. (1962). The spectral sensitivity of the consensual light reflex. *Journal of Physiology*, **164**, 478–507.

Aramant, R.B. and Seiler, M.J. (1995). Fiber and synaptic connections between embryonic retinal transplants and host retina. *Experimental Neurology*, **133**, 244–255.

Banerjee, R., Lund, R.D. and Radel, J.D. (1993). Anatomical and functional consequences of induced rejection of intracranial retinal transplants. *Neuroscience*, **56**, 939–953.

Bird, A.C. (1995). Retinal Photoreceptor Dystrophies, LI. Edward Jackson Memorial Lecture. *American Journal of Ophthalmology*, **119**, 543–562.

Björklund, A. and Stenevi, U. (1977). Reformation of the severed septohippocampal cholinergic pathway in the adult rat by transplanted septal neurons. *Cell and Tissue Research*, **185**, 289–302.

Björklund, A. and Stenevi, U. (1979). Reconstruction of the nigrostriatal dopamine pathway by intracerebral nigral transplants. *Brain Research*, **117**, 555–560.

Campbell, G. and Lieberman, A.R. (1985). The olivary pretectal nucleus: experimental anatomical studies in the rat. *Philosophical Transactions of the Royal Society of London*, **310**, 573–609.

Chan, K.M.C., Young, M.J. and Lund, R.D. (1995). Interactive events subserving the pupillary light reflex in pigmented and albino rats. *European Journal of Neuroscience*, **7**, 2053–2063.

Clarke, R.J. and Ikeda, H. (1985). Luminance and darkness detectors in the olivary pretectal nuclei and their relationship to the pupillary light reflex in the rat. I. Studies with steady luminance levels. *Experimental Brain Research*, **57**, 224–232.

Coffey, P.J., Hetherington, L., Whiteley, S.J., Litchfield, T.M., Lund, R.D. and Wright, S.R. (1995). Detection of visual patterns in dystrophic RCS rats following RPE transplants. *Society for Neuroscience Abstracts*, **21**, 1308.

Das, G.D. and Altman, J. (1971). Transplanted precursors of nerve cells: Their fate in the cerebellums of young rats. *Science*, **173**, 637–638.

David, S. and Aguayo, A.J. (1981). Axonal elongation in peripheral nervous system "bridges" after central nervous system injury in adult rats. *Science*, **214**, 931–933.

Denton, E.J. (1956). The response of the *Gekko gekko* to external light stimuli. *Journal of General Physiology*, **40**, 201–216.

Distler, C. and Hoffmann, K.-P. (1989). The pupillary light reflex in normal and innate microstrabismic cats, II: retinal and cortical input to the nucleus praetectalis olivaris. *Visual Neuroscience*, **3**, 139–153.

Dowling, J.E. and Sidman, R.L. (1962). Inherited retinal dystrophy in the rat. *Journal of Cell Biology*, **14**, 73–107.

Gamlin, P.D.R., Reiner, A., Erichsen, J.T., Karten, H.J. and Cohen, D.H. (1984). The neural substrate for the pupillary light reflex in the pigeon (*Columba livia*). *Journal of Comparative Neurology*, **226**, 523–543.

Gaze, R.M. (1970). *The Formation of Nerve Connections*. New York: Academic Press.

Gouras, P., Du, J., Kjeldbye, H., Yamamoto, S. and Zack, D.J. (1994). Long-term photoreceptor transplants in dystrophic and normal mouse retina. *Investigative Ophthalmology and Visual Science*, **35**, 3145–3153.

Itaya, S.K. and Van Hoesen, G.V. (1982). WGA-HRP as a transneuronal marker in the visual pathways of monkey and rat. *Brain Research*, **236**, 199–204.

Jaeger, R.J. and Benevento, L.A. (1980). A horseradish peroxidase study of the innervation of the internal structures of the eye. Evidence for a direct pathway. *Investigative Ophthalmology and Visual Science*, **19**, 575–583.

Jiang, L.Q. and Hamasaki, D. (1994). Corneal electrographic function rescued by normal retinal pigment epithelial grafts in retinal degenerative Royal College of Surgeons rats. *Investigative Ophthalmology and Visual Science*, **35**, 4300–4309.

Kierstead, S.A., Rasminsky, M., Fukuda, Y., Carter, D.A., Aguayo, A.J. and Vidal-Sanz, M. (1989). Electrophysiological responses in the hamster superior colliculus evoked by regenerating retinal axons. *Science*, **246**, 255–258.

Klassen, H. and Lund, R.D. (1987). Retinal transplants can drive a pupillary reflex in host rat brains. *Proceedings of the National Academy of Sciences USA*, **84**, 6958–6960.

Klassen, H. and Lund, R.D. (1988). Anatomical and behavioral correlates of a xenograft-mediated pupillary reflex. *Experimental Neurology*, **102**, 102–108.

Klassen, H. and Lund, R.D. (1990a). Retinal graft-mediated pupillary responses in rats: restoration of a reflex function in the mature mammalian brain. *Journal of Neuroscience*, **10**, 578–587.

Klassen, H. and Lund, R.D. (1990b). Parameters of retinal graft-mediated responses are related to underlying target innervation. *Brain Research*, **533**, 181–191.

Klooster, J., Beckers, H.J.M., Vrensen, G.F.J.M. and van der Want, J.J.L. (1993). The peripheral and central projections of the Edinger-Westphal nucleus in the rat. A light and electron microscopic tracing study. *Brain Research*, **632**, 260–273.

Kovalevsky, G., DiLoreto, D., Wyatt, J., del Cerro, C., Cox, C. and del Cerro, M. (1995). The intensity of the pupillary light reflex does not correlate with the number of retinal photoreceptor cells. *Experimental Neurology*, **133**, 43–49.

Lawrence, J.M., Lawson, D.D.A., Whiteley, S.J.O., Lund, R.D. and Sauve, Y. (1996). Effect of Schwann cells and macrophage inhibitory factor (MIF) on the survival of rat retinal ganglion cells (RGCs) after axotomy and peripheral nerve (PN) grafting. *Society for Neuroience Abstracts*, **22**, 320.

Lawson, D.D.A. and Lund, R.D. (1995). Synaptic basis of transplant driven pupillary light refex in neonatal and adult hosts. *Society for Neuroscience Abstracts*, **21**, 1308.

Legg, C.R. (1975). Effects of subcortical lesions on the pupillary light reflex in the rat. *Neuropsychologia*, **13**, 373–376.

Li, L.X. and Turner, J.E. (1988). Inherited retinal dystrophy in the RCS rat: prevention of photoreceptor degeneration by pigment epithelial cell transplantation. *Experimental Eye Research*, **47**, 911–917.

Li, L. and Turner, J.E. (1991). Optimal conditions for long-term photoreceptor cell rescue in RCS rats — the necessity for healthy RPE transplants. *Experimental Eye Research*, **52**, 669–679.

Loewenfeld, I.E. (1993). *The Pupil: Anatomy, Physiology, and Clinical Applications*, Detroit: Wayne State University Press.

Loewy, A.D., Saper, C.B. and Yamodis, N.D. (1978). Re-evaluation of the efferent projections of the Edinger-Westphal nucleus in the cat. *Brain Research*, **141**, 153–159.

Lopez, R., Gouras, P., Kjeldbye, H., Sullivan, B., Reppucci, V., Brittis, M., Wapner, F. and Goluboff, E. (1989). Transplanted retinal pigment epithelium modifies the retinal degeneration in the RCS rat. *Investigative Ophthalmology and Visual Science*, **30**, 586–588.

Lowenstein, O. and Loewenfeld, I.E. (1969). The pupil. In *The Eye, Vol. 3, Muscular Mechanisms*, edited by H. Davson, pp. 225–261. New York: Academic Press.

Lund, R.D. and Hauschka, S.D. (1976). Transplanted neural tissue develops connections with the host rat brain. *Science*, **193**, 582–584.

Lund, R.D., Cunningham, T.J. and Lund, J.S. (1973). Modified optic projections after unilateral eye removal in young rats. *Brain, Behavior and Evolution*, **8**, 51–72.

Lund, R.D., Rao, K.R., Hankin, M.H., Kunz, H.W. and Gill, T.J., III. (1987a). Transplantation of retina and visual cortex to rat brains of different ages; maturation, connection patterns, and immunological consequences. *Annals of the New York Academy of Sciences*, **495**, 227–241.

Lund, R.D., Rao, K.R., Hankin, M.H., Kunz, H.W. and Gill, T.J., III. (1987b). Immunogenetic aspects of neural transplantation. *Transplantation Proceedings*, **19**, 1128–1129.

Lund, R.D., Rao, K., Kunz, H.W. and Gill, T.J., III. (1988). Instability of neuronal xenografts placed in neonatal rat brains. *Transplantation*, **46**, 216–223.

Lund, R.D., Radel, J.D., Hankin, M.H., Yee, K.T., Banerjee, R., Coffey, P.J. *et al.* (1992). Intracerebral retinal transplants. In *Regeneration and Plasticity in the Mammalian Visual System*, edited by A. Zichi, pp. 125–145. Boston, MA: MIT Press

Magoun, H.W. and Ranson, S.W. (1935). The central path of the light reflex: A study of the effect of lesions. *Archives of Ophthamology*, **13**, 791–811.

Martin, X. and Dolivo, M. (1983). Neuronal and transneuronal tracing in the trigeminal system of the rat using herpes symplex virus suis. *Brain Research*, **273**, 253–276.

May, C.H. (1887). Transplantation of a rabbit's eye into the human orbit. *Archives of Ophthalmology*, **16**, 47–53.

McLoon, S.C. and Lund, R.D. (1980a). Specific projections of retina transplanted to rat brain. *Experimental Brain Research*, **40**, 273–282.

McLoon, S.C. and Lund, R.D. (1980b). Identification of cells in retinal transplants which project to host visual centers: a horseradish peroxidase study in rats. *Brain Research*, **197**, 491–495.

McLoon, S.C. and Lund, R.D. (1983). Development of fetal retina, tectum, and cortex transplanted to the superior colliculus of adult rats. *Journal of Comparative Neurology*, **217**, 376–389.

Nordtug, T. and Melф, T.B. (1992). The pupillary system of noctuid moth eyes: dynamic changes in relation to control of pupil size. *Journal of Experimental Zoology*, **262**, 16–21.

Radel, J.D., Kustra, D.J. and Lund, R.D. (1991). Rapid enhancement of transplant-mediated pupilloconstriction after elimination of competing host optic input. *Developmental Brain Research*, **60**, 275–278.

Radel, J.D., Das, S. and Lund, R.D. (1992). Development of light-activated pupilloconstriction in rats as mediated by normal and transplanted retinae. *European Journal of Neuroscience*, **4**, 603–615.

Radel, J.D., Kustra, D.J. and Lund, R.D. (1995). The pupillary light response: functional and anatomical interaction among inputs to the pretectum from transplanted retinae and host eyes. *Neuroscience*, **68**, 893–907.

Radel, J.D., Kustra, D.J., Das, S., Elton, S. and Lund, R.D. (1995). The pupillary light response: assessment of function mediated by intracranial retinal transplants. *Neuroscience*, **68**, 909–924.

Rao, K., Lund, R.D., Kunz, H.W. and Gill, T.J., III. (1989). The role of MHC and non-MHC antigens in the rejection of intracerebral allogeneic neural grafts. *Transplantation*, **48**, 1018–1021.

Sauve, Y., Sawai, H. and Rasminsky, M. (1995). Functional synaptic connections made by regenerated retinal ganglion cell axons in the superior colliculus of adult hamsters. *Journal of Neuroscience*, **15**, 665–675.

Scalia, F. (1972). The termination of retinal axons in the pretectal region of mammals. *Journal of Comparative Neurology*, **145**, 223–258.

Scalia, F. and Arango, V. (1979). Topographic organization of the projections of the retina to the pretectal region in the rat. *Journal of Comparative Neurology*, **186**, 271–292.

Schaeffel, F. and Wagner, H. (1992). Barn owls have symmetrical accomodation in both eyes, but independent pupillary responses to light. *Vision Research*, **32**, 1149–1155.

Sillito, A.M. and Zbrozyna, A.W. (1970a). The localization of pupilloconstrictor function within the midbrain of the cat. *Journal of Physiology*, **211**, 461–477.

Sillito, A.M. and Zbrozyna, A.W. (1970b). The activity characteristics of the preganglionic pupilloconstrictor neurones. *Journal of Physiology*, **211**, 767–799.

Silverman, M.S. and Hughes, S.E. (1990). Photoreceptor rescue in the RCS rat without pigment epithelium transplantation. *Current Eye Research*, **9**, 183–191.

Silverman, M.S., Hughes, S.E., Valentino, T.L. and Yao, L. (1992). Photoreceptor transplantation — anatomic, electrophysiologic, and behavioral evidence for the functional reconstruction of retinas lacking photoreceptors. *Experimental Neurology*, **115**, 87–94.

Simons, D.J. and Lund, R.D. (1985). Fetal retinae transplanted over tecta of neonatal rats respond to light and evoke patterned neuronal discharges in the host superior colliculus. *Developmental Brain Research*, **21**, 156–159.

So, K.-F. and Aguayo, A.J. (1985). Lengthy regrowth of cut axons from ganglion cells after peripheral nerve transplantation into the retina of adult rats. *Brain Research*, **328**, 349–354.

Sperry, R.W. (1945). Restoration of vision after crossing of optic nerves and after contralateral transplanation of eye. *Journal of Neurophysiology*, **7**, 57–70.

Sperry, R.W. (1963). Chemoaffinity in the orderly growth of nerve fiber patterns and connections. *Proceedings of the National Academy of Sciences USA*, **50**, 703–710.

Stone, L.S. and Zaur, I.S. (1940). Reimplantation and transplantation of adult eyes in the slamander (*Triturus viridescens*) with return of vision. *Journal of Experimental Zoology*, **85**, 243–269.

Tansley, K. (1946). The development of the rat eye in graft. *Journal of Experimental Biology*, **22**, 221–223.

Thanos, S. (1992). Adult retinofugal axons regenerating through peripheral nerve grafts can restore the light-induced pupilloconstriction reflex. *European Journal of Neuroscience*, **4**, 691–699.

Thanos, S., Mey, J. and Wild, M. (1993). Treatment of the adult retina with microglia-suppressing factors retards axotomy-induced neuronal degradation and enhances axonal regeneration *in vivo* and *in vitro*. *Journal of Neuroscience*, **13**, 455–466.

Thompson, W.G. (1890). Successful brain grafting. *New York Medical Journal*, **51**, 701–702.

Trejo, L.J. and Cicerone, C.M. (1982). Retinal sensitivity measured by the pupillary light reflex in RCS and albino rats. *Vision Research*, **22**, 1163–1171.

Trejo, L.J. and Cicerone, C.M. (1984). Cells in the pretectal olivary nucleus are in the pathway for the direct light reflex of the pupil in the rat. *Brain Research*, **300**, 49–62.

Turner, J.E. and Blair, J.R. (1986). Newborn rat retinal cells transplanted into a retinal lesion site in adult host eyes. *Developmental Brain Research*, **26**, 91–104.

Vidal-Sanz, M., Bray, G.M., Villegas-Pérez, M.P., Thanos, S. and Aguayo, A.J. (1987). Axonal regeneration and synapse formation in the superior colliculus by retinal ganglion cells in the adult rat. *Journal of Neuroscience*, **7**, 2984–2909.

Warwick, R. (1954). The ocular parasympathetic nerve supply and its mesencephalic sources. *Journal of Anatomy*, **88**, 71–93.

Wecker, J.R. and Ison, J.R. (1986). Visual function measured by reflex modification in rats with inherited retinal dystrophy. *Behavioural Neuroscience*, **100**, 679–684.

Whiteley, S.J.O., Young, M.J., Coffey, P.J. and Lund, R.D. (1993). Pupillary light reflex in RCS rats: age related changes and comparisons with visual acuity. *Society for Neurocience Abstracts*, **19**, 278.

Whiteley, S.J.O., Young, M.J., Litchfield, T.M. and Lund, R.D. (1994). Changes in the pupillary light reflex with age in RCS rats and the effect of intraretinal transplantation of retinal pigment epithelium. *Investigative Ophthalmology and Visual Science*, **34**, 1524.

Whiteley, S.J.O., Litchfield, T.M., Coffey, P.J. and Lund, R.D. (1996). Improvement of the pupillary light reflex of Royal College of Surgeons rats following RPE cell grafts. *Experimental Neurology*, **140**, 100–104.

Yamamoto, S., Du, J., Gouras, P. and Kjeldbye, H. (1993). Retinal pigment epithelial transplants and retinal function in RCS rats. *Investigative Ophthalmology and Visual Science*, **34**, 3068–3075.

Young, M.J. and Lund, R.D. (1994). The anatomical substrates subserving the pupillary light reflex in rats: origin of the consensual response. *Neuroscience*, **62**, 481–496.

Young, M.J., Rao, K. and Lund, R.D. (1989). Integrity of the blood-brain barrier in retinal xenografts is correlated with the immunological status of the host. *Journal of Comparative Neurology*, **283**, 107–117.

Young, M.J., Klassen, H.J. and Lund, R.D. (1991). Effect of lesions of the olivary pretectal nucleus on the direct and consensual pupillary light reflexes. *Society for Neuroscience Abstracts*, **17**, 558.

Young, M.J., Vidal-Sanz, M., Peinado, P. and Lund, R.D. (1994). A subpopulation of retinal ganglion cells that project to the rodent pretectum: cell size, type, and distribution. *Society for Neuroscience Abstracts*, **20**, 771.

Young, R.S.L., Han, B.C. and Wu, P.Y. (1993). Transient and sustained components of the pupillary response evoked by luminance and color. *Vision Research*, **33**, 437–446.

8 Corneal Transplantation — Basic Science and Clinical Management

Mark J. Elder

Department of Ophthalmology, Christchurch Hospital, Private Bag 4710, Christchurch, New Zealand

Corneal grafting has a slightly worse long-term outcome than solid organ transplantation dispite modern immunosuppressive drugs. Some of this result is due to co-morbidity present in the eye around the time of surgery and some due to the vast range of risk factors that influence the outcome. The clinical aspects of corneal grafting and the risk factors and outcomes associated with the surgery are reviewed along with the pathophysiology and immunology of graft rejection, and the clinical aspects of preventing and managing graft rejection. Future interventions are also discussed, focussing on the minor histocompatibility antigens and means of providing immunosuppression for these antigens and for antigens presented indirectly.

KEY WORDS: corneal graft; histocompatibility.

INTRODUCTION

Transplanted human tissue is nowadays very successful. For solid vascular organs such as kidneys and livers, the five year survival is typically 77% and 82% respectively (Landais *et al.*, 1993; Kilpe, Krakauer and Wren, 1993). Due to a complex interplay of many factors the five year survival of the cornea is similar at 72%, despite being "'immunologically privileged" (Williams *et al.*, 1993). To further understand the indications for grafting, the basis of disease and the clinical aspects of management, this treatise is divided into three broad sections. The first deals with the clinical aspects of corneal grafting and reviews the risk factors and outcomes associated with the surgery. The second surveys the pathophysiology and immunology of graft rejection. The third examines the clinical aspects of preventing and managing graft rejection. Finally, new and potential means of improving the outcome of surgery are discussed.

Correspondence: Department of Ophthalmology, Christchurch Hospital, Private Bag 4710, Christchurch, New Zealand; Tel: +64 3 364 0975; Fax: +64 3 364 0273.

TABLE 8.1
Indications for corneal grafting

Indication	Australia* 1985–92	UK** 1970–75	UK** 1985–90
Keratoconus	30%	9%	17%
Bullous keratopathy	25%	4%	8%
Failed previous graft	18%	31%	41%
Scars and opacities	11%	31%	18%
Corneal dystrophies	7%	10%	10%

* combined results from 189 surgeons (Williams *et al.*, 1993; Australian Corneal Graft Registry)
** results from a tertiary referral centre only (Sharif and Casey, 1993)

INDICATIONS FOR CORNEAL GRAFTING

There are a number of broad indications for corneal transplantation. These include; reduced visual acuity due to corneal scarring, reduced visual acuity due to irregular astigmatism (ie. keratoconus), tectonic for an impending or actual perforation, pain relief such as in bullous keratopathy and for cosmetic reasons. The proportions of patients within these subgroups is: to improve visual acuity (78%), pain (6%), tectonic (2%), cosmesis (<1%) and multiple indications (13%) (Williams *et al.*, 1995).

   In a general setting keratoconus is the commonest indication for corneal grafting (Table 8.1) (Williams *et al.*, 1993). In contrast, the commonest indications in a specialist unit in the United Kingdom has been grafting corneal scarring and for repeat grafts. This reflects tertiary referral patterns. Over twenty years, the indications for surgery have changed very little except for a reduction in the number of procedures for herpetic scarring (Table 8.1).

THE CLINICAL OUTCOMES OF CORNEAL GRAFTING

The Australian Corneal Graft Registry of 4499 grafts shows that overall graft survival is achieved in 91% at one year and in 72% at five years (Williams *et al.*, 1993). Using visual acuity as a parameter of success, 79% of eyes improved by one or more Snellen lines of vision. Of the 20% with a poor visual outcome, 50% had a failed graft and 40% had an unrelated ocular problem that caused the reduced visual acuity (Figure 8.1) (Williams *et al.*, 1993).

   Another parameter of a successful outcome is patient satisfaction. Of a cohort of 60 patients, 75% were "satisfied" with the outcome of surgery even though only 65% acheived 6/18 or better postoperatively. Satisfaction was associated with achieving a better visual acuity than the other eye, a clear graft, an improved ability to work or to enjoy

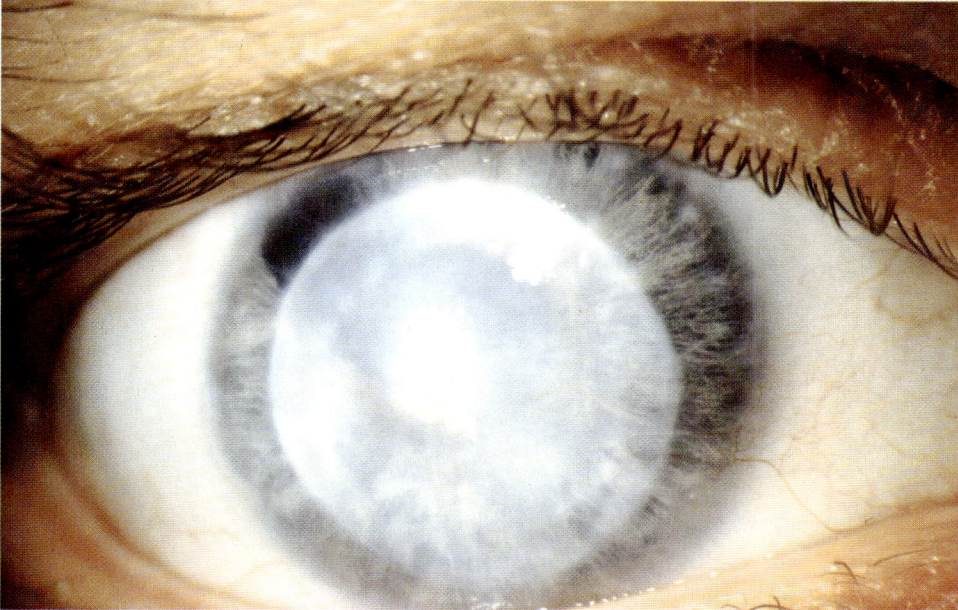

Figure 8.1   An opaque failed corneal graft.

leisure time, a lack of graft failure and not having to wear contact lenses (Williams *et al.*, 1991). The majority of patients who had grafting for pain relief or for tectonic reasons were also satisfied. Astigmatism *per se* was not associated with dissatisfaction despite a mean refractive error of five dioptres in this study.

THE REJECTION RESPONSE: THEORETICAL AND LABORATORY EVIDENCE

The cornea has long been termed "immunologically privileged" with regards to transplantation. During the 1940s, it was shown that if foreign tissue was transplanted to the cornea or anterior chamber, rejection would not occur whereas if a cornea was transplanted to the skin, a brisk rejection reaction ensued (Medawar, 1948). This illustrates that the cornea per se has antigenic properties like other tissues but that the anterior segment of the eye prevents the immune-recognition necessary for rejection. This is in keeping with other immunologically privileged sites such as the brain, the testis and the hamster cheek pouch all of which have minimal or no lymphatic drainage (Katami, 1995).

Rejection of solid vascular organs is divided into hyperacute, acute and chronic. Hyperacute rejection occurs in minutes-hours due to preformed antibodies that activate complement and cause ischaemia. This response requires prior exposure of the antigen

TABLE 8.2
Batchelor's suggested update of the transplantation laws (1995)

1. In previously non-immunised recipients, the high frequencies of T cells capable of responding (by the direct pathway of sensitisation) to MHC incompatible dendritic cells results in these cells becoming the dominant population mediating acute allograft rejection, providing that significant numbers of allogenic dendritic cells are present in the target tissue.

2. Once the passenger (allogenic) dendritic cells migrate out of the allograft, direct pathway T cells entering the graft are rendered anergic, presumably because their antigen receptor binds to alloantigen but no 'second signal' is provided. This is a major factor in the development of operational tolerance or near tolerance to organ allografts.

3. Chronic or late rejection has different clinical and pathological features to early, acute rejection, and it is predicted that the dominant T cell participating in this form of rejection are those sensitised by the indirect pathway.

4. Matching the MHC type of donor to recipient would only be expected to improve graft survival in those circumstances where graft destruction is mediated by T cells of the direct pathway. We have yet to determine the rule governing differences in the responsiveness of self-MHC-restricted T cells sensitised by the indirect pathway to peptides derived from allogeneic tissues.

5. The well know variation in susceptibility of different tissues and organs of rejection is predicted to be due to the target tissues' content of indigenous dendritic cells; this is this the most influential factor that determines the contribution made by the T cells of the direct pathway to the rejection response.

(Batchelor, 1995)

such as previous transplantation. Acute rejection occurs in the first few months, is mainly due to direct antigen presentation and is typically responsive to conventional immuno-suppression (see below). Chronic rejection is very different. It occurs months or years later, is typically unresponsive to current immunosuppression and results in organ dysfunction associated with vasculopathy and fibrosis (Friend, 1995). The predominate mechanism of corneal rejection is cell-mediated killing effected via interleukin-2 (IL-2) stimulation of CD-8+ cytotoxic killer T lymphocytes.

These phenomena of immunological rejection are governed by the "laws of transplantation". These state that the observations ascribed to "rejection" should not occur in genetically identical host and donors and that the amount and speed of rejection should be a function of the genetic differences between the tissues. These laws have been recently added to (Table 8.2) (Batchelor, 1995).

In general terms, foreign tissue can be affected by cell-mediated killing, antibody-mediated damage and inflammatory damage (Figure 8.2). All three systems are very dependent on the presentation of antigen which activates specific T helper cells, the majority of which are CD-4+ cells.

HUMORAL RESPONSE IN CORNEAL GRAFTING

Antibody-mediated damage is effected by IL-4, -5 and -6 stimulation of B cells which further process the antigen before proliferating and making memory cells and antibodies.

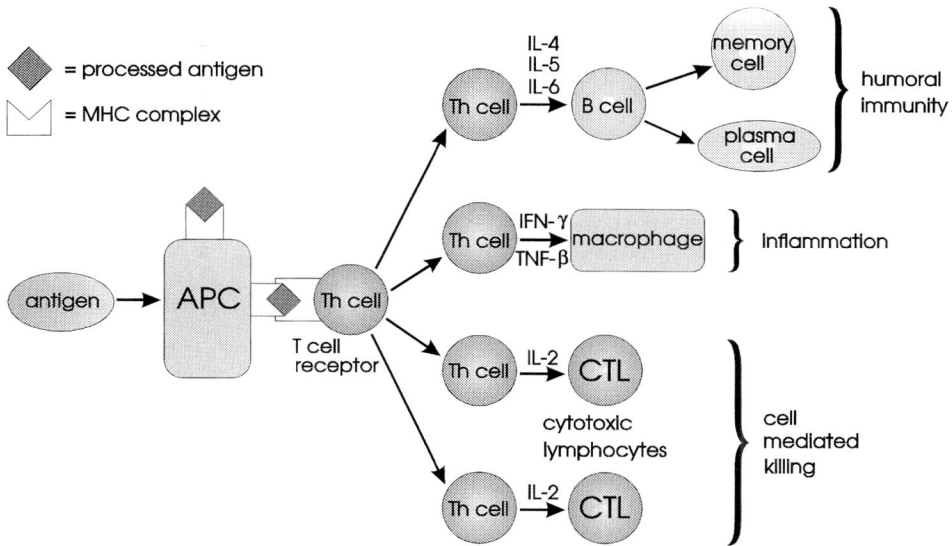

Figure 8.2 The presentation of antigen and the possible sequelae of cell-mediated, humoral and inflammatory responses. APC = antigen presenting cell; IFN-γ =interferon γ; IL-2, (-4, -5, -6) = interleukin 2, (4, 5, 6); MHC = major histocompatibility complex; Th cell = T helper cell; TNF-β = tumour necrosis factor β.

In non-ocular transplants, antibodies may be involved in hyperacute rejection due to complement activation, acute rejection or chronic rejection. For corneal grafts, there is evidence that the humoral response is minimal, particularly as the rejection reaction typically occurs before antibody production (Hutchinson, Alam and Ayliffe, 1995). However, there remains a possibility that chronic endothelial rejection may still be influenced by the humoral response (Hutchinson, Alam and Ayliffe, 1995). This aspect remains to be characterised.

INFLAMMATORY DAMAGE AND THE ROLE OF THE MACROPHAGES

Macrophages are numerous in rejecting corneal grafts (Holland *et al.*, 1991; van der Veen *et al.*, 1994). Further, the antigen presenting cell/helper cell complex is able to activate macrophages, primarily by interferon-γ and tumour necrosis factor-β. Recently, van der Veen *et al.* (1994) assessed a highly selective method of macrophage depletion by the use of liposomes containing dichloromethylene diphosphonate (Cl$_2$MDP). Compared with controls, there was a marked reduction in the rejection rate (17 *vs* >100 days) and significantly less graft vascularisation. These results are as beneficial as any immunosuppressive agents used in the rat. The exact reasons for this outcome are unclear, although there was also a reduction in graft T cells suggesting that the macrophages play a significant role in antigen presentation. Activated macrophages are also known to induce vascularisation and this is also implicated as another possible mechanism (Hunt *et al.*, 1983). Further, previous studies implicating the importance of CD-4+ T cells in rejection may have overlooked the fact that macrophages can also express CD-4 (Jefferies, Green and Williams, 1985).

ANTIGEN PRESENTING CELLS

For foreign tissue to be rejected, there must first be recognition of the foreign proteins. This process requires a heirachy of cells which present these proteins in such a manner that they are recognised by, and stimulate the appropriate efferent limb of the immune response. The dendritic cell is the main antigen presenting cell (APC) and these cells are constantly extending cytoplasmic processes to "sample the environment" (Roake, 1995). These sentinel cells phagocytose and process protein and then reexpress this on the cell surface in association with specific Class II human leucocyte antigens (HLA). These complexes are then recognised by one of 10^8 types of receptors found on specific T cells. This variation contributes to the selectivity and specificity of the immune system (Engelhard, 1994). However, foreign antigen will only be recognised by the T cells when it is presented on the cell surface as a processed peptide in association with a major histocompatibility complex molecule (MHC). Antigen alone elicites no immune response. Both Class I and II MHC molecules are involved in antigen presentation, each in slightly different ways. Once activated, the T cells proliferate and stimulate the efferent limb of the rejection response mainly by secreting IL-2 which in turn stimulates cytotoxic T cells (CTL) which destroy the appropriate tissue.

DIRECT AND INDIRECT PRESENTATION OF ANTIGEN

Antigen presentation to T cells may be classified as 'direct' or 'indirect'; the so-called dual antigen presentation theory of Lechler and Batchelor (1982) (Figure 8.3). Direct antigen presentation is where the *donor* antigen presenting cells co-present the processed foreign proteins along with HLA antigens on the cell surface. This allows specific, quiescent, host T cells to become activated, proliferate and form a population of allospecific effector cells that are not self-MHC-restricted (Williams and Coster, 1989; Batchelor, 1995). Therefore, if transplanted tissue is matched for HLA antigens, then the presentation of antigen is severely blunted and rejection is less common.

Indirect antigen presentation is where the foreign proteins are ingested and processed by the *host* APC's whereafter an allospecific self-MHC-restricted effector line of cells are generated. This response is specific to the paired presentation of foreign protein and host MHC Class II typing. Therefore, if the host and donor share the same MHC type (ie. with HLA matching) then donor rejection may be enhanced (Williams and Coster, 1989; Batchelor, 1995). Direct presentation broadly corresponds with early graft rejection whereas indirect presentation corresponds with late rejection.

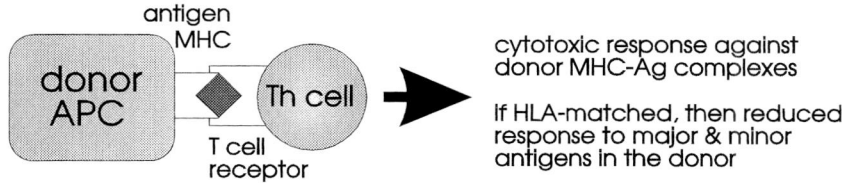

Direct antigen presentation: Acute rejection (weeks–months)

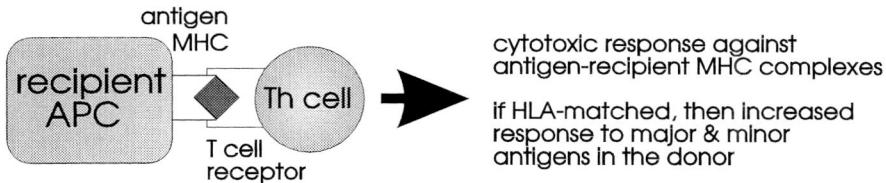

Indirect antigen presentation: Chronic rejection (months–years)

Figure 8.3   Antigen may be presented directly by donor antigen presenting cells (APC) and indirectly by recipient APC. MHC = major histocompatibility complex; Th cell = T helper cell.

THE ROLE OF LANGERHANS CELLS IN GRAFT REJECTION

The predominant APC in the cornea and limbus is the Langerhans cell which is a type of dendritic cell. These cells are almost absent in the central cornea and are most numerous in the peripheral corneal epithelium (15–20 cells/mm^2). This compares with 250–500 cells/mm^2 in the conjunctiva and skin (Rodrigues *et al.*, 1981). During inflammation or rejection the number of Langerhans cells in the cornea increases dramatically (Williams, Ash and Coster, 1985; Pepose *et al.*, 1985; Neiderkorn, Peler and Mellon, 1989).

Many investigators have considered that the number of these APC's in the donor cornea are a key factor in graft rejection (Gillette, Chandler and Greiner, 1982; Williams and Coster, 1989; Armitage, 1995). In the rat model of corneal rejection, transplantation of small areas of central cornea, which has comparatively few Langerhans cells, results in minimal rejection compared with transplantation of similar areas of peripheral cornea which has many more Langerhans cells (Katami, 1995). Depleting the corneal population of Langerhans cells by ultraviolet radiation also markedly reduces the rate of rejection in animals (Katami, 1995). In humans the majority of Langerhans cells are in the epithelium and one clinical trial demonstrated that removal of the donor epithelium just before surgery reduces the rejection rate (Tuberville, Foster and Wood, 1983; Armitage, 1995). Storage of corneas in organ culture media selectively depletes the Langerhans cells but there is no clinical difference in rejection rates compared with conventional media (Moll *et al.*, 1991).

The origin of the increased numbers of Langerhans cells in rejecting corneas is not well determined although they are probably host-derived (Pepose *et al.*, 1985). Indirect presentation of antigen via host APC's would be enhanced with graft tissue matched for HLA Class II antigens. However, the clinical evidence to support this is unclear although the Corneal Transplant Follow-up Study data implies that indirect presentation is possible (Vail *et al.*, 1994). This area requires further clarification and developments are awaited.

HISTOCOMPATIBILITY ANTIGENS

There are a variety of antigens expressed by cells and tissues that allow the determination of self and non-self. The histocompatibility antigens are composed of the MHCs and the minor complexes and these are part of the afferent limb of immune surveillance.

THE HUMAN LEUCOCYTE ANTIGEN SYSTEM

The MHC system in humans is known as the human leucocyte antigen system (HLA) and consists of a family of genes expressed at more than ten loci on the short arm of chromosome six (Taylor and Dyer, 1995). The various HLA antigens are subdivided into Class I (HLA-A, -B, -C) and Class II (HLA-D). There are many antigens in the HLA system and at present serology can define 23 HLA-A, 48 HLA-B, 9 HLA-C and 25 HLA-D antigens (Taylor and Dyer, 1995). The HLA genes are inherited in a co-dominant manner such that for each subgroup of antigens, a particular patient will express two antigens unless they are homozygous.

CLASS I HLA ANTIGENS

The main role of Class I antigens is in the presentation of cytoplasmic peptides to HLA class I restricted T cells (Townsend *et al.*, 1989). This is primarily a function of the "classical transplantation antigens" of HLA-A and HLA-B. HLA-C has very weak levels of expression compared to HLA-A and HLA-B and essentially plays no role in transplantation rejection (Bunce and Welch, 1994). Recently, other Class I antigens have been described (HLA-E, -F, -G, -H, -I, -J) but their role in transplantation is unknown (Bodmer *et al.*, 1992). Class I antigens HLA-A, -B and -C are expressed constituently by the human corneal epithelium, stroma and endothelium (Fujikawa *et al.*, 1982; Treseler, Foulks and Sanfilippo, 1984; Whitsett and Stulting, 1984; Young, Stark and Prendegast, 1985).

CLASS II HLA ANTIGENS

These antigens consist of HLA-DR, -DQ and -DP and these, together with bound antigen are selectively recognised by specific Class II restricted CD-4+ T cells (Nuchtern, Beddison and Klausner, 1990). Class II antigens are typically constitutively expressed on B-cells, monocytes, macrophages and dendritic cells. During inflammation there is upregulation of the Class II antigens by these cells and by T cells (Daar *et al.*, 1984).

Class II antigens are found on the Langerhans cells within the normal corneal epithelium and stroma but not on the endothelial cells (Fujikawa *et al.*, 1982; Treseler, Foulks and Sanfilippo, 1984; Whitsett and Stulting, 1984; Young, Stark and Prendegast, 1985). The corneal endothelium can express Class II antigens during graft rejection and after specific stimulation *in vivo* (Donnelly *et al.*, 1985; Young, Stark and Prendegast, 1985).

THE IMPORTANCE OF HLA MATCHING IN NON-OCULAR TISSUE TRANSPLANTS

The effects of HLA antigen matching for vascular organ transplants such as kidneys and livers is complex and depends on the tissue, whether it is a first or second transplant, the degree and type of HLA matching and the types and amounts of immunosuppression used. Despite this, the effects are small. For primary renal transplantation and an excellent HLA match, there is approximately a 10% increase in survival at one year compared to a total mismatch (Opelz *et al.*, 1991). It remains axiomatic that if graft survival is already very good without HLA matching, then matching can at best, have a small effect. Therefore any effect should be most noticeable for high risk grafts. For repeated renal transplants, HLA matching improves the one year survival by 14–20% (Opelz, 1989). This compares to an improvement in survival from 10% to 77% over the last fifteen years which is mainly ascribed to better and more selective immunosuppressive drugs such as cyclosporine A (Kilpe, Krakauer and Wren, 1993).

Ultimate graft survival has traditionally been the main assessment parameter but this may belie other clinical benefits. For example, for renal transplants receiving cyclosporin, azathioprine and prednisolone, there is no demonstrable effect of HLA-DR matching on graft failure (Taylor *et al.*, 1989). However, the matched group have less episodes of rejection and require less immunosuppression. This may also be true for the cornea (Taylor and Dyer, 1995).

LABORATORY TESTING OF HLA ANTIGENS FOR CLINICAL USE

If transplanted tissue needs to be HLA matched to reduce the risk of rejection, then both the donor and the recipient need to be accurately typed. This has predominately been achieved serologically by extracting lymphocytes from the blood and assaying them with various reference sera. The recipient blood can always be obtained "fresh" but for corneal harvesting, the blood obtained post-mortem is prone to inaccurate typing. A more sensitive and more specific method than serology is to amplify and then detect the appropriate HLA gene sequences using the polymerase chain reaction (PCR) (Olerup and Zetterquest, 1992). These improved techniques have shown that there is a 10–25% difference between PCR and the serological methods (Middleton *et al.*, 1988; Mytilineos, Scherer and Opelz, 1990).

Therefore, it behoves the clinician to be critical of all papers that investigate HLA antigens. Taylor and Dyer (1995) make the point that in the Collaborative Corneal Transplant Study (1992) (CCTS), too many of the HLA antigens had only one allele per class. Due to the co-dominant inheritance of HLA antigens, most individuals should have two antigens per class. Either there were an abnormally large number of homozygous patients in the CCTS study or the laboratory assessment of the antigen status was inadequate (see below).

MINOR HISTOCOMPATIBILITY ANTIGENS

The minor histocompatibility antigens (mHags) are cell surface proteins that have a role in graft rejection although they remain less well understood compared to the major group of HLA antigens (Goulmy, 1988). The mHags are named as HA-1, HA-2 etc with the exception of H-Y which is so named because it is present only on male tissue. HA-1 to -5, and H-Y are constitutively expressed by all human haemopoetic cells and HA-3 and H-Y are constitutively expressed by almost all cells (Wallny and Rammensee, 1990). The male specific H-Y antigen seems to be important in renal transplantation but only for a male donor and a female recipient (Goulmy *et al.*, 1977, 1978). In the cornea, H-Y and HA-3 are detectable (Goulmy *et al.*, 1995). Because perfectly HLA matched corneal grafts may still reject, there are clearly other antigens involved of which the mHags are candidates. Further, in humans not all MHC mismatched grafts reject and this suggests that the MHC antigens are not the sole antigens important in graft rejection. In mice, corneal grafting of various combinations of matches of MHC and mHags shows that the minor antigens are at least as important as the MHC antigens for rejection (Sonoda *et al.*, 1995). For mHag to be recognised by T cells, they must be processed by Class II MHC molecules and hence they can be only recognized via the indirect pathway of antigen presentation (Rotzscke *et al.*, 1990). Sonoder *et al.* (1995) have postulated that the dominant mechanism in corneal rejection is the mHags which constitute the majority of corneal antigens and these are presented by the indirect pathway. Recent data suggests that modern high-dose immunosuppression such as cyclosporin is unable to reduce the prevalence of rejection of HLA matched but mHag mismatched grafts (Nicholls, Bradley and Easty, 1995). If this can be confirmed, then alternative strategies will be required to improve the outcome due to mHag antigens.

CLINICAL ASPECTS OF HLA MATCHING FOR CORNEAL GRAFTS

The effects of HLA matching for corneal transplantation parallels that of other organs although the numerous clinical studies have often reached different conclusions (Allansmith, Fine and Payne, 1974; Gibbs *et al.*, 1974; Volker-Dieben *et al.*, 1982; Sanfilippo *et al.*, 1986; Boisjoly *et al.*, 1986, 1990). The two key pieces of work come from the American multicentred Collaborative Corneal Transplant Study (1992) ($n = 419$) which examined high risk grafts and the British Corneal Transplant Follow-up Study (CTFS) ($n = 2311$) (Vail *et al.*, 1994).

HLA CLASS I MATCHING

The CCTS study found no effect of matching Class I antigens. Further analysis of this data by Taylor and Dyer (1995) showed that for the subgroup of completely HLA-DR mismatched grafts, HLA-A and -B did have a significant effect on survival; 69% *vs* 59% survival at three years for complete match *vs* total mismatch of HLA-A and -B. This is consistent with Vail *et al.* (1994) who also found an independent effect for HLA-A and -B.

HLA CLASS II MATCHING

The CCTS study found no difference in graft survival for HLA-DR matching (CCTS, 1992). The study had the power to detect a 15% difference in outcome for the HLA matching. However Taylor and Dyer (1995) reanalysed this data and assessed only those cases with two alleles reported per antigen (see above). These calculations showed that HLA-DR does reduce the amount of rejection but not the amount of graft failure. This parallels the renal results. These data conflict with Vail *et al.* (1994) who showed worse survival with HLA-DR matching and this paradox is not yet resolved. The confounding variable is that early and late rejection may be stimulated by different mechanisms (direct and indirect antigen presentation) for which HLA-DR matching will worsen late, indirect presentation. Unfortunately there is no clinical distinction between the two.

THE EFFECTS OF MATCHING BLOOD GROUPS

An unusual finding of the CCTS group is that blood group matching (ABO) improved graft survival due to a reduced incidence of rejection; 69% *vs* 59% at three years for a complete ABO-match compared to a complete ABO-mismatch (CCTS, 1992). This is not in accord with previous ocular studies (Volker-Dieben *et al.*, 1982; Boisjoly *et al.*, 1986; Volker-Dieben, D'Amaro and Kok-van-Alphen, 1987) but is similar to renal, liver and cardiac transplantation (D'Amaro *et al.*, 1985; CCTS, 1992).

CLINICAL TYPES OF CORNEAL GRAFT REJECTION

In 1969, Khodadoust and Silverstein devised an elegant series of experiments in rabbits where transplants of isolated portions of cornea showed that each of the three corneal layers can independently initiate and maintain a rejection response (Khodadoust and Silverstein, 1969a,b). This was later confirmed in humans by Alldredge and Krachmer (1981). Rejection is clinically observed affecting the endothelium in 21%, the stroma in 15% and in the epithelium in 10% (Table 8.3). The prevalence of occult rejection is unknown primarily because these failed corneas are not biopsied or removed during acute or subacute failure. Overall, overt graft rejection occurs in 29% of human corneal transplants. Common parlance often equates graft rejection with endothelial rejection. This is reasonable as it is the most common and is the only type that leads to irreversible visual loss.

EPITHELIAL REJECTION

Epithelial rejection is manifest clinically as an elevated epithelial line that stains with fluoroscein or rose bengal (Figure 8.4). It typically starts at or near the host-donor junction and traverses across the cornea over days or weeks. When the rejection is complete, the

TABLE 8.3
Prevalence and types of human corneal graft rejection

Tissue	Prevalence	Mean onset	Range
Epithelium	10%	3 months	1–13 months
Stroma	15%	10 months	1.5–21 month
Endothelium	21%	8 months	0.5–29 months
Overall	29%		

(Alldredge and Krachmer, 1981)

donor epithelium will have been completely replaced by the host epithelium. The mean time of rejection is three months but it may occur up to 13 months after surgery (Alldredge and Krachmer, 1981). This implies that the donor epithelium can remain resident for at least this time and there is good evidence that it may survive for at least a year after grafting providing that the donor material was fresh (Khodadoust and Silverstein, 1969a,b; Silverstein, Rossman and Leon, 1970; Hager, Kraus-Mackiw and Schroeder, 1976). Epithelial rejection is relatively easy to halt using moderate doses of topical steroids (Smolin and Goodman, 1988; Hill, 1995).

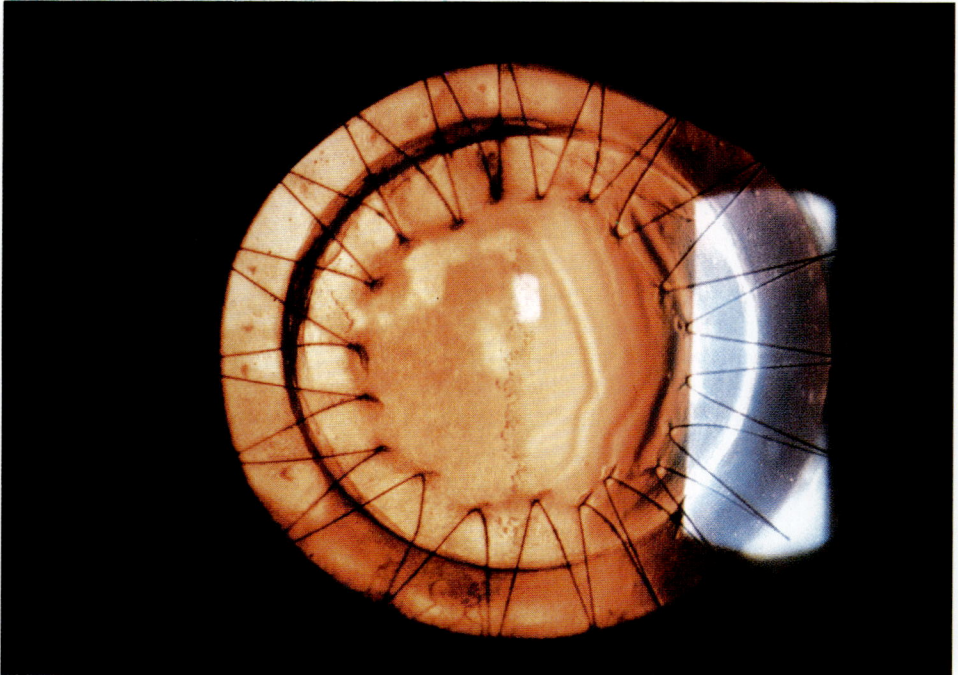

Figure 8.4   Epithelial rejection with a discrete, raised line at the junction of host and donor epithelium.

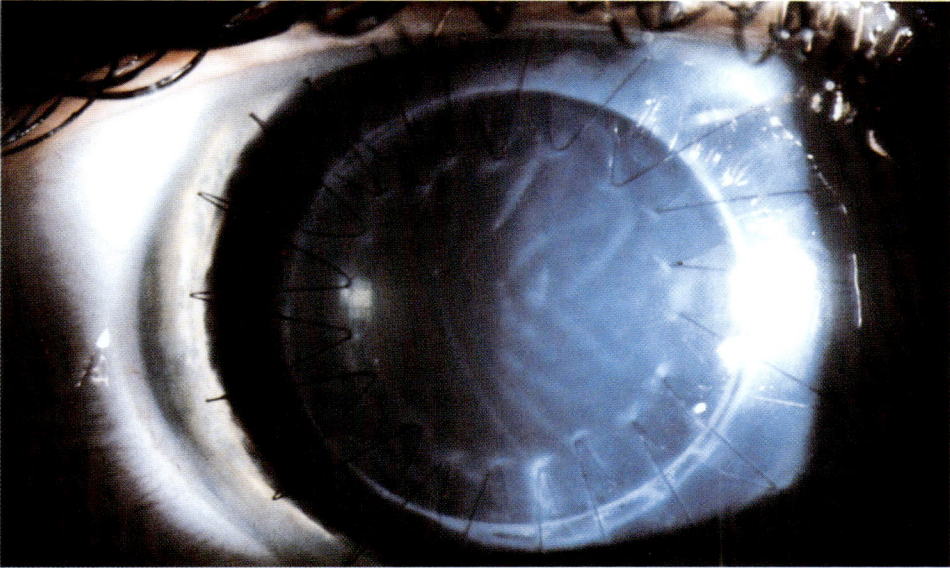

Figure 8.5   Acute endothelial rejection with corneal oedema and folds in Descemet's membrane.

STROMAL REJECTION

Stroma rejection is typically manifest as white subepithelial infiltrates in the donor anterior stroma. They resolve promptly with topical steroids and almost never contribute to graft failure (Alldredge and Krachmer, 1981).

ENDOTHELIAL REJECTION

Endothelial rejection is diagnosed clinically by the presence of keratitic precipitates on the endothelium of the donor cornea and signs of inflammation in the anterior chamber (flare and cells). This is often accompanied by oedema, particularly of the donor stroma. Occasionally, a line of keratitic precipitates is seen along the donor endothelium-Khodadoust's line (Figures 8.5 and 8.6). The line first appears near the donor-host junction and advances across the donor endothelium over days or weeks. This corresponds to a cell mediated attack of host leucocytes on the donor endothelium. This process results in donor endothelial cell death at the site of the line. If left unchecked, 'the donor endothelium is completely replaced and the graft fails permanently. The management of endothelial graft rejection is discussed later.

Patients may get rejection of one or more layers of the cornea although they usually occur almost as independent events and at different times. Endothelial rejection occurs in 72% of all cases that get rejection of any sort. Of those cases that get endothelial rejection, 28% will at some stage have epithelial rejection, 29% will get stromal rejection and 14% will get both epithelial and stromal rejection (Alldredge and Krachmer, 1981).

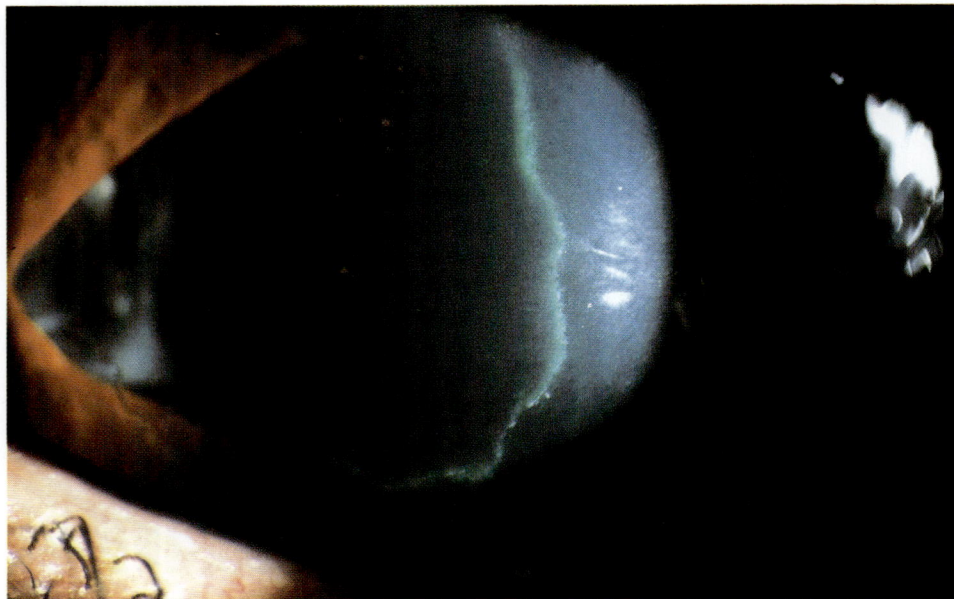

Figure 8.6   Retroillumination of acute endothelial rejection showing the rejection line of Khodadoust spanning from 6–12 o'clock (same case as Figure 8.5).

RISK FACTORS FOR REJECTION AND FAILURE

The outcome of corneal grafting can be assessed by a variety of indices and these predominantly relate to the primary indications for the surgery; visual, tectonic, cosmetic, for pain relief etc. The commonest indication for surgery is to improve the visual performance of the eye and hence graft failure has been defined as "oedema and irremediable loss of clarity in a previously thin, clear graft or as any graft that was replaced by another irrespective of clarity and for whatever reason" (Williams *et al.*, 1992a, 1993).

The main reasons for graft failure are; irreversible rejection (34%), infection (18%), glaucoma (9%), primary failure (6%) and trauma (4%). However, 30% of cases of failure are due to multifactorial causes. The features that are related to a poorer outcome with time are detailed in Table 8.4 (Williams *et al.*, 1992a).

THE CLINICAL MANAGEMENT OF CORNEAL GRAFT REJECTION

Corneal grafts can be clinically separated into three categories; low risk (avascular), medium risk (1–2 quadrant vascularisation) and high risk (3–4 quadrant vascularisation) (Hill, 1989). Other additional factors also contribute to risk (see Table 8.4). Remediable

TABLE 8.4
Risk factors and the chance of corneal graft success (n=4499)

Clinical Feature	Chance of success at 5 years
All grafts combined	72%
Number of grafts	
first	78%
second	46%
third	39%
fourth	23%
Previous graft in the contralateral eye	87% (vs 62%)
Previous blood transfusion	49% (vs 79%)
Previous inflammation	
never	91%
previously	54%
Vascularised	
no vascularisation	79%
3 quadrants	58%
Lens status	
phakic	87%
pseudophakic	61%
aphakic	36%
Keratoconus	97%
Corneal dystrophy	95%
Fuch's corneal dystrophy	76%
Suture removal	
all before 6 months	60%
after 6 months	82–90%
Microbial keratitis herpes simplex	40%
never	74%
previous disease, no recurrence	83%
recurrence post graft	22%
Raised intraocular pressure	
never	80%
only before grafting	37%
after grafting	45%
Previous rejection episode	
never	81%
1 or more	45%

(Williams *et al.*, 1992a)

clinical failure is often due to endothelial rejection and this must be adequately prevented and appropriately managed when it occurs.

PROPHYLAXIS OF ENDOTHELIAL GRAFT REJECTION

Virtually all clinicians use topical steroids during the early postoperative period even if the graft is "low risk". Typically, topical dexamethasone is administered 4 times/day for low risk cases and 7 times/day for high risk cases (Rinne and Stulting, 1992). The CCTS study (1992) had excellent results for high risk patients and this has been ascribed to the intensive topical steroids and good follow up. While topical steroids are undoubtably very effective and have virtually no systemic morbidity, they are often insufficient for prophylaxis in very high risk cases. The addition of systemic steroids (25 mg/day for a month, then 10 mg/day) does not provide any extra prophylaxis, nor does topical cyclosporin (Hill, 1989, 1995). However systemic cyclosporin at 3–4 mg/kg/day does reduce the risk of rejection (Hill, 1986, 1994; Miller *et al.*, 1988). A better outcome was achieved with 12 months cyclosporin compared with 4–6 months (Hill, 1994). Hill (1995) recommends this regime for "high-risk" cases that have bilateral disease or only one eye.

MANAGEMENT OF ESTABLISHED ENDOTHELIAL REJECTION

During endothelial rejection the non-replicating donor endothelium is destroyed by cytotoxic lymphocytes. If sufficient cells are destroyed, then the graft fails (Brooks, Grant and Gillies, 1989). Therefore endothelial rejection requires urgent, optimal management to minimise graft failure and this requires that the patients present to an ophthalmologist promptly if there is any suspicion of rejection. Clinically, rejection is classified as "*definite*" (corneal oedema and a Khodadoust rejection line in a previously clear graft), "*probable*" (corneal oedema and signs of corneal or anterior chamber inflammation without a rejection line in a previously clear graft), "*possible*" (corneal oedema with no signs of inflammation) (Stulting *et al.*, 1988).

The commonest management for endothelial rejection is to give intensive topical steroids alone, typically hourly dexamethasone (Larkin, 1994). However much more effective is an adjunctive single injection of intravenous methylprednisolone 125–500 mg (Hill, Maske and Watson, 1991; Hill and Ivey, 1994). This single injection is at least as effective as a two week course of oral steroids at 60–80 mg/day. A second pulse of intravenous methyprednisolone 1–2 days later confirs no extra benefit. It is possible that this regime may reduce the risk of further rejection episodes. Patients who are taking oral cyclosporin at the time of a rejection episode are also more likely to experience reversal of the event (Hill, 1995).

MANAGEMENT OF OTHER RISK FACTORS

Viral infection, mainly herpetic disease, is a significant cause of a poor outcome. This could be improved by avoiding surgery during active infection and preventing recurrence of the infection after surgery. This may be achieved with oral or topical antiviral agents such as acyclovir (Foster and Barnes, 1992; Moyes *et al.*, 1994). For patients with previous herpes simplex keratitis, topical acyclovir 4–5 times/day for one year postoperatively significantly reduces the incidence of herpetic recurrence, graft rejection and the risk of failure (Moyes *et al.*, 1994).

FUTURE TREATMENTS

Antibodies are potentially able to selectively reduce both immune-recognition and its subsequent response and this has had some benefits in solid organ transplants. In rabbits, intravenous administration of antibodies non-specifically directed against all T cells partially prevents corneal rejection (Smolin, 1969a; Waltman, Faulkner and Burde, 1969).

Unfortunately these antibodies are ineffective at preventing rejection when used topically or subconjunctivally (Smolin, 1969b; Polack, Townsend and Waltman, 1972). Intriguingly, intravenous administration of anti-CD-4 antibodies significantly prevents graft rejection although anti-CD-8 antibodies do not (He, Ross and Niederkorn, 1991). This implies that CD-8+ cells are not the sole or main cause of rejection. Topically applied anti-CD-4 antibodies contained within liposomes also reduces the amount of corneal graft rejection in the rat (Pleyer et al., 1995). The effects on modulation of the T cells have been mainly studied on the prevention of rejection although acute rejection can be reversed in rabbits using anti-T cell or anti-myeloid cell antibodies but not with anti-HLA Class II antigens (Williams et al., 1992b).

Leucocyte adhesion to various tissue components is greatly enhanced by the presence of specialised adhesion molecules. These include the intercellular adhesion molecules (ICAM), the vascular adhesion molecules (VCAM) and the neural adhesion molecules (NCAM). NCAM is constitutively expressed by normal corneal endothelium whereas VCAM and ICAM are upregulated in graft rejection and inflammation (Foets et al., 1992; Philipp and Gottinger, 1993; Whitcup et al., 1993; Claesson, Larkin and Elder, 1995). During rejection, endothelial damage is primarily lymphocyte mediated and therefore modulation of the adhesion molecules could theoretically reduce the prevalence of rejection and/or reduce its severity. Pavilack and Chaves (1994) have shown that using antibodies to ICAM-1, or its ligand LFA-1 leads to improved graft viability in high risk rat corneal grafts. Future therapies may prevent the upregulation of the adhesion molecules or impair their interaction with leucocytes.

Gene therapy is the concept that therapeutically useful genetic material may be able to be inserted into specific cell lines and thereby effect the behaviour of that cell. The genes are typically inserted into the cell using retrovirus vectors or adenovirus vectors which themselves have been rendered "harmless". Cell specificity is achieved by either isolating the tissue during the inoculation or by using a cell-specific vector. The nucleotide sequence of many proteins is now known and the synthesis of specific sequences of DNA is comparatively easily with recombinant techniques. Therefore the possibilities of therapeutic intervention are almost endless and these have widespread implications to all tissues and diseases (Mulligan, 1993).

Human corneal epithelium has been successfully transfected with T antigen (ts58 SV40) and in this instance, the cells retain their normal epithelial characteristics but become immortalised (Okamoto et al., 1995). Murine corneal endothelium has been successfully transfected with a replication-deficient adenovirus and the associated marker was expressed for two weeks during which no toxicity was apparent (Budenz et al., 1995). At present, there are still formidable problems in ensuring safety and specificity. Further, the adenovirus vectors provide only transient expression of the gene products but these early works suggest that gene therapy is a viable therapeutic possibility in corneal disease.

CONCLUSIONS

Corneal grafting has a slightly worse long-term outcome than solid organ transplantation dispite modern immunosuppressive drugs. Some of this result is due to co-morbidity present in the eye around the time of surgery and some due to the vast range of risk factors that influence the outcome. In this regard, the eye is different from a kidney in that the entire globe must be working well for good vision to occur whereas a kidney functions 'alone'. Low risk grafts such as keratoconus and corneal dystrophies have an excellent outcome of 97% at 5 years. This contrasts with the very poor outcome of repeated grafts or very vascular corneas. The CCTS (1992) ultimately concluded that for high risk corneal grafts, the best outcome would be achieved with "high dose postoperative steroids therapy, good compliance and close-follow up". Their steroid regime was initially two hourly, reducing to 4 times/day for weeks 2–8 and reducing to once a day at 16 weeks. There is also a role for oral cyclosporin for very high risk cases.

HLA matching for Class I antigens is beneficial but matching for Class II antigens is of no proven benefit and may even be deleterious. ABO matching has a clinically significant effect and is easy and cost effective compared with HLA antigen matching. Future interventions will focus on the minor histocompatibility antigens and means of providing immunosuppression for these antigens and for antigens presented indirectly.

REFERENCES

Allansmith, M.R., Fine, M. and Payne, R. (1974). Histocompatibility typing and corneal transplantation. *Transactions of the American Academy of Ophthalmology and Otology*, **78**, 445–460.

Alldredge O.C. and Krachmer, J.H. (1981). Clinical types of corneal transplant rejection. Their manifestations, frequency, preoperative correlates and treatment. *Archives of Ophthalmology*, **99**, 599–604.

Armitage, W.J. (1995). The effects of storage of corneal tissue on Langerhans cells. *Eye*, **9**, 228–232.

Batchelor, J.R. (1995). The laws of transplantation: A modern perspective. *Eye*, **9**, 152–154.

Bodmer, J.G., Marsh, S.G., Albert, E.D., Bodmer, F., Dupont, B., Erlich, H.A., Mach, B., Mayr, W.R., Parham, P. and Sasazuki, T. (1992). Nomenclature for factors of the HLA system, 1991. *Tissue Antigens*, **39**, 161–173.

Boisjoly, H.M., Roy, R., Dube, I., Laughrea, P.A., Michaud, R., Douville, P. and Heebert, J. (1986). HLA-A, -B and -DR matching in corneal transplantation. *Ophthalmology*, **93**, 1290–1297.

Boisjoly, H.M., Roy, R., Bernard, P.M., Dube, I., Laughrea, P. and Bazin, R. (1990). Association between corneal allograft reactions and HLA compatibility. *Ophthalmology*, **97**, 1689–1698.

Brooks, A.M., Grant, G. and Gillies W.E. (1989). Assessment of the corneal endothelium following kerato-plasty. *Australian and New Zealand Journal of Ophthalmology*, **17**, 379–385.

Budenz, D.L., Bennett, J., Alonso, L. and Maguire, A. (1995). *In vivo* transfer into murine trabecular meshwork and corneal endothelial cells. *Investigative Ophthalmology and Visual Science (Suppl)*, **36**, 846.

Bunce, M. and Welch, K.I. (1994). Rapid DNA typing for HLA-C using sequence-specific primers (PCR-SSP): identification of serological and non-serologically defined HLA-A alleles including several new alleles. *Tissue Antigens*, **43**, 7–17.

Claesson, M., Larkin, D.F. and Elder, M. (1995). Expression of adhesion and MHC molecules by human corneal endothelium on a flat mount preparation. *Investigative Ophthalmology and Visual Science (Suppl)*, **35**, 301.

Collaborative Corneal Transplantation Studies Research Group. (1992). The collaborative corneal transplantation studies (CCTS). Effectiveness of histocompatibility matching in high-risk corneal transplantation. *Archives of Ophthalmology*, **110**, 1392–1403.

D'Amaro, J., Hendricks, G.F., Persijin, G.G. and van Rood, J.J. (1985). The influence of sex and ABO blood group on HLA-A, -B and DR matching in renal transplantation. *Transplantation Proceedings*, **17**, 759–760.

Daar, A.S., Fuggle, S.V., Fabre, J.W., Ting, A. and Morris, P.J. (1984). The detailed distribution of MHC Class II antigens in normal human organs. *Transplantation*, **38**, 293–298.

Donnelly, J.J., Li, W., Rockey, J.H. and Prendergast, R.A. (1985). Induction of Class II (Ia) alloantigen expression on corneal endothelium *in vivo* and *in vitro*. *Investigative Ophthalmology and Visual Science*, **26**, 575–580.

Engelhard, V.H. (1994). How cells process antigens. *Scientific American*, **271**, 44–51.

Foets, B.J., van den Oord, J.J., Volpes, R. and Missotten, L. (1992). *In situ* immunohistochemistry of cell adhesion molecules on human corneal endothelial cells. *British Journal of Ophthalmology*, **76**, 205–9.

Foster, C.S. and Barnes, N.P. (1992). Systemic acyclovir and penetrating keratoplasty for herpes simplex keratitis. *Documenta Ophthalmologica*, **80**, 363–369.

Friend, P.J. (1995). Rejection reactions to different organ transplants. *Eye*, **9**, 190–191.

Fujikawa, L.S., Colvin, R.B., Bhan, A.K., Fuller, T.C. and Foster, C.S. (1982). Expression of HLA-A/B/C and-DR locus antigens on epithelial, stromal and endothelial cells of the human cornea. *Cornea*, **1**, 213–22.

Gibbs, D.C., Batchelor, J.R., Werb, A., Schlesinger, W. and Casey, T.A. (1974). The influence of tissue-type compatibility on the fate of full thickness corneal grafts. *Transactions of the Ophthalmological Societies of the UK*, **94**, 101–126.

Gillette, T.E., Chandler, J.W. and Greiner, J.V. (1982). Langerhans cells of the ocular surface. *Ophthalmology*, **89**, 700–710.

Goulmy, E. (1988). Minor histocompatibility antigens in man and their role in transplantation. *Transplantation Reviews*, **2**, 29–53.

Goulmy, E., Termijtelen, A., Bradley, B.A. and Van Rood, J.J. (1977). Y-antigen killing by T cells of women is restricted by HLA. *Nature*, **266**, 544–545.

Goulmy, E., Bradley, B.A., Lansbergen, Q. and Van Rood, J.J. (1978). The importance of H-Y incompatibility in human organ transplantation. *Transplantation*, **25**, 315–319.

Goulmy, E., Pool, J., van Lochem, E. and Volker-Dieben, H. (1995). The role of human minor histocompatibility antigens in graft failure: a mini review. *Eye*, **9**, 180–184.

Hager, H.D., Kraus-Mackiw, E. and Schroeder, T.M. (1976). Sex chromatin in a corneal graft in adult cystinosis. *Klinische Monatsblatter für Augenheilkunde*, **169**, 234–239.

He, Y.G., Ross, J. and Niederkorn, J.Y. (1991). Promotion of murine orthotopic corneal allograft survival by systemic administration of an anti-CD4 monclonal antibody. *Investigative Ophthalmology and Visual Science*, **32**, 2723–2728.

Hill, J.C. (1986). The use of systemic cyclosporin in human corneal transplantation: a preliminary report. *Documenta Ophthalmologica*, **62**, 337–344.

Hill, J.C. (1989). The use of cyclosporin in high-risk keratoplasty. *American Journal of Ophthalmology*, **107**, 506–510.

Hill, J.C. (1994). Systemic cyclosporin in high-risk keratoplasty: short versus long-term therapy. *Ophthalmology*, **101**, 128–133.

Hill, J.C. (1995). Immunosuppression in corneal transplantation. *Eye*, **9**, 247–253.

Hill, J.C. and Ivey, A. (1994). Corticosteroids in corneal graft rejection: double versus single pulse therapy. *Cornea*, **13**, 383–384.

Hill, J.C., Maske, R. and Watson, P.G. (1991). The use of a single pulse of intravenous methylprednisolone in the treatment of corneal graft rejection: a preliminary report. *Eye*, **5**, 420–424.

Holland, E.J., Cha, C.C., Wetzig, R.P., Palestine, A.G. and Nussenblatt, R.B. (1991). Clinical and immunohistochemical studies of corneal rejection in the rat penetrating keratoplasty model. *Cornea*, **10**, 374–380.

Hunt, T.K., Knighton, D.R., Thakral, K.K., Goodson, W.H. and Andrews, W.S. (1983). Studies on inflammation and wound healing: Angiogenesis and collagen synthesis stimulated *in vivo* by resident and activated wound macrophages. *Surgery*, **96**, 48–54.

Hutchinson, I.V., Alam, Y. and Ayliffe, W. (1995). The humoral response to an allograft. *Eye*, **9**, 155–160.

Jefferies, W.A., Green, J.R. and Williams, A.F. (1985). Authentic T helper CD4 (W3/25) antigen on rat peritoneal macrophages. *Journal of Experimental Medicine*, **162**, 117–127.

Katami, M. (1995). The mechanisms of corneal graft failure in the rat. *Eye*, **9**, 197–207.

Khodadoust, A.A. and Silverstein, A.M. (1969a). The survival and rejection of epithelium in experimental corneal transplants. *Investigative Ophthalmology*, **8**, 169–179.

Khodadoust, A.A. and Silverstein, A.M. (1969b). Transplantation and rejection of individual cell layers of the cornea. *Investigative Ophthalmology*, **8**, 180–195.

Kilpe, V.E., Krakauer, H. and Wren, R.E. (1993). An analysis of liver transplant experience from 37 transplant centres reported to Medicare. *Transplantation*, **56**, 554–561.

Landais, P., Jais, J.P., Margreiter, R., Tufveson, G., Brunner, F., Selwood, N. and Mallick, N. (1993). Modelling long-term survival in 52315 first cadaveric grafts: the European experience. *Transplantation Proceedings*, **25**, 1316–1317.

Larkin, D.F.P. (1994). Corneal allograft rejection. *British Journal of Ophthalmology*, **78**, 649–652.

Lechler, R.I. and Batchelor, J.R. (1982). Restoration of immunogenicity to passenger cell-depleted kidney allografts by the addition of donor strain dendritic cells. *Journal of Experimental Medicine*, **155**, 31–41.

Medawar, P.B. (1948). Immunity to homologous grafted skin. III. The fate of skin homografts transplanted to the brain, to subcutaneous tissue, and to the anterior chamber of the eye. *British Journal of Experimental Pathology*, **29**, 58–69.

Middleton, D., Savage, D.A., Cullen, C. and Martin, J. (1988). Discrepancies in serological tissue typing revealed by DNA techniques. *Transplant International*, **1**, 161–164.

Miller, K., Huber, C., Neiderweiser, D. and Gottinger, W. (1988). Successful engraftment of high risk corneal allografts with short-term immunosuppression with cyclosporin. *Transplantation*, **45**, 651–653.

Moll, A.C., van Rij, G., Beekhuis, W.H., Reneradel de Lavalette, J.H., Hermans, J., Pels, E. and Rinkel-van Driel, E. (1991). Effect of donor cornea preservation in tissue culture and in McCarey-Kaufman medium on corneal graft rejection and visual acuity. *Documenta Ophthalmologica*, **78**, 272–278.

Moyes, A.L., Sugar, A., Musch, D.C. and Barnes, R.D. (1994). Antiviral therapy after penetrating keratoplasty for herpes simplex keratitis. *Archives of Ophthalmology*, **112**, 601–607.

Mulligan, R.C. (1993). The basic science of gene therapy. *Science*, **260**, 926–932.

Mytilineos, J., Scherer, S. and Opelz, G. (1990). Comparison of RFLP-DR beta and serological HLA-DR typing in 1500 individuals. *Transplantation*, **50**, 870–873.

Neiderkorn, J.K., Peler, J.S. and Mellon, J. (1989). Phagocytosis of particulate antigens by corneal epithelial cells stimulates interleukin-1 secretion and migration of Langerhans cells into the cornea. *Regional Immunology*, **2**, 83–90.

Nicholls, S.M., Bradley, B.A. and Easty, D.L. (1995). Non-MHC antigens and their relative resistance to immunosuppression after corneal transplantation. *Eye*, **9**, 208–214.

Nuchtern, J.G., Beddison, W.E. and Klausner, R.D. (1990). Class II MHC molecules can use the endogenous pathway of antigen presentation. *Nature*, **343**, 74–76.

Okamoto, S., Kosaku, K., SunderRaj, N., Hassell, J.R., Pipas, J.M. and Kao, W.W. (1995). Characterisation of immortalized human corneal epithelial cells transfected with a temperature-sensitive T antigen. *Investigative Ophthalmology and Visual Science (Suppl)*, **36**, 608.

Olerup, O. and Zetterquest, H. (1992). HLA-DR typing by PCR amplification with sequence specific primers (PCR-SSP) in 2 hours: an alternative to serological DR typing in clinical practise including donor-recipient matching in cadaveric renal transplantation. *Tissue Antigens*, **39**, 225–235.

Opelz, G. (1989). Influence of HLA matching on survival of second kidney transplants in cyclosporin-treated recipients. *Transplantation*, **47**, 823–827.

Opelz, G., Schwarz, V., Englemann, A., Back, D., Wilk, D. and Keppel, E. (1991). Long-term impact of HLA matching on kidney graft survival in cyclosporin treated patients. *Transplantation Proceedings*, **23**, 373–375.

Pavilack, M.A. and Chaves, H.V. (1994). Antibodies to ICAM-1 and LFA-1 in high-risk corneal transplantation. *Investigative Ophthalmology and Visual Science (Suppl)*, **35**, 1896.

Pepose, J.S, Gardner, K.M., Nestor, M.S., Foos, R.Y. and Pettit, T.H.E. (1985). Detection of HLA class I and II antigens in rejected human allografts. *Ophthalmology*, **92**, 1480–1484.

Philipp, W. and Gottinger, W. (1993). Leucocyte adhesion molecules in diseased corneas. *Investigative Ophthalmology and Visual Science*, **34**, 2449–2459.

Pleyer, U., Milani, J.K, Dukes, A., Chou, J., Lutz, S., Ruckert, D., Theil, H.J. and Mondino, B.J. (1995). Effects of topically applied anti-CD4 monoclonal antibodies on orthotopic corneal allografts in a rat model. *Investigative Ophthalmology and Visual Science*, **36**, 52–61.

Polack, F.M., Townsend, W.M. and Waltman, S. (1972). Antilymphocyte serum and corneal graft rejection. *American Journal of Ophthalmology*, **73**, 52–55.

Rinne, J.R. and Stulting, R.D. (1992). Current practices in the prevention and treatment of corneal graft rejection. *Cornea*, **11**, 326–328.

Roake, J.A. (1995). Pathways of dendritic cell differentiation and development. *Eye*, **9**, 161–166.

Rodrigues, M.M., Rowen, G., Hackett, J. and Bakos, I. (1981). Langerhans cell in the normal conjunctiva and peripheral cornea of selected species. *Investigative Ophthalmology and Visual Science*, **21**, 759–765.

Rotzscke, O., Falk, K., Wallny, H.J., Faath, S. and Rammensee, H.G. (1990). Characterization of naturally occuring minor histocompatibility peptides including H-4 and H-Y. *Science*, **249**, 283–287.

Sanfilippo, F., MacQueen, J.M., Vaughn, W.K. and Foulks, G.N. (1986). Reduced graft rejection with good HLA-A and -B matching in high-risk corneal transplantation. *New England Journal of Medicine*, **315**, 29–35.

Sharif, K.W. and Casey, T.A. (1993). Changing indications for penetrating keratoplasty, 1971–1990. *Eye*, **7**, 485–488.

Silverstein, A.M., Rossman, A.M. and Leon, A.S. (1970). Survival of donor epithelium in experimental corneal xenografts. *American Journal of Ophthalmology*, **69**, 448–453.

Smolin, G. (1969a). Suppression of the corneal homograft by antilymphocyte serum. *Archives of Ophthalmology*, **81**, 571–576.

Smolin, G. (1969b). Corneal homograft reaction following subconjunctival antilyphocyte serum. *American Journal of Ophthalmology*, **67**, 137–139.

Smolin, G. and Goodman, D. (1988). Corneal graft reaction. *International Ophthalmology Clinics*, **28**, 30–36.

Sonoda, Y., Sano, Y., Ksander B. and Streilein, J.W. (1995). Characterisation of cell-mediated immune responses elicited by orthotopic corneal allografts in mice. *Investigative Ophthalmology and Visual Science*, **36**, 427–34.

Stulting, R.D., Waring, G.O., Bridges, W.Z. and Cavanagh, H.D. (1988). Effect of donor epithelium on corneal transplant survival. *Ophthalmology*, **95**, 803–812.

Taylor, C.J. and Dyer, P.A. (1995). Histocompatibility antigens. *Eye*, **9**, 173–179.

Taylor, C.J., Chapman, J.R., Ting, A. and Morris, P.J. (1989). Characterization of lymphocytotoxic antibodies causing a positive crossmatch in renal transplantation. Relationship to primary and regraft outcome. *Transplantation* , **48**, 953–958.

Taylor, C.J., Welsh, K.I., Gray, C.M., Bunce, M., Bayne, A. and Sutton, P.M. (1994).. Clinical and socio-economic benefits of serological HLA-DR matching for renal transplantation over three eras of immunosuppression regimens in a single unit. In: *Clinical Transplants 1993*, edited by P.I. Terasaki and J.M. Cecka, pp. 233–241. Los Angeles, CA: UCLA Tissue Typing Laboratory.

Townsend, A., Ohlem, C., Bastin, J., Ljunggren, H.G., Foster, L. and Karre, K. (1989). Association of class I major histocompatibility heavy and light chains induced by viral peptides. *Nature*, **340**, 443–448.

Treseler, P.A., Foulks, G.N. and Sanfilippo, F. (1984). The expression of HLA antigen by cells in the human cornea. *American Journal of Ophthalmology*, **98**, 763–772.

Tuberville, A.W., Foster, C.S. and Wood, T.O. (1983). The effect of donor corneal epithelium removal on the incidence of allograft rejection reactions. *Ophthalmology*, **90**, 1351–1356.

Vail, A., Gore, S.M., Bradley, B.A., Easty, D.L., Rogers, C.A. and Armitage, W.J. (1994), on behalf of the Corneal Transplant Follow-up Study Collaborators. Influence of donor histocompatibility factors on corneal graft outcome. *Transplantation*, **58**, 1210–1217.

van der Veen, G., Broersma, L., Dijkstra, C.D., van Rooijen, N., van Rij, G. and van der Gaag, R. (1994). Prevention of corneal allograft rejection in rats treated with subconjunctival injections of liposomes containing dichloromethylene diphosphonate. *Investigative Ophthalmology and Visual Science*, **35**, 3505–3515.

Volker-Dieben, H.J., Kok-van-Alphen, C.C., Lansbergen, Q. and Persin, G.G. (1982). The effect of prospective HLA-A and -B matching on corneal graft survival. *Acta Ophthalmologica*, **60**, 203–212.

Volker-Dieben, H.J., D'Amaro, J. and Kok-van-Alphen, C.C. (1987). Hierarchy of prognostic factors for corneal allograft survival. *Australian and New Zealand Journal of Ophthalmology*, **15**, 11–18.

Wallny, H.J. and Rammensee, H.G. (1990). Identification of classical minor histocompatibility antigen as cell-derived peptide. *Nature*, **343**, 275–278.

Waltman, S.R., Faulkner, W. and Burde, R.M. (1969). Modification of the ocular immune response. I. Use of antilymphocytic serum to prevent immune rejection of penetrating corneal homografts. *Investigative Ophthalmology and Visual Science*, **8**, 196–200.

Whitcup, S.M., Nussenblatt, R.B., Price, F.W. and Chao, C. (1993). Expression of cell adhesion molecules in corneal graft failure. *Cornea*, **12**, 475–480.

Whitsett, C.F. and Stulting, R.D. (1984). The distribution of HLA antigens on human corneal tissue. *Investigative Ophthalmology and Visual Science*, **25**, 519–524.

Williams, K.A. and Coster, D.J. (1989). The role of the limbus in corneal allograft rejection. *Eye*, **3**, 158–166.

Williams, K.A., Ash, J.K. and Coster, D.J. (1985). Histocompatibility antigen and passenger cell content of normal and diseased human cornea. *Transplantation*, **39**, 265–269.

Williams, K.A., Ash, J.K., Parajasegaram, P., Harris, S. and Coster, D.J. (1991). Long-term outcome after corneal transplantation: visual result and patient perception of success. *Ophthalmology*, **98**, 651–657.

Williams, K.A., Roder, D., Esterman, A., Muehlberg, S.M. and Coster, D.J. (1992a). Factors predictive of corneal graft survival: report from the Australian Corneal Graft Registry. *Ophthalmology*, **99**, 403–414.

Williams, K.A., Standfield, S.D., Wing, S.J., Barras, C.W., Mills, R.A. and Comacchio, R.M. (1992b). Patterns of corneal graft rejection in the rabbit and reversal of rejection with monoclonal antibodies. *Transplantation*, **54**, 38–43.

Williams, K.A., Muehlberg, S.M., Win, S.J. and Coster, D.J. (1993), on behalf of all contributors. The Australian Corneal Graft Registry, 1990–92 report. *Australian and New Zealand Journal of Ophthalmology*, **21** (Suppl.), 1–48.

Williams, K.A., Muehlberg, S.M., Lewis, R.F. and Coster, D.J. (1995). How successful is the Australian Corneal Graft Register. *Eye*, **9**, 219–227.

Young, E., Stark, W.J. and Prendegast, R.A. (1985). Immunology of corneal allograft rejection: HLA-DR antigens on human corneal cells. *Investigative Ophthalmology and Visual Science*, **26**, 571–574.

9 Tears and the Dry Eye

Mark J. Elder[1] and John K.G. Dart[2]

[1]*Department of Ophthalmology, Christchurch Hospital, Private Bag 4710, Christchurch, New Zealand*
[2]*Corneal & External Disease Service, Moorfields Eye Hospital, City Road, London EC1V 2PD, UK*

The ocular surface is a specialised structure that protects and maintains a smooth and transparent cornea and this ensures that light can be adequately refracted and focused on to the retina. To provide these unique functions, the corneal epithelium must be constantly moistened, desiccation can lead to damage within minutes. The tear film ensures that the epithelium remains transparent and healthy, smooths the ocular surface and provides protection against infection. Because of the nature of the tear film and its interaction with the cornea, in humans it is refreshed every few seconds with blinking. During sleep the lids are able to completely cover the ocular surface thereby providing further protection. To ensure that blinking and ocular movement can occur without corneal trauma, the inside of the lids and the surface of the globe is covered with mucous membrane, the conjunctiva, and this too is covered by the tear film. Adequate production of the tear film is assured by an intrinsic basal level of secretion and various reflexes that can increase tear production, increase the blink rate, close the eye or enact protective behavioural changes. This chapter will discuss the anatomy and physiology of the tear film and the ocular surface, outline the innervation and neural control mechanisms of the ocular surface and discuss the diseases and the management of tear film disorders.

ANATOMY OF THE OCULAR SURFACE

The "ocular surface" is defined as the continuous sheet of tissue that extends from the grey line of the eyelid margin and includes the posterior lid surface, the fornix, the anterior surface of the globe and the cornea and is composed of the epithelium of the conjunctiva and cornea (Thoft and Friend, 1977). The normal corneal epithelium is a specialised, stratified squamous epithelium. It functions to provide an optically smooth, transparent anterior surface to the cornea, aids in the maintenance of relative stromal dehydration and provides a barrier against trauma and microbes. To ensure survival of the cornea and hence the animal there is usually a rapid healing response that does not impair transparency.

Correspondence: J.K.G. Dart, Corneal and External Disease Service, Moorfields Eye Hospital, City Road, London EC1V 2PD, UK; Tel: +44 171 566 2320; Fax: +44 171 566 2019.

ANATOMY OF THE CORNEAL EPITHELIUM

The corneal epithelium covers the entire anterior corneal surface and extends from limbus to limbus. It is a stratified, non-keratinized, non-secretory epithelium five to seven cells thick. There are three layers of cells; basal cells, wing cells and superficial cells. The basal cells are a single, cuboidal layer that rest on a 50 nm thick epithelial basement membrane and then on Bowman's membrane. Mitotic activity may be seen and there are abundant mitochondria, Golgi apparati, tonofilaments, actin filaments and glycogen. The wing cells are a layer 2–3 cells thick that have "wing like" projections laterally. They are more differentiated than the basal cells and have an abundance of cytokeratin tonofilaments. The superficial cells are flatter and more differentiated than the wing cells and may be classified into light and dark cells. The smaller, light cells are "younger" and have more microvilli than the older, darker cells that are the cells ready to slough off. The microvilli are covered by a glycocalyx and then a mucin layer. The cells adhere to each other by a variety of junctional structures. The superficial cells are joined by zonula occludens that make the epithelium a semi-permeable membrane. The wing cells and basal cells adhere using desmosomes and gap junctions that allow much freer fluid and ion exchange. The basal cells also adhere to the basement membrane by hemi-desmosomes and anchoring filaments that extend into the stroma (Jakobiec, 1983; Gipson, Spurr-Michaud and Tisdale, 1987). This enhances epithelial adhesion.

CONJUNCTIVAL AND LIMBAL EPITHELIAL ANATOMY

The conjunctiva is a mucous membrane that lines the posterior surface of the upper and lower lids, the fornices, and covers the globe up to the limbus. The conjunctiva is subdivided by regions into the tarsal, forniceal and bulbar conjunctiva. All conjunctiva consists of a stratified squamous epithelium that rests on the vascular substantia propria, unlike the corneal epithelium that rests directly on Bowman's membrane. Both types of epithelia have surface microvilli, attach to adjacent cells via desmosomes and attach to the underlying basement membrane by hemi-desmosomes (Jakobiec, 1983). However, conjunctival epithelial cells have cytoplasmic tonofilaments organised into dense bundles, there are more mitochondria, there are wider intercellular spaces and it is normal to find lymphocytes, neutrophils, masts cells, Langerhans cells and melanocytes within both the epithelium and the substantia propria. The conjunctival stroma contains blood vessels, lymphatics and nerves and rests on the episclera.

The limbus may be defined anatomically, surgically or clinically. The clinical limbus is the 1 mm wide region immediately posterior to the clear cornea and hence the limbal epithelium is a transition zone between conjunctival and corneal epithelia (Wanko, Lloyd and Mathews, 1964). The layers increase to 10–15 cells thick and goblet cells are absent. Langerhans cells, mast cells, lymphocytes and melanocytes are present. The limbal epithelium is smooth on its superficial aspect but dips deep into radially arranged furrows — the Palisades of Vogt (Gipson, 1989). It has been postulated that this provides a larger surface area for the diffusion of nutrients from the avascular bed to the basal cells (Gipson, 1989). The basal cell membrane is markedly infolded with a thick basement membrane-like layer and anchoring filaments (Iwamoto and Smelser, 1965).

CORNEAL EPITHELIAL EMBRYOLOGY

The corneal and conjunctival epithelium and the ocular adnexal elements all derive from the surface ectoderm that overlies the optic vesicle. The ocular surface epithelium is present from five weeks gestation and is bathed in amniotic fluid from seven weeks. At five weeks gestation there are two cell layers of corneal epithelium and this increases to seven by 36 weeks (Hay, 1980).

CORNEAL EPITHELIAL PHYSIOLOGY

The cornea is avascular and hence oxygen is supplied mainly from the tears. Lesser amounts are contributed from the aqueous component and limbal vasculature; the oxygen tension of tears is 150 mmHg, blood is 95 mmHg and aqueous is 40 mmHg (Holly and Lemp, 1977). When the eye is closed, the oxygen is mainly supplied from the superior palpebral conjunctival vessels (Holden and Sweeney, 1985). The epithelium consumes ten times more oxygen than the stroma (Riley, 1969).

While much of the oxygen is derived from the air, the glucose and amino acids are derived mainly from the aqueous. Smaller amounts come from the tears and the limbal vessels. The epithelium metabolises glucose via the hexose monophosphate shunt (aerobic and anaerobic conditions) and the anaerobic Embden-Meyerhof pathway. This latter pathway produces pyruvate and lactate which can be oxidised via the Krebs Cycle. Under hypoxic conditions such as contact lens wear these agents cause osmotic oedema (Klyce, 1981). This may be slow to resolve because the superficial location of the tight junctions within the epithelium means that pyruvate and lactate must diffuse through the stroma to the aqueous.

A major function of the epithelium is to act as a barrier. This keeps the stroma dehydrated and therefore transparent and it resists infection. The barrier is due to the tight junctions between adjacent cells. Physiologically, this is measured as the resting membrane potential (RMP). The RMP of the wing cells is -30 mV and this compares to the RMP of the basal cells of -15 mV (Klyce, 1972). These potentials are due to active transport of sodium into the stroma via Na^+/K^+ ATPase pumps and this facilitates chloride transport into the tears via chloride channels (Klyce, 1985). Together, these mechanisms result in the osmotic transport of water from the stroma into the tears thereby promoting stromal dehydration. This is in addition to the osmotic effects of tear evaporation and the endothelial pumping mechanism.

THE ORIGIN OF THE CORNEAL EPITHELIUM

The corneal and conjunctival epithelium are derived from different sources although both are able to modulate their phenotype and behaviour to mimic each other. The corneal epithelium has been proposed to be derived from stem cells that reside at the basal limbus. These produce transient amplifying cells that populate the entire basal corneal epithelial layer. These basal cells then replicate and produce normal corneal epithelium that is constantly being lost through desquamation (Thoft and Friend, 1983; Schermer, Galvin and Sun, 1986; Cotsarelis et al., 1989; Chen and Tseng, 1990; Kruse et al., 1990). The entire epithelium, except for the basal cells, is renewed every seven days in humans (Hanna, Bicknell and O'Brien, 1961).

RESPONSE TO CORNEAL EPITHELIAL LOSS

The loss of an area of cells due to trauma causes the cells at the wound edge to retract and lose their hemi-desmosomal attachments to the underlying basement membrane. The cells then enlarge and migrate by an amoeboid action to cover the defect, initially as a one cell thick layer. After wound closure, basal mitosis resumes and the additional layers are produced (Pfister, 1975). Lesions in the superior cornea heal faster than inferior lesions and the superior edge advances more rapidly than the inferior edge; Bron (1973) suggests that this is due to the enhanced ability of the superior limbus to provide new basal cells from stem-cell precursors.

THE TEAR FILM

The pre-corneal tear-film is a complex three-layered structure composed mainly of water, dissolved ions and two other layers to ensure adequate wetting and prevent excessive evaporation (Table 9.1) (Holly and Lemp, 1977). The basal layer of the tear film is mainly mucin and this adheres to the microvilli of the underlying corneal epithelium which has been said to be intrinsically hydrophobic (Holly and Lemp, 1977). However the methods used to establish this proposal have been shown to have produced artefactual non-wetting (Cope *et al.*, 1986). Careful removal of tear film mucin does not affect the wetting of the ocular surface epithelium probably because of the presence of a mucin glycocalyx secreted by both the corneal and conjunctival epithelial cells that maintains the ability of these cells to wet (Dilly and Mackie, 1981; Greiner *et al.*, 1985; Nichols, Chiappino and Dawson, 1985; Gipson, 1994). The ocular surface mucin is therefore a bilayer composed of glycocalyx secreted from the epithelial cells and superficial mucin layer secreted mainly by the goblet cells. Traditionally, the mucin layer was thought to be about 0.02–0.05 μm thick but it appears that processing artefacts caused shrinkage and it is about 1 μm thick over the cornea and 2–7.0 μm thick over the conjunctiva (Nichols, Chiappino and Dawson, 1985). This layer allows the rest of the tear film to be evenly distributed over the ocular surface for up to 30 s without blinking. Because blinking is more frequent than this, there is a degree of reserve in maintaining a continuous tear film. Some ocular surface diseases prevent the tear film remaining continuous for more than 10 s unless refreshed by blinking and this may lead to a compensatory increased blink rate and intermittent corneal desiccation.

The middle aqueous layer is about 7 μm thick and contains ions, proteins and immunoglobulins. It is produced by the lacrimal glands and the accessory lacrimal glands at about 1.2 μl/min (Mishima, 1965; Jordan and Baum, 1980). Clinically the volume of aqueous production is assessed by Schirmer's test while its osmolarity, protein content, lysozyme, lactoferrin content, IgA and IgE levels can also be determined and have some clinical relevance.

The superficial layer is 0.1 μm thick and consists mainly of waxy and cholesterol lipids. These are derived from the meibomian glands (McDonald, 1968; Brauninger, Shah and

TABLE 9.1
Normal tear film characteristics

Tear film thickness	6.6 μm
lipid layer	0.1 μm
aqueous layer	6.5 μm
mucin layer	1.0–7.0 μm
Total tear film volume	6.2 ± 2.0 μl
Marginal strip volume	2.9 μl
Pre-ocular tear film volume	1.1 μl
Cul-de-sac	4.5 μl
Flow rate	1.2 μl per minute
Evaporation rate	0.085 μl per minute
Osmolarity	300–304 mOsm/litre
Break up time	> 10 s
Schirmer I test	> 5.5–8 mm

Kaufman, 1972). This layer is oily and hence decreases evaporation and aids stabilisation of the less viscous aqueous layer. Evaporation accounts for about 25% of tear film loss but measurement of this is not yet a useful clinical tool (Mishima, 1961; Lemp and Wolfley, 1992).

TEAR FILM PHYSIOLOGY

The aqueous layer is constantly refreshed by the addition of new aqueous from the lacrimal gland located in the superotemporal quadrant of the orbit. The aqueous flows across the globe and drains via the lacrimal puncti located in the medial canthal region. In the fornices, a rolled thread of mucus traps unwanted particles. Mucin, lipid, excess mucus and entrapped particles also drain via the puncti. This is aided by the dynamics of blinking that squeezes the tear film medially. Particles and tears are actively sucked into the puncti by the "lacrimal pump" (Jones, 1966). These mechanisms allow for constant renewal and maintenance of the tear film and for the removal of unwanted particles, lipid contaminated mucin and organisms.

The tear film is constantly changing. With blinking, the palpebral aperture closes completely and this compresses the lipid layer 1000-fold and increases the volume of the meniscus which acts as a storage area. When the eyes open, the aqueous thins and the lipid layer is redistributed over the surface. This is achieved despite very rapid eyelid opening and closing speeds. Between blinks the tear film remains continuous because of the interplay of surface tensions between the epithelium, the mucin coating and the lipid layers (Holly and Lemp, 1977). When the eyes are continually open, the aqueous thins and the lipid is able to contact the mucin layer. This alters the surface tension and causes the formation of dry spots which are clinically significant (Holly, 1973, 1985).

All patients with dry eyes of any cause have some impairment of tear film stability. In health, there are two mechanisms that contribute to this — the balance of surface tensions and the viscosity. The hydrophobicity of corneal epithelium, at least when the glycocalyx is damaged, is similar to plastics such as polyethylene. For a film of water to be continuous on such a surface, it must be thicker than 200 μm (Holly, 1985). A normal tear film is about 6–8 μm thick and this illustrates the importance of the systems that encourage wetting and a low surface tension. The major hydrophilic component is the mucin system while the lipid layer has a minor effect. Normally, the lipid layer is separated from the mucin layer by the thickness of the aqueous and blinking clears lipid within the aqueous and mucin layers. If the aqueous layer is thin, the mucin layer can become more easily contaminated with lipid and the surface tension rises dramatically. Therefore, tear film stability can be affected by the mucin layer, a reduced aqueous layer, or a reduced, abnormal or contaminated lipid layer (Holly, 1985).

Water and simple ionic solutions have the same viscosity at both low and high flow rates (shear rates). This contrasts to tears which have a very high viscosity when the flow rate is low such as between blinks and a very low viscosity during high flow rates such as during blinking. This "non-Newtonian" behaviour is similar to tomato sauce, undried paint and wet concrete. The ocular advantages are that the low viscosity reduces the drag on the lids during blinking and the high viscosity aids tear film stability (Bron, 1985).

ANATOMY OF THE ADNEXAL ELEMENTS THAT CONTRIBUTE TO THE TEAR FILM

The glands that contribute to the tear film are the meibomian glands, the glands of Moll and Zeis, the main lacrimal glands and the accessory lacrimal glands of Kruse and Wolfring. The meibomian glands constitute about 25 glands in the upper lid and 20 in the lower lid and are specialised sebaceous glands that are orientated vertically within the tarsal plates of the eyelids. Each gland consists of 30–40 acini that open into a central canal. This duct is lined by stratified squamous epithelium which drains to the posterior lid margin (Wolff, 1954). The mouths of the glands are easily visible with a slitlamp microscope.

The glands of Zeis are sebaceous glands associated with the eyelash follicles and serve to keep the lashes supple. The glands of Moll are sweat glands within the eyelids and empty between the eyelashes on the anterior lid margin. Neither the glands of Zeis or Moll contribute significantly to the tear film.

THE GOBLET CELLS

The goblet cells are located solely within the conjunctival epithelium and are distributed mainly medially and in the inferior fornix (Kessing, 1968). They provide the majority of the tear film mucin and do so by discharging their contents directly onto the conjunctival surface.

THE LACRIMAL GLANDS

The lacrimal glands consist of the main lacrimal gland and the accessory glands of Wolfring and Krause. The main lacrimal gland is a tubuloacinar gland located in the superotemporal quadrant of the orbit. The acinine tissue communicates with ducts and the aqueous discharges into the superotemporal conjunctival fornix via 5–7 ducts. These can be destroyed after trauma, chemical burns or surgery in this area. There are 2–4 glands of Wolfring at the superotemporal border of the tarsal plate of the upper lid and 1–2 at the inferotemporal border of the lower lid. The glands of Krause are situated subconjunctivally in the fornices of the eyelids and number 42 in the upper lid and 6–8 in the lower lid. Both sets of these exocrine glands contribute to the aqueous component of the tear film.

INNERVATION OF THE OCULAR SURFACE AND ADNEXA

SENSORY INNERVATION

The innervation to the ocular surface and adnexa is sensory, motor, sympathetic and parasympathetic. The majority of the sensory innervation of the cornea, conjunctiva and eyelids is via the ophthalmic division of the trigeminal nerve which divides into the frontal, lacrimal and nasociliary branches which enter the orbit through the superior orbital fissure. The frontal nerve divides into the supratrochlear nerve which supplies sensation to the medial aspect of the upper eyelid and adjacent brow and the bridge of the nose and into the supraorbital nerve which supplies the medial aspect of the eyelid, corresponding tarsal conjunctiva, frontal sinus and forehead as far as the vertex. The lacrimal nerve supplies the lateral aspect of the upper eyelid and tarsal conjunctiva and a small area of adjacent skin. The nasociliary nerve courses medially from the superior orbital fissure, passes through the ciliary ganglion approximately 1 cm behind the globe, gives off the short and long ciliary nerves to the posterior globe, gives off the posterior ethmoidal nerve and courses anteriorly to divide into the anterior ethmoidal and infratrochlear nerves about 1 cm behind the medial canthus. These nerves carry sensory input from the cornea, bulbar conjunctiva, ethmoid sinuses, the nasal cavity and skin on the tip of the nose. The maxillary division of the trigeminal nerve provides sensory fibres to the inferior and inferotemporal eyelid and tarsal conjunctiva via the zygomatic nerve which enters the orbit through the inferior orbital fissure and divides into the zygomaticotemporal and infraorbital nerves (Wolff, 1954).

The cornea is highly innervated. The nerves enter the corneal stroma radially and immediately lose their myelination to aid transparency. The bare axons pierce Bowman's membrane and innervate the epithelium directly. In contrast, the conjunctiva is mainly innervated around blood vessels. There is no innervation of goblet cells and minimal innervation of the epithelium (Ruskell, 1985). The eyelids are innervated by plexi that lie deep to the orbicularis muscle, from which fibres penetrate the subepithelial and epithelial tissue and terminate in a wide variety of nerve endings (Burton, 1992). Corneal sensitivity can be quantified with an instrument such as a Cochet-Bonnet aesthesiometer where a fine nylon thread is pressed against the eye. The central cornea is 3 times more sensitive than its periphery and 9 times more sensitive than the conjunctiva (Burton, 1992). Corneal sensitivity is reduced in patients with brown irises, contact lens wearers and during the last half of the menstrual cycle (Miller, 1985).

MOTOR INNERVATION

The motor system provides control over the size of the palpebral aperture, enables blinking and ensures that it is closed during sleep. The upper lid is opened and elevated by the levator palpebrae superioris which is innervated by the third cranial nerve. When the upper lid is open, its exact position is fine tuned by Muller's muscle which is innervated by the sympathetic nervous system. With excessive sympathetic drive, such as occurs with anxiety or fear, the upper lid becomes over-elevated. During blinking, the eyelids are closed by the palpebral portion of the orbicularis oculi while both the palpebral and orbital portions are used during forced closure or blepharospasm. The orbicularis oculi is innervated by the zygomatic and temporal branches of the seventh cranial nerve (Wolff, 1954).

PARASYMPATHETIC INNERVATION

The parasympathetic nervous system is responsible for pupillary constriction and stimulation of lacrimal gland secretion. The preganglionic neurons are located in the lacrimal nucleus in the tegmental pons and their axons travel with the nervus intermedius to join the facial nerve until the petrous temporal bone where they separate to become the greater superficial petrous temporal nerve which synapses in the sphenopalatine ganglion in the pterygopalatine fossa. The postganglionic fibres join the maxillary division of the trigeminal nerve, enters the orbit via the inferior orbital fissure and reaches the lacrimal gland within the zygomatic branch (Wolff, 1954).

SYMPATHETIC INNERVATION

The sympathetic nervous system is responsible for pupillary dilatation and vasoconstriction of the conjunctiva and lacrimal gland. The nerves reach the orbit by leaving the spinal cord at the level of first three thoracic vertebrae, synapse in the superior cervical ganglion and travel with the internal carotid artery. Fibres run forwards with the ophthalmic artery and various divisions of the sensory nerves (Wolff, 1954).

REGULATION OF THE TEAR FILM

The regulation of the tear film is complex and not completely understood. Unlike the aqueous components, the mucin and lipid layers are not under direct neural control. The goblet cells increase secretion in the presence of prostaglandins, histamines, directly irritant chemicals and parasympathetic agonists (Lemp and Wolfley, 1992). Therefore increased mucin production often accompanies conjunctival inflammation of any cause and this is seen clinically as an increased discharge. Non-goblet cell mucin production seems to behave in a similar manner (Greiner *et al.*, 1985). The control of meibomian gland production is not understood although control of blinking may play a role. Aqueous secretion has traditionally been divided into "basal" and "reflex". Recent evidence suggests that the basal secretion is also driven by the reflex feedback loops and that the distinction is artificial (Jordan and Baum, 1980).

NEURAL REGULATION OF THE TEAR FILM AND ADNEXA; BLINKING AND LACRIMATION

The ocular surface and tear film is protected and controlled by spontaneous and persistent blinking and a variety of reflexes including the blink reflex and the trigeminal-lacrimal reflex. Spontaneous blinking is essentially independant of external factors and occurs approximately every 4 s in females and every 2.8 s in males (Hart, 1992). The rate is intrinsic to the individual and the species and is influenced by the "background levels" of afferent signal from the ocular surface. Reflex blinking is a physiological protective response to specfic tactile, visual or auditory stimuli.

The main tactile input is from stimulation of the ophthalmic division of the trigeminal nerve, particularly the corneal surface, although any painful stimulus from any dermatome can lead to reflex blinking. Input from the visual system may elicit the blink reflex, the dazzle reflex or the menace reflex depending on the stimuli. In a similar manner, blinking is also induced by very loud noises. Blepharospasm may be invoked if any of the types of stimulation is severe.

Lacrimal gland secretion is controlled by a similar set of tactile reflexes to blinking, using the afferent signal from the ophthalmic division of the trigeminal nerve and the efferent signal from the parasympathetic nervous system. This feedback loop controls the amount of aqueous tears secreted. Stimulation over and above the normal signal levels leads to a reflex increase in lacrimal gland secretion. Lacrimation can also be produced by severe stimulation of any part of the body or by emotional distress (psychogenic tearing) (Hart, 1992). The sympathetic nervous system has a minimal role in tear regulation (Miller, 1985).

DISEASES THAT AFFECT THE TEAR FILM

Clinically significant diseases of the tear film can occur due to a deficiency in any or all of its layers or from the addition of toxic components. The commonest diseases are due to keratoconjunctivitis sicca, Sjogren's syndrome and lid margin disease and these account for approximately 90% of dry eyes. The term "dry eye" is strictly a deficiency of one or more of the tear film layers and clinically it is useful to use this as a classification and management concept. However, the term has often been used in the past to describe aqueous deficiency alone. The expression "dry eye" must be distinguished from that of "ocular surface disease" which is a term that has been recently introduced to describe "a group of disorders, of diverse pathogenesis, in which disease results from the failure of the anatomical or physiological mechanisms responsible for maintaining a healthy ocular surface". Tear film deficiencies may cause or be associated with ocular surface disease and vice versa. The key distinction is that one term describes changes in the tear film and the other in the cellular covering of the eye surface.

AQUEOUS DEFICIENT DRY EYES

A unilateral aqueous deficient dry eye is uncommon but may be due to lacrimal gland inflammation (dacryoadenitis), trauma to the lacrimal gland from surgery, chemical burns or irradiation, inadequate parasympathetic innervation from a VII nerve palsy or abnormal congenital development such as in anhydrotic ectodermal dysplasia. Bilateral deficiency may be due to keratoconjunctivitis sicca, Sjogren's syndrome, sarcoidosis, Riley-Day syndrome and congenital alacrima (Bron, 1985). The most common disease in a general population is a mild aqueous deficiency due to senile degeneration of the lacrimal gland. It is often accompanied by mild lid margin disease that exacerbates the symptoms.

KERATOCONJUNCTIVITIS SICCA AND SJOGREN'S SYNDROME

Keratoconjunctivitis sicca (KCS) is a consequence of lacrimal gland atrophy. The disease is more common in women and has been associated with hormonal imbalance. KCS *per se* is an isolated ocular disease and is diagnosed as an abnormal result in two of three tests; Schirmer's test, Rose bengal staining and tear break-up time (Farris *et al.*, 1983; Whitcher, 1987). Primary Sjogren's syndrome consists of patients with KCS and evidence of the same disease process in the glands servicing the mouth, pharynx, bronchi and vagina. Sixty percent of patients with autoimmune diseases such as rheumatoid arthritis, systemic lupus erythematosus, polyarteritis nodosa, coeliac disease, thyroid disease and graft-*vs*-host disease are associated with Sjogren's syndrome and this is called secondary Sjogren's syndrome (Manthorpe *et al.*, 1981; Baum, 1985). Rheumatoid arthritis is the most common single association (30%). The ocular signs and symptoms are those of aqueous deficiency and the spectrum of disease varies from very mild to severe (*vide infra*). The natural history of the disease is that the signs and symptoms may progress slowly or plateau at any stage (Wilson *et al.*, 1991).

Histologically, there is conjunctival epithelial acanthosis and loss of goblet cells combined with stromal infiltration with lymphocytes. There may be infiltration of acute inflammatory cells. These conjunctival features are common to all aqueous deficiency, regardless of the cause. The lacrimal gland may become clinically enlarged and be focally or diffusely infiltrated with lymphocytes and plasma cells. Initially the gland retains its architecture but later the acinar parenchyma becomes atrophic (Damato *et al.*, 1984; Spencer and Zimmerman, 1985).

CONGENITAL ALACRIMA

This may unilateral or bilateral and is occasionally associated with ectodermal dysplasia or Mobius syndrome. Keratoconjunctivitis sicca occurs in approximately 36% of cases (Spencer and Zimmerman, 1985).

ERYTHEMA MULTIFORME

Erythema multiforme is an acute, self-limited mucocutaneous disease that may be divided into major and minor forms. The minor form is clinically less severe and affects skin and, at most, one mucosal surface. In contrast, the major form (Stevens-Johnson syndrome) maybe life threatening, affects skin, two or more mucosal surfaces and is associated with systemic toxicity. It affects children and young adults and the acute episode lasts about six weeks. Precipitating factors are mainly drugs (approximately 50%) or infections (Yetiv, Bianchine and Owen, 1980). Common triggers include sulphonamides, barbiturates, *Mycoplasma pneumoniae* or *Herpes simplex* (Shelley, 1967; Huff, Weston and Tonnesen, 1983). Acute pseudomembranous or membranous conjunctivitis may develop and lasts about three weeks before spontaneously resolving. The sequelae are variable and range from no damage to conjunctival cicatrisation that is non-progressive (Mondino, 1990). Histologically, skin and mucous membrane both show an acute vasculitis with lymphocytic infiltration and there is an associated acute inflammatory infiltrate and epithelial necrosis (Buchner, Lozada and Silverman, 1980). The chronic severe manifestations involve cicatrisation and loss of goblet cells. This results in reduced aqueous and mucin production.

OCULAR CICATRICIAL PEMPHIGOID

Ocular cicatricial pemphigoid (OCP) is an autoimmune disease directed against mucosal basement membrane (Bernauer, Itin and Kirtschig, 1997). The conjunctiva gradually fibroses, typically associated with episodes of inflammation. These eyes are prone to lid margin disease which affects the lipid layer. Aqueous deficiency develops only very late in the disease and much of what is ascribed to dryness is secondary to unrelated ocular surface disease (Foster, 1986; Roat, Sossi and Thoft, 1989; Mondino, 1990; Elder, 1997).

SARCOIDOSIS

Sarcoidosis can cause lacrimal gland enlargement and aqueous deficient dry eyes. The gland may be enlarged due to infiltration by sarcoid granulomas and if untreated may become fibrotic leading to a permanently dry eye.

ANHYDROTIC ECTODERMAL DYSPLASIA

These patients have a congenital reduction or absence of sweat glands throughout the body. The eyes have an aqueous deficiency due to lacrimal gland involvement and absence of the meibomian glands. Patients are unable to sweat, have minimal body hair, abnormal nails, prematurely aged skin that is dry and wrinkled and have characteristic facies showing a sunken cheeks, a saddle nose, large ears and thick lips. Ninety percent are male. A few patients have a cleft lip and palate, lobster claw hands and feet (ectodactyly) and abnormal teeth (Baum and Bull, 1974; Baum, 1985).

DRUGS CAUSING AQUEOUS DEFICIENCY

Any drug that has a parasympatholytic or sympathomimetic effect can decrease the neural stimulation to the lacrimal gland and hence provoke aqueous deficiency. This includes atropine-like agents, nasal decongestants, antihistamines, tricyclic antidepressants, some antihypertensives, sleeping tablets, diuretics and topical β-blockers (Baum, 1985; McMonnies, 1986; McMonnies and Ho, 1987a,b; Kruppens et al., 1992).

NEUROLOGICAL DISEASES AFFECTING THE TEAR FILM

Of the three components of the tear film, only the aqueous portion is under neural control. The afferent pathways are the sensory nerves of the trigeminal nerve and from supranuclear sources whereas the efferent system is that of the parasympathetic nervous system. The tear film can be affected by a lesion at any site of the neural pathway; afferent, supranuclear, nuclear and efferent.

AFFERENT DEFECTS

Decreased corneal and conjunctival sensation may result in functional dryness because there is no capability for reflex tearing. This may be congenital, due to diabetes, herpes zoster or any lesion of the trigeminal nerve. The lack of the neurotrophic influences also leads to decreased corneal epithelial adhesion and punctate epitheliopathy.

SUPRANUCLEAR AND NUCLEAR LESIONS

Excess lacrimation occurs in pseudobulbar palsy where patients appear to be inappropriately emotionally distressed and weep. Excess tearing also occurs in meningitis and under light general anaesthesia. Nuclear and brainstem lesions are very rare causes of altered lacrimation (Miller, 1985).

PERIPHERAL NERVOUS SYSTEM LESIONS

Any lesion along the course of the parasympathetic supply to the lacrimal gland may result in altered tearing. This includes acoustic neuromas, lesions involving the petrous temporal bone, meningeal tumours, nasopharyngeal carcinoma, sphenoid sinus disease, carotid artery aneurysms, sphenopalatine lesions and herpes zoster. It is not possible to clinically distinguish preganglionic from postganglionic lesions. The parasympathetic supply runs with the facial nerve until the petrous temporal bone and a lesion in this region will lead to a unilateral facial palsy and a dry eye such as may occur with Bell's palsy. The parasympathetic supply may also be affected where it runs with the maxillary division of the trigeminal nerve anterior to the inferior orbital fissure.

CROCODILE TEARS

Crocodile tears refer to the abnormal production of tears during eating. This may be congenital, may appear a few weeks after an acute facial palsy or may appear several months after a facial palsy such as Bell's. It results from re-innervation of the lacrimal gland by misdirected fibres originally destined for the salivary gland (Miller, 1985).

RILEY-DAY SYNDROME

Riley-Day syndrome (familial dysautonomia) is a congenital disease where there are aberrant autonomic nervous system responses. This results in markedly reduced lacrimal gland secretion although this becomes almost normal when stimulated with parasympathomimetic drugs. The pupil does not dilate with topical adrenaline and there may be altered corneal sensitivity, strabismus, myopia and retinal vascular tortuosity. The cornea often has inferior punctate staining and recurrent epithelial defects similar to the neuroparalytic keratitis and the neural component of Riley-Day has been additionally implicated in the pathogenesis of the corneal manifestations (Spencer and Zimmerman, 1985). Systemically, there is excessive sweating and salivation, postural hypotension, skin blotching, diarrhoea, recurrent respiratory infection and emotional instability. It is an autosomally inherited disease that affects almost solely Ashekenazic Jews and most patients die before age fifty due to intercurrent infection (Riley *et al.*, 1949; Goldberg, Payne and Brunt, 1968; Baum, 1985).

MUCIN DEFICIENCY

Mucin is produced by the goblet cells and the conjunctival and corneal epithelial cells. Therefore clinical deficiency of mucin occurs when there is less surface area of conjunctiva such as with cicatrisation or after surgical excision and with loss of goblet cells such as occurs in chemical burns, cicatrisation, vitamin A deficiency and KCS (Ralph, 1975; Kinoshita *et al.*, 1983). The goblet cell density can be assessed clinically using impression cytology (Tseng *et al.*, 1985; Tseng and Farazdaghi, 1988). When there is a major loss of goblet cells, the ocular mucin levels are reduced by approximately 50% (Dohlman *et al.*, 1976).

VITAMIN A DEFICIENCY

Dietary vitamin A deficiency causing xerosis is a major cause of acquired blindness in the Third World (Sandford-Smith and Whittle, 1979; Olurin, 1970). The conjunctiva is initially and mainly affected, typically by a loss of goblet cells, acanthosis, keratinisation and stromal inflammation. Clinically, there are reduced levels of mucin and mucus and the conjunctiva becomes inflamed and develops focal white plaques in the interpalpebral fissure, Bitots spots. Corneal vascularisation, scarring and ulceration occur due to bacterial infection, measles or herpes simplex. Treatment with oral Vitamin A results in a resolution of the conjunctival changes within days (Spencer and Zimmerman, 1985).

LIPID DEFICIENCY AND DISEASE

Abnormality of the lipid layer occurs due to disease of the meibomian glands or contamination of the lipid layer. The commonest cause is "blepharitis" which is a "catch-all" term for a variety of eyelid margin diseases, all of which can cause ocular surface disease and tear film disease. Chronic blepharitis can be subdivided into meibomian gland disease, seborrheic blepharitis and staphylococcal blepharitis (McCulley, Dougherty and Denean, 1982).

MEIBOMIAN GLAND DISEASE

Meibomian gland disease (MGD) results in blockage of the glands due to ductal thickening and changes in the lipid chemistry and this leads to a reduction in the lipid layer of the tear film. The lipid has an increased melting point and an increase in free fatty acids which are directly toxic to the ocular surface and can destabilise the tear film. The glands may be inflamed due to the inspissated secretions and there are clinical signs of inflammation around the gland orifices. The secretions may solidify and this contributes to enlargement of the glands. These changes are visible through the tarsal conjunctiva and may be further delineated using transillumination of the everted lid (Robin et al., 1985). Histologically, there is obstruction and dilation of the ducts, acini enlargement degeneration, squamous metaplasia and cystic degeneration, chronic granulomatous inflammation and abnormal keratinisation of the ductal epithelium which extends from deep in the body of the gland to its orifice. This is similar to the pathogenesis of acne vulgaris and it has been postulated that MGD is due to abnormalities of the keratinisation process (Knutson, 1974; Jester, Nicolaides and Smith, 1981; Gutgesell, Stern and Hood, 1982).

By squeezing the glands gently, the percentage of patent ducts, the secretion and its viscosity can be observed. In health approximately 40% of glands are patent at any given time and the secretions have the consistency of olive oil (Hom et al., 1990). In MGD, fewer ducts are patent and the secretions are thicker, typically like toothpaste. MGD is associated with seborrheic blepharitis in 36%, aqueous deficiency in 35% and with acne rosacea but not with resident populations of abnormal bacteria per se (McCulley and Sciallis, 1977; McCulley, Dougherty and Denean, 1982). The symptoms are those of dry eyes in general and of burning in the morning. The cornerstones of treatment are lid hygiene, lid massage and a course of oral tetracycline. The lids should be cleaned with a weak solution of bicarbonate of soda or saline initially twice daily and reduced to once daily after a clinical improvement. Baby shampoo is often recommended but is frequently irritant. This removes desquamating skin and rancid skin lipids and reduces the resident bacterial numbers. The upper and lower lids need to be massaged towards the lid margin such that the glands are "milked" and emptied and this is best performed after applications of heat using a moist flannel. Oral tetracycline, 250 mg twice daily initially for two months, is indicated for severe disease that is associated with acne rosacea and failure of these other measures (Salamon, 1985).

SEBORRHOEIC BLEPHARITIS

These patients are characterised by greasy crusting of the eyelid margin and eyelashes and of eyelid inflammation. It is associated with seborrhoeic dermatitis in 95%, particularly mild dermatitis of the skin of the brow, nasolabial folds, scalp, behind the ears and the sternum. It is also associated with MGD (35%) and KCS (33%). The management is to ensure adequate lid margin hygiene. Topical steroid ointment twice a day to the lid skin and lid margin for three weeks may also be helpful (McCulley, Dougherty and Denean, 1982).

STAPHYLOCOCCAL BLEPHARITIS

Of all lid margin disease, staphylococcal blepharitis tends to have the most lid inflammation, especially of the anterior lid margin, the flakes of skin are dry rather than greasy and there are collarettes of inflammatory exudate around the bases of the lashes. Compared to seborrhoeic blepharitis, these patients are younger (mean age 40 *vs* 50 years) and often female (80% *vs* 44%). The bacteria involved are *Staphylococcus aureus* (50%) and *Staphylococcus epidermidis* (100%). It is associated with KCS (50%), follicular or papillary conjunctivitis (15%), atopy (9%) but not seborrhoeic dermatitis (McCulley, Dougherty and Denean, 1982). The management is an appropriate course of topical antibiotic, such as chloramphenicol. Resistant disease may need systemic antibiotics.

ABNORMAL PRODUCTS IN THE TEAR FILM OF DISEASED CONJUNCTIVA

A variety of inflammatory conjunctival diseases have abnormal products in the tear film. In vernal keratoconjunctivitis there are elevated tear-film levels of major basic protein, Charcot-Leyden crystal protein, prostaglandins PGF and HETE, complement C3, factor B and C3 anaphylotoxin (Dhir *et al.*, 1979; Udell *et al.*, 1981; Ehlers and Donshik, 1992). Eosinophil major basic protein has been identified in the base and mucous plug of two corneal shield ulcers and is implicated in their aetiology (Trocme, Raizman and Bartley, 1992). Prostaglandin PGF has been identified in trachoma (Dhir *et al.*, 1979) and leukotriene B_4 (LTB_4) has been identified in acute allergic eye disease (Bisgaard *et al.*, 1985). Compared to controls, OCP patients have statistically elevated levels of thromboxane A_2, PGE_2, 6-keto-$PGF_{1\alpha}$, S-HETE, LTB_4 and leukotriene C_4 (LTC_4). Thromboxane A_2 causes platelet aggregation, PGF and HETE augment mediator release, LTB_4 causes chemotaxis of eosinophils and neutrophils and LTC_4 causes vasodilatation. PGE_2 causes vasodilatation and may also act as a negative feedback inhibitor of T-cell production, lymphokine production and macrophage activity (Goodwin and Ceuppens, 1983). It is probable that similar mediators of inflammation are found in many inflammatory conjunctival conditions including KCS. These mediators may contribute to the toxicity of the ocular surface and the effects may be worse if there is aqueous deficiency or a reduction in the protective mucin layer.

THE DIAGNOSTIC APPROACH TO DRY EYES

The clinician makes an assessment of tear film and ocular surface disease based on the interpretation of data collected by history taking, clinical examination, special tests and laboratory tests. Each piece of data makes its contribution to the final diagnosis by its own sensitivity and specificity. The history is most important, followed by the clinical examination. In patients with tear film disorders and ocular surface disease, laboratory investigations are often of minimal help in confirming a diagnosis or planning management. It is the history that puts the diagnosis and its severity into perspective and is vital in planning short-term and long-term management goals. The dilemma of tear film and ocular surface disease is that the correlation between clinical signs and symptoms is variable, both at any given time and from day to day. Therefore, the following groups of patients may be defined.

1. Symptoms of a dry eye but with no corresponding clinical signs.
2. Symptoms of a dry eye but with minimal corresponding clinical signs.
3. Symptoms of a dry eye and corresponding clinical signs.
4. No symptoms of a dry eye but with clinical signs.
5. A wet or watery eye as a manifestation of ocular surface disease.
6. The intermittently dry eye.
    (after Snyder, 1994)

PATIENT HISTORY

The history is the only tool that allows the clinician to determine the relevance of any clinical signs and to determine the effect of the signs and symptoms on the patients lifestyle. Further, it is potentially the most sensitive tool for diagnosing dry eyes. McMonnies and co-workers have devised a formal series of questions that has a sensitivity of 87% and a specificity of 87% (Table 9.2) and this performance is approximately 10% better than any single or combination of clinical signs or tests (McMonnies, 1986; McMonnies and Ho, 1987a,b; Golding and Brennan, 1993). Compared to clinical signs, the features in the history also have a significantly lower inter-observer reliability rating (Anderson et al., 1972).

The history needs to assess the patient's current symptoms, secondary symptoms and exacerbating factors, risk factors for dry eyes, previous ocular treatments, presence of associated systemic or ocular diseases, contraindications to possible treatment and the relationship with the patient's socio-economic status. The differential diagnosis must also be considered. The common diagnoses with similar symptoms are the family of lid margin diseases ("blepharitis") (McCulley, Dougherty and Denean, 1982). Less common is rosacea keratitis, allergic eye disease and ocular exposure from ectropion (Lui and Stasior, 1983; Friedlaender, Ohashi and Kelly, 1984; Browning and Proia, 1986). However numerous other disorders, amongst the many causes of ocular surface diseases, may present with a dry feeling, burning or uncomfortable eyes and be misdiagnosed as tear deficient disorders.

Aspects of the history that are statistically associated with dry eyes are: a previous history of eyedrop use for dry eyes, symptoms of soreness, scratchiness, dryness, grittiness, burning, particularly if "often or constant", an unusual sensitivity to smoke, smog, air conditioning or central heating, eyes that become very red and irritated when swimming or the day after alcohol, the use of certain medication, arthritis or thyroid abnormalities, symptoms of dryness of the nose, mouth, throat, chest or vagina, sleeping with the eyes partly open and ocular irritation on waking (McMonnies and Ho, 1987a,b).

TABLE 9.2
McMonnies' Questionnaire for Dry Eyes (McMonnies, 1986)

Age:
    under 25   25–45   over 45 years

Currently wearing:
    no contact lenses   hard lenses   soft lenses

1. Have you ever had drops prescribed or other treatment for dry eyes?
    yes   no   uncertain

2. Do you ever experience any of the following (underline those that apply to you)
    soreness   scratchiness   dryness   grittiness   burning

3. How often do your eyes have these symptoms
    never   sometimes   often   constantly

4. Are your eyes unusually sensitive to smoke, smog air conditioning or central heating?
    yes   no   sometimes

5. Do your eyes easily become very red and irritated when swimming?
    not applicable   yes   no   sometimes

6. Are your eyes dry and irritable the day after drinking alcohol?
    not applicable   yes   no   sometimes

7. Do you take (please underline) antihistamines tablets or use antihistamine eyedrops, diuretics (fluid tablets), sleeping tablets, tranquillisers, oral contraceptives, medication for duodenal ulcer or digestion, or for high blood pressure or
    ....?
    (write in any medication that you are taking which is not listed).

8. Do you suffer from arthritis?
    yes   no   uncertain

9. Do you experience dryness of the nose, mouth, throat, chest or vagina?
    never   sometimes   often   constantly

10. Do you suffer from thyroid abnormalities?
    yes   no   uncertain

11. Are you known to sleep with your eyes partly open?
    yes   no   sometimes

12. Do you have eye irritation as you wake from sleep?
    yes   no   sometimes

CLINICAL TESTS FOR ASSESSING POTENTIALLY DRY EYES

The clinical tests are summarised in Figure 9.1, and may be classified in the following way:

1.   Tear film biomicroscopy
2.   Schirmer's tests
3.   Osmolarity
4.   Lactoferrin testing
5.   Dynamic film tests
6.   Ocular surface tests

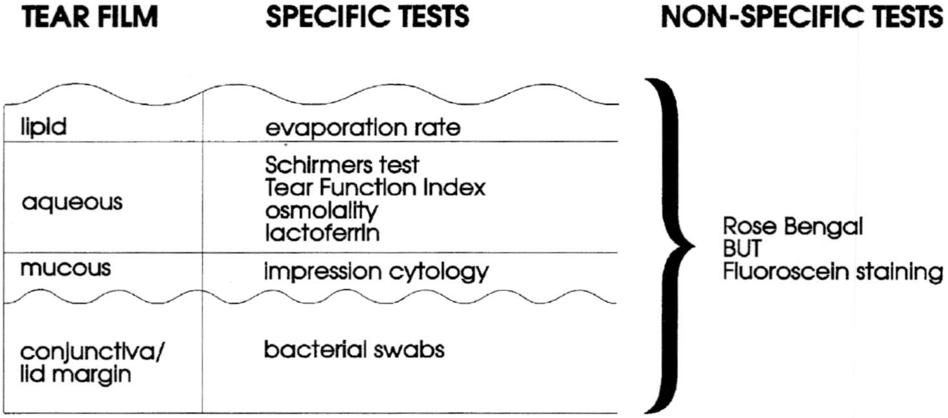

TEAR FILM   **SPECIFIC TESTS**   **NON-SPECIFIC TESTS**

lipid	evaporation rate	
aqueous	Schirmers test Tear Function Index osmolality lactoferrin	Rose Bengal BUT Fluoroscein staining
mucous	impression cytology	
conjunctiva/ lid margin	bacterial swabs	

Figure 9.1   A summary of the specific and non specific tests available for the assessment of the different tear film components (BUT: Tear break up time).

Tear film biomicroscopy

The tear film and the ocular adnexa can be easily viewed with the slit-lamp biomicroscope. The tear film forms a meniscus at the lid margins, the height of which is 0.2 mm or greater in 94% of normal eyes (Lamberts, Foster and Perry, 1979). This height may be reduced in aqueous deficient dry eyes (Baum, 1985; Whitcher, 1987). Unfortunately, the meniscus height varies from 0.1–0.6 mm in normal patients and there is a poor correlation between this sign, Schirmer tests and aqueous deficiency (Scherz, Doane and Dohman, 1974; Lamberts, Foster and Perry, 1979).

The tear film may contain particulate matter and mucous strands and both of these components are more common and are in greater amounts in aqueous deficient dry eyes (Baum, 1985; Whitcher, 1987; Mackie and Seal, 1981). Because of internal reflection and destructive interference within the thin superficial lipid layer, its thickness can be estimated from its colour when it is illuminated by diffuse oblique white light. For example: no colour, <100 nm; yellow, 100-140 nm; red, 190–205 nm; purple, 210 nm; blue, 250 nm; green, 280–310 nm (Serdarevic and Koester, 1985). This is similar to the Newton's rings seen within an oil droplet on water. In normal eyes the layer may be too thin to have a visible colour however narrowing the palpebral aperture thickens the layer and makes it visible. The layer is increased in thickness in dry eyes (163 *vs* 102 nm), blepharitis (129 *vs* 102 nm) and in inflammatory conditions of the conjunctiva (McDonald, 1969; Norn, 1979).

The ocular adnexa require evaluation for any patient with suspected dry eyes. In health, the majority of the meibomian gland orifices should be open. The body of the glands are visible through the tarsal conjunctiva and should be parallel like piano keys rather than dilated. Other signs of lid margin disease include dry flaky skin around the base of the lashes, lid notching and follicular conjunctivitis due to bacterial exotoxins (McCulley, Dougherty and Denean, 1982). Follicular conjunctivitis suggests either chronic toxicity from topical drugs or their preservatives or the presence of pathogenic lid margin bacteria (McCulley, Dougherty and Denean, 1982).

Schirmer's tests

Schirmer's tests assess the volume of the aqueous component of the tear film by using standardised strips of blotting paper to wick tears from the conjunctival sac. A strip of Whatman No. 41 filter paper 5 × 35 mm is folded 5 mm from its end and is placed into the lower fornix for 5 minutes after which the length of wet paper is measured. Normal wetting is regarded as 10 mm or more and is typically 20–35 mm in normal young adults. The "standard" and most commonly performed Schirmer I test does not use topical anaesthesia. The addition of anaesthesia (Schirmer II test) reduces the amount of wetting by 40% by abolishing any additional reflex tearing that may be caused by the presence of the paper strips. It is said to more accurately, but not completely, represent basal tear production (Lamberts, Foster and Perry, 1979; Jordan and Baum, 1980; Clinch *et al.*, 1983). However, an adequate time period is required between the topical anaesthesia and the test to allow the diluted tear film to return to normal. This test may be of little value if the hypothesis that basal tear secretion may not exist is accepted — all tear secretion being the result of reflex tear secretion resulting from environmental stimuli (Jordan and Baum, 1980). The reduction in the volume of tears measured that results from the use of topical anaesthesia (Schirmer II test) is then explained by suggesting that some of these environmental stimuli are non-ocular. The Schirmer III test aims to assess the maximal tear secretion by deliberately causing a noxious stimulus, such as direct stimulation of the nasal mucosa. Therefore, these tests have the potential to distinguish between the basal secretion of tear film aqueous component, if this exists, and that produced secondary to varying degrees of stimulation.

For the Schirmer I test, wetting of less than 5–5.5 mm is indicative of significant aqueous deficiency (van Bijsterveld, 1969; Lamberts, Foster and Perry, 1979). The test may vary from day-to-day and the precision is improved by averaging three recordings taken on separate occasions. A Schirmer I test of less than 5.5 mm is 83% sensitive for KCS and 85% specific while a test result of less than 8 mm is 60–74% sensitive and 77% specific (van Bijsterveld, 1969; Goren and Goren, 1988). The test is slightly less specific in males (72% *vs* 80%) (Lamberts, Foster and Perry, 1979).

Tear osmolarity

The normal tear film osmolarity is 300–304 mOsmol/l whereas in aqueous deficient dry eyes it is 343–63 mOsmol/l (Gilbard, Farris and Santamaria, 1978; Farris *et al.*, 1983; Gilbard, 1985; Tiffany, Chew and Bron, 1994). In one series, no normal patient had an osmolarity greater than 310 mOsmol/l and no patient with aqueous deficiency had an osmolarity less than 312 mOsmol/l (Gilbard, Farris and Santamaria, 1978). Both aqueous deficient dry eyes and meibomian gland disease result in a significant elevation of the osmolarity ($P < 0.01$, $P < 0.01$) (Tiffany, Chew and Bron, 1994). In aqueous deficient dry eyes, the altered osmolarity may entirely explain the patients symptoms and pathological features although the clinical importance of this is unknown (Balik, 1952; Gilbard, Farris and Santamaria, 1978; Gilbard, 1985). The environmental humidity dramatically alters the osmolarity. An increase of humidity from 30–45% to 65–75% reduces the tear film osmolarity from 327 to 295 mOsmol/litre in the same patients (Simmons, Craig and Tomlinson, 1993). When testing the osmolarity, it is important that only 0.1–0.4 µl is sampled by capillary tube to prevent artefactual changes. These small volumes can then be assessed using freezing-point depression or vapour-pressure osmometry (Gilbard, Farris and Santamaria, 1978; Tiffany, Chew and Bron, 1994). The latter instrument is able to be used in outpatients and gives an "instant answer".

Tear film lactoferrin levels

Lactoferrin is a protein found in normal tears at 1.4 mg/ml and arises predominant from its secretion from the lacrimal gland. It makes up 25% of all tear film proteins, is antibacterial and acts to inhibit complement (Kijlstra and Jeurisse, 1982; Janssen and van Bijsterveld, 1983; Kijlstra, Jeurisse and Koning, 1983). It can be measured using commercially available radial immunodiffusion test kits (Lactoplate, Eagle Vision Inc.), although the results take at least three days (Mancini, Carbonarra and Heremens, 1965; Janssen and van Bijsterveld, 1983; Kijlstra, Jeurisse and Koning, 1983). Levels of lactoferrin below 1 mg/ml are 68–71% sensitive and 90% specific for aqueous deficient dry eyes and Goren and Goren (1988) claim that it is the best single test of aqueous deficiency.

Tear function index

Recently, Xu *et al.* (1995) have described a tear function index (TFI) that has a specificity of 92% and a sensitivity of 79%. This is calculated as the value of the Schirmer test with anaesthesia divided by the tear clearance rate (TCR). The test involves staining the tear film with 10 µl of 0.5% fluoroscein and 0.4% oxybuprocaine, waiting 5 minutes and then performing a Schirmer test for 5 minutes. The residual colour of the blotting paper is then compared with standard published colour charts (Zappia and Milder, 1972; Xu *et al.*, 1995).

Tear film tests

Break-up times   The complex tear film is refreshed and redistributed with every blink. If blinking is prevented, then the tear film thins and after a time ceases to be continuous across the ocular surface due to surface hydrophobicity. Tear film break-up time (TBUT) is defined as the time from a complete blink to the first randomly distributed dry spot and is greater than ten seconds in normal eyes. The TBUT is affected by lid holding and topical anaesthesia and neither are recommended. Localised surface disease may consistently cause overlying tear film break up and therefore a true TBUT requires that the break-up be randomly distributed. For any individual, the TBUT is moderately repeatable, the standard deviation is typically 1–4 s (Lemp and Hamill, 1973; Vanley, Leopold and Gregg, 1977). A reduced TBUT is associated with a reduction in any of the components of the tear film (Lemp and Hamill, 1973). In meibomian gland disease the TBUT is reduced (7.2 s *vs* 28.2 s) but returns to normal with expression of the meibomian glands (McCulley and Sciallis, 1983). The TBUT is increased after the application of topical lubrication and decreased by 0.01% benzalkonium chloride (Norn, 1985). To avoid any confounding effects of the addition of fluoroscein, a non-invasive TBUT (NIBUT) determines the break-up when the image of a reflected grid becomes distorted. This test claims specificity of 86% and sensitivity of 82% in dry eyes using normal as greater than

10 s (Mengher, Pandher and Bron, 1986). Commercial instruments are now available for the measurement of NIBUT (Craig *et al.*, 1995).

Tear evaporation rates The tear evaporation rate (TER) is measured by assessing the humidity inside goggles that allow blinking but seal the eye from the environment. The normal value is 4×10^{-7} g/cm^2/s and this may be increased up to 5-fold if there is a reduction in the tear film lipid due to meibomian gland disease or in KCS (Rolando, Refojo and Kenyon, 1983; Foster, 1986; Tsubota and Yamada, 1992). Unfortunately, the test requires complex equipment and therefore is not routinely available.

Ocular surface tests

Fluorescein staining Fluorescein is a water soluble vital dye that stains the ocular surface cells where there is disruption of the cell-cell junctions (Feenstra and Tseng, 1992). It is administered as 0.25% eyedrops or by application of fluorescein-impregnated paper and fluoresces well with cobalt-blue illumination. Fluorescein is non-toxic and diffuses freely into the corneal stroma and aqueous. The normal ocular surface does not stain and it is cleared from the tear film within 5–10 min. The patterns of fluoroscein staining have been helpful in the diagnosis of punctate and gross epithelial lesions, contact lens fitting and tear volume and clearance (Coster, 1988). In very dry eyes, there is often staining of the ocular surface in a punctate pattern (Whitcher, 1987; Feenstra and Tseng, 1992).

Rose Bengal staining Rose Bengal (tetrachloro-tetriodo fluoroscein sodium) is a red coloured, vital dye that is manufactured as 1% eyedrops or impregnated into blotting paper for diagnostic purposes only. Traditionally, it has been believed that this dye stains only mucus and dead or devitalised cells and therefore it does not stain the conjunctiva or corneal epithelium of normal eyes (Sjogren, 1933; Forster, 1951; van Biijsterveld, 1969; Norn, 1969, 1970). However, recent work has suggested that it stains any normal ocular surface cells that lack the physiological mucin coating (Feenstra and Tseng, 1992). Staining can be prevented by coating the cells with mucin, albumin or carboxycellulose.

Positive staining of the ocular surface is particularly sensitive and specific for dry eyes (93% and 93%) (Golding and Brennan, 1993) and is useful for distinguishing it from meibomian gland disease (Plugfelder *et al.*, 1994). Severe disease is associated with more staining over a larger area and several grading systems exist to document this (van Biijsterveld, 1969; Laroche and Campbell, 1988; Whitcher, 1987). For example, van Biijsterveld (1969) divides the conjunctiva into temporal, central and nasal thirds and grades the staining intensity from 0–3 giving a maximum score of '9' (see Figure 9.2). Xu *et al.* (1995) grades the staining as follows: 0, no staining; 1, scattered minute; 2, moderate spotty; 3, diffuse blotchy staining.

The pattern of Rose Bengal staining is helpful in distinguishing various ocular surface diseases. For example, in dry eyes there is diffuse staining which may be more marked inferiorly, lid margin disease may cause localised bulbar conjunctival staining directly adjacent to the affected lid margin, hard contact lens wearers will often have characteristic staining of the peripheral cornea at "3 & 9 o'clock or at 4 & 8 o'clock" due to epithelial dehydration at the lens edge (Businger, Treiber and Flury, 1989). Because of the sensitivity of the test artefactual changes are easily induced by a Schirmer test, lid eversion or applanation tonometry, each of which lead to characteristic staining patterns. Patients

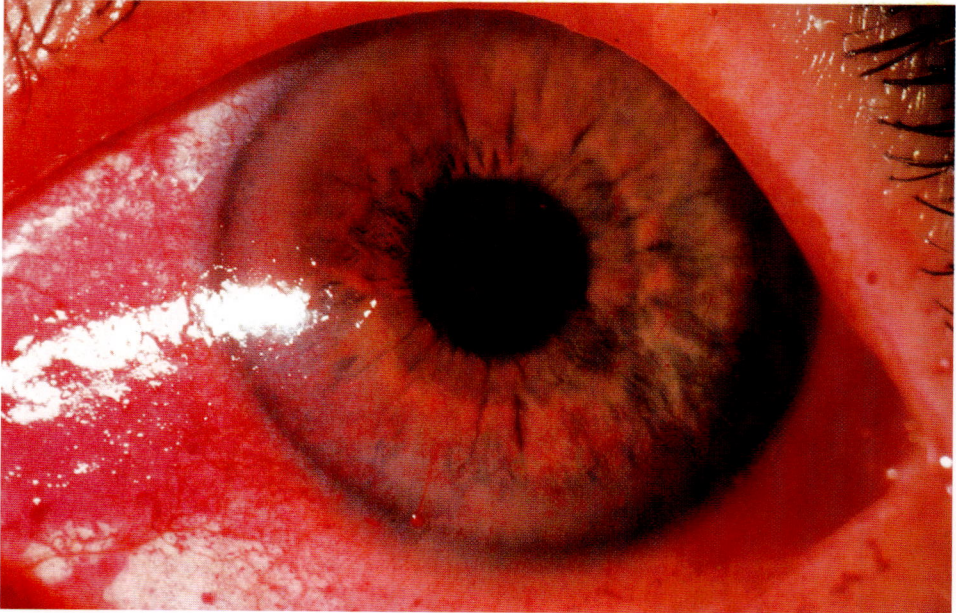

Figure 9.2   Rose bengal staining of the cornea and conjunctiva in dry eye showing the most severe grade. (Courtesy of Peter Wright).

using topical acetylcysteine drops will also have a generalised increase in the staining because of the mucolytic effect (Feenstra and Tseng, 1992; Thermes, Molon-Noblet and Grove, 1991).

Rose Bengal is toxic to any ocular surface cell that is not covered in mucin. Cell death results at concentrations of 0.01% and there is an additional photodynamic effect. This makes it particularly irritating to very dry eyes and some patients may have discomfort for several days after exposure (Feenstra and Tseng, 1992).

Impression cytology   Impression cytology is the technique whereby cells from the ocular surface are removed using cellulose acetate filter paper and then examined cytologically. This technique can determine cell morphology, goblet cell density and epithelial mucin expression all of which can be altered in dry eyes (Tseng *et al.*, 1985; Tseng and Farazdaghi, 1988; Pflugfelder *et al.*, 1994).

THE GENERAL OCULAR EXAMINATION

A general ocular examination and adnexal examination is mandatory in ocular surface disease. Corneal sensation and a normal blink reflex should be determined before topical anaesthesia is instilled. Conjunctival and corneal sensitivity can be quantified using a Cochet-Bonnet aesthesiometer (Burton, 1992).

SYSTEMIC EXAMINATION

A general medical examination may be rewarding in dry eyes. Patients may have a dry mouth in Sjogren's syndrome and this is confirmed by the absence of pooling of saliva under the tongue and there may be skin or joint manifestations of the collagen diseases.

BLOOD TESTS

Blood tests may be useful in confirming the presence of systemic diseases that have minimal overt clinical signs. For example, antibodies SS-A (anti-Ro) and SS-B (anti-La) in Sjogren's syndrome, antinuclear antibody (ANA) in the collagen diseases, rheumatoid factor in rheumatoid arthritis and fasting glucose in suspected diabetes (Wilson *et al.*, 1991).

THE CLINICAL APPROACH TO DIAGNOSIS

The clinician has a wide variety of means of collecting data about the ocular surface but clearly not all methods are particularly sensitive or specific to any given disease process. To make a diagnosis of dry eyes, the most important aspect is the history and the importance of the various aspects can be garnered from McMonnies' questionnaire (Table 9.2). Next, all patients need a complete clinical examination of the anterior and posterior segments of the globe and the ocular adnexa. Thereafter, a variety of clinical tests are required. For the diagnosis of aqueous deficient dry eyes, the best clinical tests are given in Table 9.3. It is clear that there is a clinical trade-off between a test being sensitive enough to diagnose the condition most of the time and being specific enough not to make a false diagnosis. The use of several tests is often but not always helpful. Goren and Goren (1988) conclude that the best test is to rely only on tear film lactoferrin. In their study, combining lactoferrin with any other test reduces the specificity of the outcome from 90% to 69% or less (Table 9.3). Unfortunately, lactoferrin assays are often not routinely available. Furthermore, in their study, any Rose Bengal staining was considered abnormal whereas the this test becomes very sensitive and specific when an eye scores more than 3.5 of a maximum of 9 points on the scale of van Bijsterveld (1969). Golding and Brennan (1993) conclude that the combination of McMonnies' questionnaire, TBUT <10 s and Rose Bengal staining are the best single and combination of tests. Dry eyes can be distinguished from MGD by an abnormal Schirmer test, rapid TBUT, irregular keratoscopic mires, greater Rose Bengal and fluoroscein staining and reduced goblet cells on impression cytology (Pflugfelder *et al.*, 1994).

THE CLINICAL MANAGEMENT OF DRY EYES

DEFINING ACHIEVABLE OBJECTIVES

The diagnosis of "dry eyes" is a catch-all for many diseases of varying severity. The disease is typically chronic, the response to treatment is variable and for those patients with severe lacrimal gland atrophy, no cure is possible. Further, there is considerable

TABLE 9.3
Clinical Tests for the Diagnosis of Dry Eyes

Test	Sensitivity	Specificity
McMonnies' questionnaire[1]	87%	87%
Schirmer I test		
< 8 mm[2]	60–74%	77%
< 5.5 mm[3]	83%	85%
Rose Bengal staining		
any staining[2]	28–36%	90%
score >3.5/9[3]	95%	96%
Lactoferrin assay[2]	68–71%	90%
Break up time (TBUT)		
abnormal < 10 s[1]	73%	83%
abnormal < 8 s[2]	54–60%	72%
Non-invasive break up time (NIBUT)[4]	82%	86%
Fluoroscein staining[1]	67%	100%
Schirmer test plus Rose bengal[2]	72–82%	49%
Schirmer test plus TBUT[2]	72–83%	56%
Schirmer's test plus lactoferrin assay[2]	76–85%	69%
Tear function index[5]	79%	92%

[1] Golding and Brennan, 1993
[2] Goren and Goren, 1988
[3] van Bijsterveld, 1969
[4] Mengher, Pandher and Bron, 1986
[5] Xu *et al.*, 1995

disparity between patients symptoms and their clinical signs and tests. Therefore it is very important that before treatment is initiated the management objectives are clearly defined, achievable and satisfy the patient's reasons for consultation. Some patients will be content with no treatment other than the advice that they do not have any blinding disease. Other patients will be able to manipulate their environment to adequately control their symptoms while some will require medical and surgical intervention. The principles of management are to maximise the visual potential and minimise the symptoms so that there is no impairment of the patients occupational, economic or social life.

Management principles

Management aims to minimise tear loss by evaporation, provide replacement tears, treat inflammation as required, prevent iatrogenic disease, manage lid margin disease and manage any ocular complications. Most important is to first do no harm, *"primum non nocere"*.

ENVIRONMENTAL ASPECTS OF DRY EYES

The maintenance of an adequate tear film requires a balance between its production, drainage via the nasolacrimal apparatus and evaporation. Evaporative losses make up 25% of total losses and the environment has a major influence on this proportion. Air humidity, temperature, wind and direct radiant heating from sunshine all increase evaporation. These are enhanced by a larger palpebral aperture, more prominent eyes and a reduced blink rate. Patients need to be aware of this physiology so that they can manipulate their environment and hence minimise their symptoms. For example, air conditioned rooms may be adjusted to increase the humidity, car ventilation systems should avoid a direct draft of air to the eyes and sunglasses and hats may reduce solar and wind-related drying. Users video display units (VDUs) often have symptoms of dry eyes mainly due to reduced blinking and hence they should be encouraged to actively blink throughout the day or to punctuate their work with other activities (Nakamori et al., 1993, 1994). The daily and seasonal weather changes often causes significant variations in symptoms. In climates with damp winters, many patients may be symptom-free during this period. An occupational aspect of the disease may be noted if the symptoms stop during a holiday and similar changes may be noted due to geographic relocation.

Of all the aspects of the management of dry eyes, the environmental are most important. It allows the patient to understand their disease and to be able to control some or much of their symptoms without medication. This "handing back" of the control of a disease to the patient is essential for any chronic condition.

TEAR FILM SUPPLEMENTS

The mainstay of management is tear film supplements. These include saline drops, hypotonic solutions and drops containing polymers. The addition of various polymers such as the polyethers, the polyvinyls, the dextrans or the cellulose derivatives attempt to either prolong their ocular retention time or improve the wetting characteristics (Gilbard, 1985). There are many over-the-counter type products available, each claiming various advantages. The clinical differences between each are small and there is typically more patient-patient variation than eyedrop-eyedrop variation. This idiosyncratic response means that a given patient should be given a choice of several types of eyedrop if they do not respond to the physicians first recommendation. Some patients may also prefer hypotonic drops although these are of limited commercial availability (Gilbard, 1985).

Patients with intermittent symptoms will usually only require tear supplementation during these times. Patients with continual symptoms should be instructed to install the drops every two hours for the first four days. This allows the patient to determine the maximum benefit from the drops. Thereafter, the medication frequency can be reduced according to symptoms, typically to 4–6 times/day. With severe disease, patients may require the drops every ten minutes. In these cases, saline drops should be tried as they have a lower viscosity and hence give clearer vision sooner after installing the drops. Installing drops may be difficult for elderly patients or those with arthritis of the hands such as those with rheumatoid arthritis. This must be evaluated and the patients should be observed putting their drops in. A variety of commercial devices can assist this task and it should be noted that some brands of eyedrop bottles are very rigid and difficult to "squeeze" the drops from. A formal assessment by Occupational Health personnel may be very valuable.

Ointment lasts much longer than eyedrops but impairs the visual acuity because it adds an irregular greasy film to the ocular surface. It is a most useful adjunct to eyedrops if given just before bedtime or for those patients with symptoms maximal after awaking. Very severe disease may get no benefit from eyedrops and require ointment up to two hourly. Other topical medication should be installed before the ointment as it may act as a barrier to penetration. Carbomer gels are intemediate in viscosity between ointment and drops.

Slow-release tear film supplements

When a solid rod of hydoxylpropylcellulose is placed into the lower conjunctival fornix, the subsequent water absorbtion leads to dispersal of this lubricant over approximately 24 h (Lacrisert, Merke Sharpe and Dohme). This system offers no ocular benefit over conventional drops and may dislodge although some patients, with enough tears to be absorbed by the lubricant, may find it convenient (Lamberts, 1980).

TOXICITY OF TOPICAL MEDICATION

Almost all commercially available eyedrops contain preservatives, typically benzalkonium chloride. When used chronically, these agents are toxic to the corneal and conjunctival epithelium and cause chronic inflammatory cell infiltration of the conjunctival stroma (Wilson, Duncan and Jay, 1975; Burstein, 1985; Broadway et al., 1994). Patients using eyedrops every hour are also administering preservatives every hour and this effect is additive with multiple medications. This may provide a vicious cycle whereby the dry-eye symptoms are exacerbated by the preservatives but soothed by the lubricants. The alternatives are to either stop all topical treatment or to switch to non-preserved medication as a trial. These are usually commercially available although are often expensive and potentially carry a small increased risk of infection. The over-riding principle of topical medication is to maximise the benefit while minimising the side-effects (Burstein, 1985).

ANTI-INFLAMMATORY AGENTS

A significant proportion of patients with dry eyes may have clinical features of conjunctival inflammation. This may be a consequence of the mechanical effects of the lack of lubrication, the effects of topical medication or those of lid margin disease. Often the inflammation will resolve with adequate management of these areas but there is a small group of patients for whom the inflammation *per se* exacerbates the surface disease. These patients will often benefit from a short course of topical steroids such as dexamethasone 0.1% or prednisolone acetate 0.3% four times a day. If there is no clinical benefit after one month, then this medication should be stopped. If there is benefit, then it should be tapered off and stopped within 2–3 months. These patients often report significant symptomatic benefit from the steroids and want to continue with them for prolonged periods but because of the chronicity of the disease, the patients are at a high risk of cataracts and glaucoma. Only in extreme circumstances is long term steroids warranted for severe dry eyes, and then only at very low doses and frequencies such as prednisolone acetate 0.1% once a day. There has been recent interest in the use of topical cyclosporine for treatment of dry eye syndromes in animals (Kaswan, 1994; Jabs *et al.*, 1996) that has identified the potential for the use of this drug in humans. Cyclosporin, when used topically, has no apparent serious side effects, in particular those of glaucoma and cataract that are associated with steroid use. However most preparations have been difficult to tolerate. Several large clinical trials are currently underway in humans and results are awaited with great interest. Some small trials have been reported which suggest that it may be effective in humans although the mechanism is possibly more complex than a reduction in the level of inflammatory activity in the conjunctiva and lacrimal gland (Gunduz and Ozdemir, 1994).

PRESERVATION OF THE EXISTING TEARS

The existing tear film can be preserved by blocking the lacrimal puncti and hence preventing nasolacrimal drainage although this risks a watery eye, either constantly or intermittently. Patients most suited to occlusion of the puncti have a reduced tear film and reduced reflex tearing. This is evidenced by a Schirmer test of <5.5 mm and an inability to produce excess tears such as when peeling an onion or watching a sad movie. This subgroup should have the puncti temporarily occluded either with commercially available punctal plugs, gelatine rods or a 10/0 nylon suture. If the patient does not have epiphora over the subsequent one week, then the puncti should be permanently occluded. This can be achieved with hot-wire cautery which is inserted unheated about 6 mm into the dilated puncti, heated and slowly withdrawn. Silicone punctum plugs can be left in to achieve semi-permanent or permanent occlusion. During the temporary occlusion it is unimportant whether the patient notices symptomatic improvement or requires less tear supplements as the ocular surface may take weeks to improve (Wright, 1985a). The puncti can occasionally recanalise and this must be assessed if symptoms subsequently worsen.

Figure 9.3 summarises the principal management options in patients diagnosed with aqueous deficient dry eye.

MANAGEMENT OF COMPLICATIONS

Punctate keratopathy

Punctate keratopathy (superficial punctate keratitis, SPK, or punctate epithelial keratopathy, PEK) reflects damage to the ocular surface from any the disease processes involved in dry eyes. Each problem needs managing in turn and the pattern of the staining is a clue to the major cause: inferior — lid margin disease, exposure due to ectropion, lagophthalmos, topical drug toxicity; superior — superior limbic keratitis with or without thyroid disease; "3 & 9 o'clock" — contact lens desiccation; diffuse — aqueous tear film deficiency (Coster, 1988).

Excess mucous

Dry eyes frequently accumulate excess mucus both because there is a reduced volume of aqueous to wash it away and because it is produced in excess secondary to conjunctival

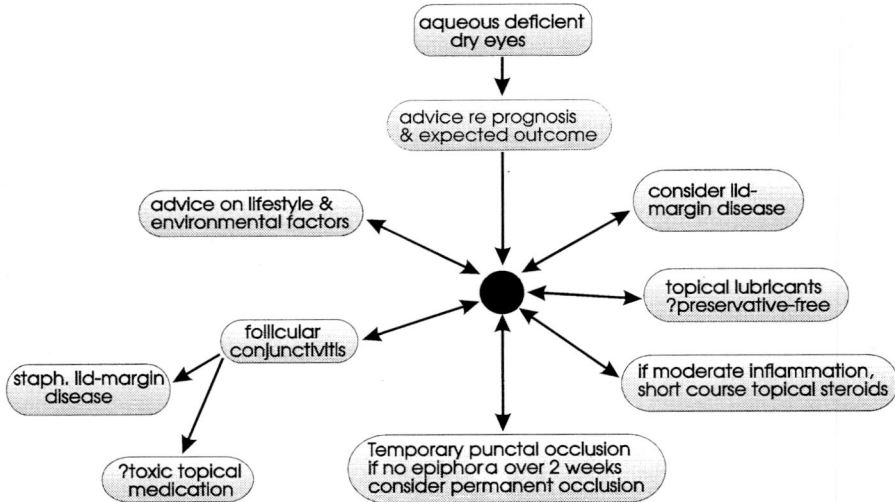

Figure 9.3   A summary of the management options for the aqueous deficient dry eye.

inflammation. Adequate lubrication is required; inflammation may require treating with topical steroids and the excess mucous can be dissolved with topical acetylcysteine 5–10% 2–4 times per day.

Filamentary keratitis

Filamentary keratitis results from the adhesion of small strands of mucous or epithelial cells which become adherent to the corneal surface (see Figure 9.4). They are easily seen with the slitlamp microscope and pull on the cornea with blinking, resulting in detachment of the adherent corneal epithelium, thereby causing discomfort. Individual filaments can be removed with forceps. Their presence usually indicates more severe disease, often with conjunctival inflammation. The initial management is to provide adequate lubrication, resolve any lid margin disease and to treat any conjunctival inflammation. Thereafter, topical acetylcysteine 5–10% may be helpful in dissolving the filaments or preventing their development (Jones and Coop, 1965). In appropriate cases therapeutic contact lenses may be very helpful in relieving symptoms (see below).

Persistent epithelial defect

A macroscopic corneal epithelial defect is usually associated with ocular exposure, severe surface disease (see Figure 9.5) or herpetic viral infection (an indolent ulcer, often in a geographic pattern). Bacterial and viral infection should be excluded even if the surrounding cornea is transparent with no clinical signs of infection as is commonly the case in patients being treated with steroids that suppress the signs of infection. Lubricating ointments are required every 2–3 hours. The eye may benefit from padding. If a defect returns after padding and intensive lubrication, then either a temporary surgical tarsorrhaphy or temporary ptosis, induced with a botulinum toxin injection lasting for about 6 weeks, is required (Kirkness et al., 1988). Exposure must be sought and treated aggressively as any epithelial defect dramatically increases the risk of infectious keratitis

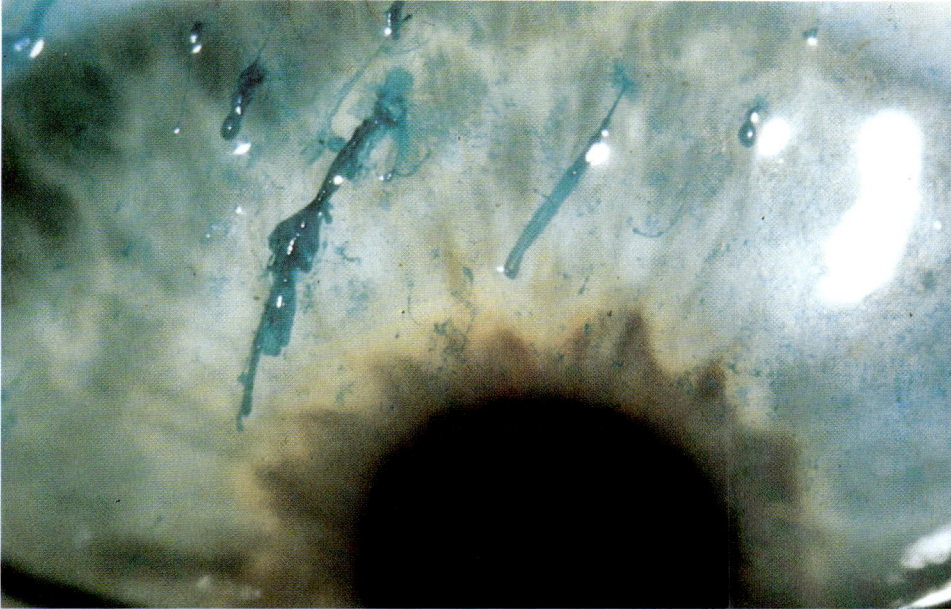

Figure 9.4   Filamentary keratitis stained with lissamine green. (Courtesy of Peter Wright).

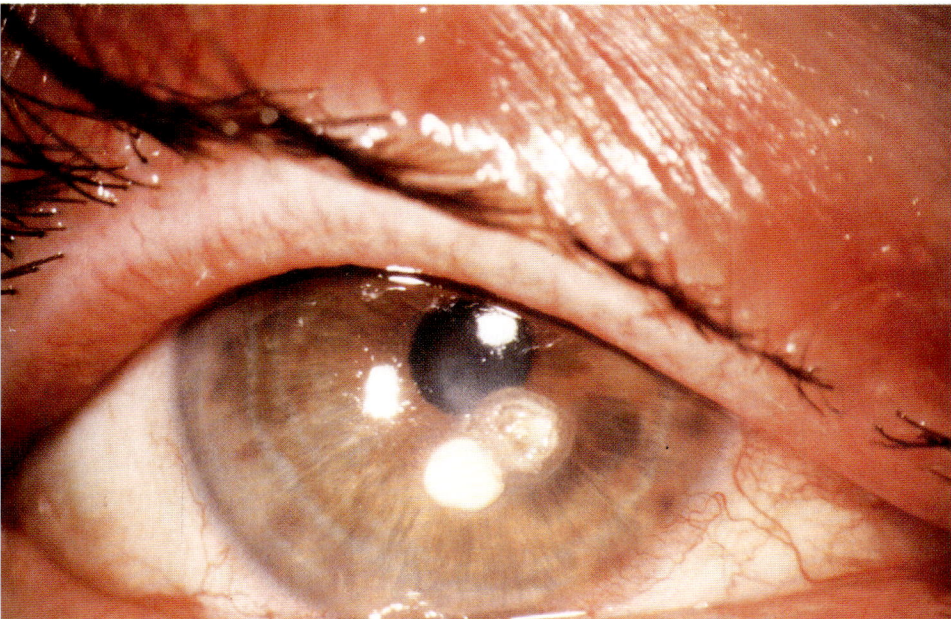

Figure 9.5   A persistent epithelial defect due to severe surface disease in Sjogren's syndrome. The persistent epithelial defect has been present for some months and has a mucous plaque in the base.

and patients should be on prophylactic antibiotics until the defect has healed. In the short term, the lids may be taped together overnight. Persistent exposure may require a surgical tarsorrhaphy, a botulinum-induced ptosis or corrective plastic surgery for ectropion.

MEANS OF IMPROVING ENDOGENOUS TEAR PRODUCTION

It is often possible to stimulate additional lacrimal gland production even in dry eyes and this has been used as the basis for therapy. Parasympathomimetic drugs such as intravenous mecholyl stimulate tear production in patients with Riley-Day syndrome and with KCS (Bron, 1985). Other secretogogues such as oral bromhexine, physalaemin, eledoisin, evening primrose oil and topical pilocarpine all increase the lacrimal gland production of tears but as yet there is no clinical role for any of these agents (Bron, 1985).

THE ROLE OF CONTACT LENSES

Contact lenses have a minor therapeutic role in dry eyes. In the short term, soft (hydrogel) contact lenses may be used as a bandage to aid healing of epithelial defects. Unfortunately lens wear increases evaporation from an ocular surface that is already tear deficient, added to which these lenses require hydration. Because of these conflicting factors hydrogel lenses often dehydrate in the dry eye resulting in reduced oxygen transmission through the lens causing hypoxia and altered lens conformation, tightening or displacement. Even when fully hydrated the lenses always cause some epithelial hypoxia which may also contribute to surface damage and an increased risk of bacterial keratitis. For these reasons epithelial defects are usually best treated by other means in severely dry eyes, although in eyes with only moderate aqueous tear deficiency hydrogel lenses can be useful, particularly for treating filamentary keratopathy that is resistant to medical therapy. However, other types of therapeutic lenses, in particular silicone rubber (Bacon, Astin and Dart, 1994) and rigid gas permeable or scleral lenses (Foss, Trodd and Dart, 1994) can be used as they are not dependant on the tear film for hydration. Silicone rubber lenses which have excellent oxygen transmission, require minimal lubrication and are rigid enough to provide a regular anterior refracting surface. They are indicated for persistent epithelial defects where the eye is required to see, *vis-a-vis* tarsorrhaphy (see Figure 9.6). The lenses may also improve the visual acuity if the underlying cornea is irregularly shaped. These lenses are not widely available and are difficult to fit but may be the only means of improving visual function in severe diseases such as Stevens Johnson syndrome.

   Some patients with dry eyes want to wear contact lenses rather than glasses for cosmetic reasons. These patients may have reduced wearing times and have increased risks of infection but this may be acceptable, especially for patients with very large refractive errors. For soft lenses, but not for hard lenses, preserved tear film supplements are contraindicated while are being worn as the lens absorbs the drug and its preservatives.

RETINOIC ACID

Patients with very dry eyes may develop areas of ocular surface squamous metaplasia or keratinisation which may cause symptoms of discomfort. Topical retinoic acid, 0.05%

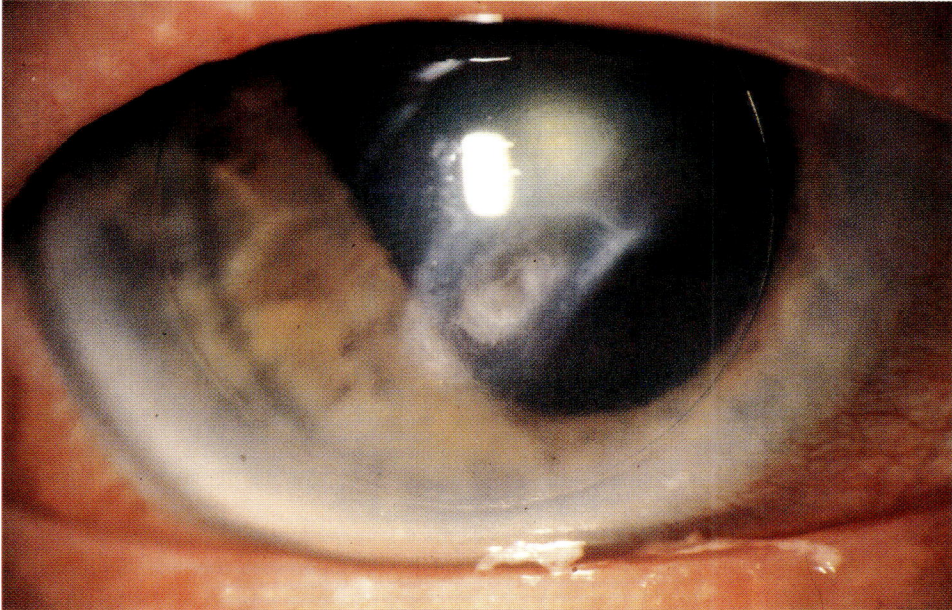

Figure 9.6   A silicone rubber contact lens used to treat a persistent epithelial defect, that has lead to a micro-perforation of the cornea, in Sjogren's syndrome.

once a day, is sometimes able to reverse this metaplasia both macroscopically and cytologically (Soong *et al.*, 1988; Tseng and Farazdaghi, 1988). Unfortunately there is usually minimal symptomatic improvement and the drops can be very irritating to the ocular surface (Tseng *et al.*, 1985; Wright, 1985b).

SURGICAL MANAGEMENT OF DRY EYES

Surgical management can be used to normalise lid architecture, as an adjunct to the medical treatment of complications, to increase ocular wetting, to minimise tear film loss and to improve the visual acuity. Patients with ectropion, chronic exposure or trichiasis need appropriate surgical procedures to provide normal lid-globe contact with a normal palpebral aperture that closes fully. A permanent lateral or medial tarsorrhaphy may result in a reduced palpebral aperture consistent with a satisfactory long-term outcome. Corneal scarring that reduces the visual acuity can often be corrected with a penetrating keratoplasty although the risks of failure are higher.

Various motorised pumps can provide supplementary tears to the eyes and while they are not yet clinically useful, this may change (Doane, 1980). Historically, various operations have been described that allow parotid duct secretions to wash over the eye but the flow is typically unmetered and too great and the procedure is no longer performed (Bennet and Bailey, 1957). Recently, submandibular gland transplantion to the temporal fossa, with a microvascular anastomosis, has been performed to solve the problems associated with parotid duct transplantation (Kumar *et al.*, 1990; MacLeod and Robbins, 1992). The submandibular gland produces a mixed seromucinous secretion, transplantation of the gland abolishes the gustatory reflex that proved a problem for parotid gland transplantation and any excess secretion can be managed by a gland reduction. Early results have shown that that the procedure has the potential to reduce, or eliminate, the requirement for pharmaceutical lubrication and it is hoped that it may result in an improvement in the ocular surface such that keratoplasty can be carried out successfully, a procedure that is at present contraindicated in the very dry eye (Geerling *et al.*, 1998).

OVERVIEW OF THE MANAGEMENT OF DRY EYES

The key features of managing patients with dry eyes are to allow the patient to improve their own environment, to use supplementary lubricants appropriately, to manage the lid margin diseases and the complications adequately and to avoid iatrogenic disease. Sometimes unusual medical and surgical options are required. Patients' problems are dynamic and change with time and with the seasons.

In a five year longitudinal study of 98 patients with severe dry eyes, all patients were initially prescribed the alkaline topical lubricant BJ6 four times a day and instructed to bathe their eyes in saline solution twice a day and to remove excess mucus (Jones and Coop, 1965; Williamson *et al.*, 1974). Fifty percent of patients maintained improvement of their signs and symptoms over five years. Patients that did not, had punctal occlusion and of these 70% maintained improvement of their signs and symptoms over the next three years. The 20 patients that failed to respond to topical lubricants and punctal occlusion were treated with topical acetylcysteine 5% and of these, 30% maintained improvement in signs and symptoms over one year. Recurrent infection, particularly *Staphylococcus* occurred in 21% and corneal ulceration occurred in 7% over the five years (Williamson *et al.*, 1974). This study clearly illustrates the long-term outcomes for patients with dry eye.

Tears and the dry eye remain challenging fields for both the laboratory and clinical scientist. This review has shown that there are many unanswered questions about the homeostatic mechanisms affecting the production and regulation of the normal tear film. The role of the ocular surface epithelium in the maintenance of a stable tear film is only now becoming understood. The pathogenesis of tear film disorders and their treatment has been a stimulating field with the prospect of new developments in the understanding of this group of disorders and their management.

REFERENCES

Anderson J.A., Whaley, K., Williamson, J. and Buchanan, W.W. (1972). A statistical aid to the diagnosis of KCS. *Quarterly Journal of Medicine New Series XLI*, **162**, 175–189.
Bacon, A.S., Astin, C. and Dart, J.K.G. (1994). Silicone rubber contact lenses for the compromised cornea. *Cornea*, **13**, 422–428.
Balik, J. (1952). The lacrimal fluid in keratoconjunctivitis sicca: A quantitative and qualitative investigation. *American Journal of Ophthalmology*, **35**, 773–782.

Baum, J. (1985). Clinical manifestations of dry eye states. *Transactions of the Ophthalmological Society UK*, **104**, 415–423.

Baum, J. and Bull, M.J. (1974). Ocular manifestations of the ectodactyl ectodermal dysplasia, cleft lip-palate syndrome. *American Journal of Ophthalmology*, **78**, 211–16.

Bennett, J.E. and Bailey, A.L. (1957). Surgical transplanatation of total xerophthalmia. transplanation of the parotid duct to the inferior cul-de-sac with report of a case. *Archives of Ophthalmology*, **58**, 372–374.

Bernauer, W., Itin, P.H. and Kirtschig, G. (1997). Cicatricial pemphigoid. *Documenta Ophthalmologica*, **28**, 46–63.

Bisgaard, H., Ford-Hutchinson, A.W., Charleson, S. and Taudorf, E. (1985). Production of leukotrienes in human skin and conjunctival mucosa after specific allergen challenge. *Allergy*, **40**, 417–423.

Brauninger, G.E., Shah, D.O. and Kaufman, H.E. (1972). Direct physical demonstration of the oil layer on the tear film surface. *American Journal of Ophthalmology*, **73**, 132–134.

Broadway, D.C., Grierson, I., O'Brien, C. and Hitchings, R.A. (1994). Adverse effects of topical antiglaucoma medication. I. The conjunctival cell profile. *Archives of Ophthalmology*, **112**, 1437–1445.

Bron, A.J. (1973). Vortex patterns of the corneal epithelium. *Transactions of the Ophthalmological Society UK*, **93**, 455–472.

Bron, A.J. (1985). Prospects for the dry eye. *Transactions of the Ophthalmological Society UK*, **104**, 801–826.

Browning, D.J. and Proia, A.D. (1983). Ocular rosacea. *Survey of Ophthalmology*, **31**, 145–158.

Buchner, A., Lozada, F. and Silverman, S. (1980). Histopathologic spectrum of oral erythema multiforme. *Oral Surgery, Oral Medicine, Oral Pathology*, **49**, 221–228.

Burstein, N.L. (1985). The effects of topical drugs and preservatives on the tears and corneal epithelium in dry eye. *Transactions of the Ophthalmological Society UK*, **104**, 402–409.

Burton, H. (1992). Somatic sensations from the eye. In *Adler's Physiology of the Eye: Clinical Application, 9th edition*, edited by W.M. Hart, pp. 71–100, St Louis: Mosby.

Businger, U., Trieber, A. and Flury, C. (1989). The aetiology and management of three and nine o'clock staining. *International Contact Lens Clinics*, **16**, 136–139.

Chen, J.J.Y. and Tseng, S.C.G. (1990). Corneal epithelial wound healing in partial limbal deficiency. *Investigative Ophthalmology and Visual Science*, **31**, 1301–1314.

Clinch T.E., Benedetto, D.A., Felberg, N.T. and Laibson, P.R. (1983). Schirmer's test. A closer look. *Archives of Ophthalmology*, **101**, 1383–1386.

Cope, C., Dilly, P.N., Kaura, R. and Tiffany, J.M. (1986). Wettability of the corneal surface: a reappraisal. *Current Eye Research*, **5**, 777–785.

Coster, D.J. (1988). Superficial keratopathy. In *Clinical Ophthalmology, Volume 4*, edited by T.D. Duane and E.A. Jaeger, pp. 1–8,, Philadelphia: JB Lippincott Company.

Cotsarelis, G., Cheng, S.-Z., Dong, G., Sun,T.-T. and Lavker, R.M. (1989). Existence of slow-cycling limbal epithelial basal cells that can be preferentially stimulated to proliferate:Implications on epithelial stem cells. *Cell*, **57**, 201–208.

Craig, J.P., Blades, K., Patel, S. and Sturrock, R.D. (1995). Tear lipid layer structure and stability following expression of the meibomian glands. *Ophthalmic and Physiological Optics*, **15**, 569–574.

Damato, B.E., Allan, D., Murray, S.B. and Lee, W.R. (1984). Senile atrophy of the lacrimal gland: the contribution of chronic inflammatory disease. *British Journal of Ophthalmology*, **68**, 674–80.

Dhir, S.P., Garg, S.K., Sharma, Y.R. and Lath, N.K. (1979). Prostaglandins in human tears. *American Journal Ophthalmology*, **87**, 403–404.

Dilly, P.N. and Mackie, I.A. (1981). Surface changes in the anaesthetic conjunctiva in man with special reference to the production of mucus from a non-goblet-cell source. *British Journal of Ophthalmology*, **65**, 833–842.

Doane, M.G. (1980). Methods of ophthalmic fluid delivery. *International Ophthalmology Clinics*, **20**, 93–101.

Dohlman, C.H., Friend, J., Kalevar,.E., Yagonda, D. and Balazs, E. (1976). The glycoprotein (mucus) content of tear from normals and dry eye patients. *Experimental Eye Research*, **22**, 359–365.

Ehlers, W.H. and Donshik, P.C. (1992). Allergic ocular disorders: a spectrum of diseases. *The CLAO Journal*, **18**, 117–124.

Elder, M.J. (1997). Keratopathy in chronic progressive conjunctival cicatrisation. *Documenta Ophthalmologica*, **28**, 182–191.

Farris, R.L., Gilbard, J.P., Stuchell, R.N. and Mandell, I.D. (1983). Diagnostic tests in keratoconjunctivitis sica. *The CLAO Journal*, **9**, 23–28.

Feenstra, R.P. and Tseng, S.C. (1992). Comparison of fluoroscein and rose bengal staining. *Ophthalmology*, **99**, 605–617.

Forster, H.W. Jr. (1951). Rose bengal test in diagnosis of defiecient tear formation. *Archives of Ophthalmology*, **45**, 419–424.

Foss, A.J.E., Trodd, T.C. and Dart, J.K.G. (1994). Current indications for scleral contact lenses. *The CLAO Journal*, **20**, 115–118.

Foster, C.S. (1986). Cicatricial pemphigoid. *Transactions of the American Ophthalmological Society*, **84**, 527–663.

Friedlaender, M.H., Ohashi, Y. and Kelley, J. (1984). Diagnosis of allergic conjunctivitis. *Archives of Ophthalmology*, **102**, 1198–1199.

Geerling, G., Sieg, P., Bastian, G.-O. and Laqua, H. (1998). Transplantaion of the autologous submandibular gland for most severe cases of keratoconjunctivitis sicca. *Ophthalmology*, **105**, 327–335.

Gilbard, J.P. (1985). Topical therapy for dry eyes. *Transactions of the Ophthalmological Society UK*, **104**, 484–488.

Gilbard, J.P., Farris, R.L. and Santamaria, J. (1978). Osmolarity of tear microvolumes in keratoconjunctivitis sicca. *Archives of Ophthalmology*, **96**, 677–681.

Gipson, I.K. (1989). The epithelial basement membrane zone of the limbus. *Eye*, **3**, 132–140.

Gipson, I.K. (1994). Evidence that the entire ocular surface epithelium produces mucins for the tear film. *Investigative Ophthalmology and Visual Science (Suppl)*, **35**, 2589.

Gipson, I.K., Spurr-Michaud, S. and Tisdale, A.S. (1987). Anchoring fibrils form a complex network in human and rabbit cornea. *Investigative Ophthalmology and Visual Science*, **28**, 212–220.

Goldberg, M.F., Payne, J.W. and Brunt, P.W. (1968). Opthalmic studies of familial dysautonomia: The Riley-Day syndrome. *Archives of Ophthalmology*, **80**, 732–743.

Golding, T.R. and Brennan, N.A. (1993). Diagnostic accuracy and intercorrelation of clinical tests for dry eye. *Investigative Ophthalmology and Visual Science (Suppl)*, **34**, 605.

Goodwin, J.S. and Ceuppens, J. (1983). Regulation of the immune response by prostaglandins. *Journal of Clinical Immunology*, **3**, 295–315.

Goren, M.B. and Goren, S.B. (1988). Diagnostic tests in patients with symptoms of keratoconjunctivitis sicca. *American Journal of Ophthalmology*, **106**, 570–574.

Greiner, J.V., Weidman, T.A., Korb, D.R. and Allansmith, M.R. (1985). Histochemical analysis of secretory vesicles in nongoblet conjunctival epithelial cells. *Acta Ophthalmologica*, **63**, 89–92.

Gunduz, K. and Ozdemir, O. (1994). Topical cyclosporin treatment of keratoconjunctivitis sicca in secondary Sjogrens syndrome. *Acta Ophthalmologica*, **72**, 438–442.

Gutgesell, V.J., Stern, G.A. and Hood, C.I. (1982). Histopathology of meibomian gland dysfunction. *American Journal of Opthalmology*, **94**, 383–387.

Hanna, C., Bicknell, D.S. and O'Brien, J.E. (1961). Cell turnover in the adult human eye. *Archives of Ophthalmology*, **65**, 695–698.

Hart WM. (1992). The eyelids. In *Adler's Physiology of the Eye: Clinical Application*, 9th edition, edited by W.M. Hart, pp. 1–17, St Louis: Mosby.

Hay, E.D. (1980). Development of the vertebrate cornea. *International Review of Cytology*, **63**, 263–267.

Holden, B.A. and Sweeney, D.F. (1985). The oxygen tension and temperature of the superior palpebral conjunctiva. *Acta Ophthalmologica*, **63**, 100–103.

Holly, F.J. (1973). Formation and rupture of the tear film. *Experimental Eye Research*, **15**, 515–525.

Holly, F.J. (1985). Physical chemistry of the normal and diseased tear film. *Transactions of the Ophthalmological Society UK*, **104**, 374–380.

Holly, F.J. and Lemp M.A. (1977). Tear physiology and dry eyes. *Survey of Ophthalmology*, **22**, 69–87.

Hom, M.M., Martinson, J., Knapp, L. and Paugh, J.R. (1990). Prevalence of meibomian gland dysfunction. *Optometry and Vision Science*, **67**, 710–712.

Huff, J.C., Weston, W.L. and Tonnesen, M.G. (1983). Erythema multiforme: a critical review of characteristics, diagnostic criteria, and causes. *Journal of the American Academy of Dermatology*, **8**, 763–775.

Iwamoto, T. and Smelser, G.K. (1965). Electron microscope studies on the mast cells and the blood and lymphatic capillaries of the human corneal limbus. *Investigative Ophthalmology*, **4**, 815–834.

Jabs, D.A, Lee, B., Burek, C.L., Saboori, A.M. and Prendergast, R.A. (1996). Cyclosporine therapy supresses ocular and lacrimal gland diseasse in MRL/Mp-lpr/lpr mice. *Investigative Ophthalmology and Visual Science*, **37**, 377–383.

Jakobiec, F.A. (Ed) (1983). *Ocular Anatomy, Embryology and Teratology*, pp. 591, 592, 733–760. Philadelphia: Harper & Row.

Janssen, P.T. and van Bijsterveld, O.P. (1983). A simple test for lacrimal gland function. A tear lactoferrin assay by radial immunodiffusion. *Graefes Archive for Clinical and Experimental Ophthalmology*, **220**, 171–174.

Jester, J., Nicolaides, N. and Smith, R.E. (1981). Meibomium gland studies: histologic and ultrastructural investigations. *Investigative Ophthalmology and Visual Science*, **20**, 537–547.

Jones, L.T. (1966). The lacrimal secretory system and its treatment. *American Journal of Ophthalmology*, **62**, 47–60.

Jones, B.R. and Coop, H.V. (1965). The management of keratoconjuctivitis sicca. *Transactions of the Ophthalmological Societiesof the UK*, **85**, 379–390.

Jordan, A. and Baum, J.L. (1980). Basic tear flow, does it exist? *Ophthalmology*, **87**, 920–930.

Kaswan, R. (1994). Characteristics of a canine model of KCS: effective treatment with topical cyclosporine. *Advances in Experimental Medicine and Biology*, **350**, 583–594.

Kessing, S.V. (1968). Mucous gland system of the conjunctiva. A quantitative anatomical study. *Acta Ophthalmologica Supplement*, **95**, 1.

Kijlstra, A. and Jeurissen, S.H. (1982). Modulation of classical C3 convertase of complement by tear lactoferrin. *Immunology*, **47**, 263–270.

Kijlstra, A., Jeurissen, S.H. and Koning, K.M. (1983). Lactoferrin levels in human tears. *British Journal of Ophthalmology*, **67**, 199–202.

Kinoshita, S., Kiorpes, T.C., Friend, J. and Thoft, R.A. (1983). Goblet cell density in ocular surface disease. *Archives of Ophthalmology*, **101**, 1284–1287.

Kirkness, C.M., Adams, G.G., Dilly, P.N. and Lee, J.P. (1988). Botulinum toxin A induced ptosis in corneal disease. *Ophthalmology*, **95**, 473–480.

Klyce, S.D. (1972). Electrical profiles in the corneal epithelium. *Journal of Physiology*, **226**, 407–426.

Klyce, S.D. (1981). Stromal lactate accumulation can account for corneal oedema osmotically following epithelial hypoxia in the rabbit. *Journal of Physiology*, **321**, 49–64.

Klyce, S.D. and Crosson, C.E. (1985). Transport processes across the rabbit corneal epithelium: a review. *Current Eye Research*, **4**, 323–331.

Knutson, D.D. (1974). Ultrastructural observations in acne vulgaris: the normal sebaceous follicle and acne lesions. *Journal of Investigative Dermatology*, **62**, 288–307.

Kruppens, E.V., Stolwijk, T.R., de Keizer, R.J. and van Best, J.A. (1992). Basal tear turnover and topical timolol in glaucoma patients and health controls using fluorophotometry. *Investigative Ophthalmology and Visual Science*, **33**, 3442–3448.

Kruse, F.E., Chen, J.K., Tsai, R.J.F. and Tseng, S.C.G. (1990). Conjunctival transdifferentiation is due to the incomplete removal of limbal basal epithelium. *Investigative Ophthalmology and Visual Science*, **31**, 1903–1913.

Kumar, P.A.V., MacLeod, A.M., O'Brien, B. McC., Hickey, M.J. and Knight, K.R. (1990). Microvascular submandibular gland transfer for the management of xerophthalmia. An experimental study. *British Journal of Plastic Surgery*, **43**, 431–436.

Laroche, R.R. and Campbell, R.C. (1988). Quantitative Rose Bengal staining technique for external ocular diseases. *Annals of Ophthalmology*, **20**, 274–276.

Lamberts, D.W. (1980). Solid delivery devices. *International Ophthalmology Clinics*, **20**, 63–77.

Lamberts, D.W, Foster, C.S. and Perry, H.D. (1979). Schirmer test after topical anaesthesia and the tear meniscus height in normal eyes. *Archives of Ophthalmology*, **97**, 1082–1085.

Lemp, M.A. and Hamill, J.R. (1973). Factors affecting tear film break up time in normal eyes. *Archives of Ophthalmology*, **89**, 103–105.

Lemp, M.A. and Wolfley, D.E. (1992). The lacrimal apparatus. In *Adler's Physiology of the Eye: Clinical Application, 9th edition*, edited by W.M. Hart, pp. 18–28, St Louis: Mosby.

Lui, D. and Stasior, O.G. (1983). Lower eyelid laxity and ocular symptoms. *American Journal of Ophthalmology*, **95**, 545–551.

Mackie, I.A. and Seal, D.V. (1981). The questionably dry eye. *British Journal of Ophthalmology*, **65**, 2–9.

MacLeod, A.M. and Robbins, S.P. (1992). Microvascular submandibular gland transfer. An alternative approach for total xerophthalmia. *Australian and New Zealand Journal of Ophthalmology*, **20**, 99–103.

Mancini, G., Carbonara, A.O. and Heremans, J.F. (1965). Immuno-chemical quantification of antigens by radial immunodiffusion. *Journal of Immunochemistry*, **2**, 235–254.

Manthorpe, R., Frost-Larsen, K., Isaser, H. and Prause, J.U. (1981). Sjogrens syndrome. A review with emphasis on immunological features. *Allergy*, **36**, 139–153.

McCulley, J.P. and Sciallis, G.F. (1977). Meibomian keratoconjunctivitis. *American Journal of Ophthalmology*, **84**, 788–793.

McCulley, J.P. and Sciallis, G.F. (1983). Meibomian keratoconjunctivitis oculo-dermal correlates. *The CLAO Journal*, **9**, 130–132.

McCulley, J.P., Dougherty, J.M. and Denean, D.G. (1982). Classification of chronic blepharitis. *Ophthalmology*, **89**, 1173–1180.

McDonald, J.E. (1968). Surface phenomena of tear films. *Transactions ot the American Ophthalmological Society*, **66**, 905–939.

McDonald, J.E. (1969). Surface phenomena of the tear film. *American Journal of Ophthalmology*, **67**, 56–65.

McMonnies, C.W. and Ho, A. (1987a). Patient history screening for dry eye conditions. *Journal of the American Optometric Association*, **58**, 296–301.

McMonnies, C.W. and Ho, A. (1987a). Responses to a dry eye questionaire from a normal population. *Journal of the American Optometric Association*, **58**, 588–591.

McMonnies, C.W. (1986). Key questions in a dry eye history. *Journal of the American Optometric Association*, **57**, 512–517.

Mengher, L.S., Pandher, K.S. and Bron, A.J. (1986). Non-invasive tear film break-up time: sensitivity and specificity. *Acta Ophthalmologica*, **64**, 441–444.

Miller, N.R. (Ed) (1985) *Walsh and Hoyt's Clinical Neuro-Ophthalmology, Volume 2*. Baltimore: Williams and Wilkins.

Mishima, S. (1961). The oily layer of the tear film and evaporation from the corneal surface. *Experimental Eye Research*, **1**, 39–45.

Mishima, S. (1965). Some physiological aspects of the precorneal tear film. *Archives of Ophthalmology*, **73**, 233–241.

Mondino, B.J. (1990). Cicatricial pemphigoid and erythema multiforme. *Ophthalmology*, **97**, 939–952.

Nakamori, K., Nakajima, T., Odawara, M., Yoshida, T. and Tsubota, K. (1993). Dry eyes in VDT work. *Investigative Ophthalmology and Visual Science*, **34**, 3806.

Nakamori, K., Odawara, M., Nakajima, T., Yamada, K., Mizutani, T., Yoshida T. *et al.* (1994). Dry eyes and ocular fatique in VDT work. *Investigative Ophthalmology and Visual Science*, **35**, 2045.

Nichols, B.A., Chiappino, M.L. and Dawson, C.R. (1985). Demonstration of mucus layer tear film by electron microscopy. *Investigative Ophthalmology and Visual Science*, **26**, 464–473.

Norn, M.S. (1969). Dead, degenerated and living cells in conjunctival fluid and mucous thread. *Acta Ophthalmologica*, **47**, 1102–1115.

Norn, M.S. (1970). Rose bengal vital staining. *Acta Ophthalmologica*, **48**, 546–559.

Norn, M.S. (1979). Semiquantitative interference study of fatty layer of precorneal film. *Acta Ophthalmologica*, **57**, 766–774.

Norn, M.S. (1985). The effects of drugs on the tear flow. *Transactions of the Ophthalmological Society UK*, **104**, 410–414.

Olurin, O. (1970). Aetiology of Blindness in Nigerian Children. *American Journal of Ophthalmology*, **70**, 533–540.

Pfister, R.R. (1975). The healing of corneal abrasions in the rabbit: a scanning electron microscope study. *Investigative Ophthalmology and Visual Science*, **14**, 648–661.

Pflugfelder, S.C., Tseng, S.C., Sanabria, O., Kell, H., Garcia, C.G., Felix, C. *et al.* (1994). Diagnostic classification of conditions causing ocular irritation. *Investigative Ophthalmology and Visual Science (Suppl)*, **35**, 2034.

Ralph, R.A. (1975). Conjunctival goblet cell density in normal subjects and in dry eyes. *Investigative Ophthalmology and Visual Science*, **14**, 299–302.

Riley, M.V. (1969). Glucose and oxygen utilization by the rabbit cornea. *Experimental Eye Research*, **8**, 193–200.

Riley, C.M., Day, R.L, Greeley, D. and Langford W.C. (1949). Central autonomic dysfunction with defective lacrimation. I. Report of five cases. *Pediatrics*, **3**, 468–478.

Roat, M.I., Sossi G. and Thoft, R.A. (1989). Hyperproliferation of conjunctival fibroblasts from patients with cicatricial pemphigoid. *Archives of Ophthalmology*, **107**, 1064–1067.

Robin, J., Jester, J., Noble, J., Nicolaides, N. and Smith, R.E. (1985). *In vivo* transillumination biomicroscopy and photography of meibomian gland dysfunction. A clinical study. *Ophthalmology* **92**, 1423–1426.

Rolando, M., Refojo, M.F. and Kenyon, K.R. (1983). Increased tear evaporation in eyes with keratoconjunctivitis sicca. *Archives of Ophthalmology*, **101**, 557–558.

Ruskell, G.L. (1985). Innervation of the conjunctiva. *Transactions of the Ophthalmological Society UK*, **104**, 390–395.

Salamon, S.M. (1985). Tetracyclines in ophthalmology. *Survey of Ophthalmology*, **29**, 265–275.

Sandford-Smith, J.H. and Whittle, H.C. (1979). Corneal ulceration following measles in Nigerian children. *British Journal of Ophthalmology*, **63**, 720–724.

Schermer, A., Galvin, S. and Sun, T.-T. (1986). Differentiation-related expression of a major 64K corneal keratin *in vivo* and in culture suggests limbal location of corneal stem cells. *Journal of Cell Biology*, **103**, 49–62.

Scherz, W., Doane, M.G. and Dohlman, C.H. (1974). Tear volume in normal eyes and keratoconjunctivitis sicca. *Albrecht von Graefes Archiv für Klinische und Experimentelle Ophthalmoogie*, **192**, 141–150.

Serdarevic, O.N. and Koester, C.J. (1985). Colour wide field specular microscopic investigation of corneal surface disorders. *Transactions of the Ophthalmological Society UK*, **104**, 439–445.

Shelley, W.B. (1967). Herpes simplex virus as a cause of erthyema multiforme. *Journal of the American Medical Association*, **201**, 153–156.

Simmons, P.A., Craig, J.P. and Tomlinson, A. (1993). Effect of atmospheric humidity on human tear osmolarity. *Investigative Ophthalmology and Visual Science*, **34**, 3814.

Sjogren, H. (1933). Zur kenntnis der keratoconjunctivitis sicca (keratitis filiformis) bei hypofunktion der tranendrusen. *Acta Ophthamologica (kobenharn)*, **11**, 1–151.

Snyder, C. (1994). Anomalies of the tears and the preocular tear film. In *Anterior segment complications of contact lens wear*, edited by J. Silbert, pp. 2. New York, Edinburgh: Churchill Livingston Inc.

Soong, H.K., Martin, N.F., Wagoner, M.D., Alfonso, E., Mandelbaum, S.H., Laibson, P.R., Smith, R.E. and Udell, I. (1988). Topical retinoid therapy for squamous metaplasia of various ocular surface disorders. *Ophthalmology*, **95**, 1442–1446.

Spencer, W.H. and Zimmerman, L.E. (1985). The conjunctiva. In *Ophthalmic Pathology, An Atlas and Textbook, 3rd Edition*, edited by W.B. Spencer, pp. 169–174, Philadelphia: WB Saunders Company.

Thermes, F., Molon-Noblot, S. and Grove J. (1991). Effects of acetylcysteine on rabbit conjunctival and corneal surfaces. *Investigative Ophthalmology and Visual Science*, **32**, 2958–2963.

Thoft, R.A. and Friend, J. (1977). Biochemical transformation of regenerating ocular surface epithelium. *Investigative Ophthalmology and Visual Science*, **16**, 14–20.

Thoft, R.A. and Friend, J. (1983). The X, Y, Z hypothesis of corneal epithelial maintenance. *Investigative Ophthalmology and Visual Science*, **24**, 1442–1443.

Tiffany, J.M., Chew, C.K. and Bron, A.J. (1994). Vapour-pressure osmometry of human tears. *Investigative Ophthalmology and Visual Science*, **35**, 2031.

Trocme, S.D., Raizman, M.B. and Bartley, G.B. (1992). Medical therapy for ocular allergy. *Mayo Clinic Proceedings*, **67**, 557–565.

Tseng, S.C. and Farazdaghi, M. (1988). Reversal of conjunctival transdifferentiation by topical retinoic acid. *Cornea*, **7**, 273–279.

Tseng, S.C., Maumenee, A.E., Stark,W.J., Maumenee, I.H., Jensen, A.D., Green, W.R. and Kenyon, K.R. (1985). Topical retinoid treatment for various dry-eye disorders. *Ophthalmology*, **92**, 717–727.

Tsubota, K. and Yamada, M. (1992). Tear evaporation from the ocular surface. *Investigative Ophthalmology and Visual Science*, **33**, 2942–2950.

Udell, I.J., Gleich, G.J., Allansmith, M.R., Ackerman, S.J. and Abelson, M.B. (1981). Eosinophil granule major basic protein and Charcot-Leyden crystal protein in human tears. *American Journal of Ophthalmology*, **92**, 824–828.

van Biijsterveld. (1969). Diagnostic tests in the sicca syndrome. *Archives of Ophthalmology*, **82**, 10–14.

Vanley, G.T., Leopold, I.H. and Gregg, T.H. (1977). Interpretation of tear film breakup. *Archives of Ophthalmology*, **95**, 445–448.

Wanko, T., Lloyd, B.L. and Mathews, J. (1964). The fine structure of human conjunctiva in the perilimbal zone. *Investigative Ophthalmology*, **3**, 285–301.

Whitcher, J.P. (1987). Clinical diagnosis of the dry eye. *International Ophthalmology Clinics*, **27**, 7–24.

Williamson, J., Doig, W.M., Forrester, J.V., Tham, M.H., Whaley, K. and Carson Dick, W. (1974). Management of the dry eye in Sjogren's syndrome. *British Journal of Ophthalmology*, **58**, 798–805.

Wilson, W.S., Duncan, A.J. and Jay, J.L. (1975). Effects of benzalkonium chloride on the stability of the precorneal tearfilm in rabbit and man. *British Journal of Ophthalmology*, **59**, 667–669.

Wilson, J.D., Braunveld, E., Isselbacker, K.J., Petersdorf, R.G., Martin, J.B., Fauci, A.S. and Root, R.K. (Eds). (1991). *Harrison's Principles of Internal Medicine, 12th edition.*, pp. 1432–1449. New York: McGraw Hill.

Wolff, E. (1954). *Anatomy of the Eye and Orbit, 4th edition.* New York: Blakiston.

Wright, P. (1985a). Other forms of treatment of dry eyes. *Transactions of the Ophthalmological Society UK*, **104**, 497–498.

Wright, P. (1985b). Topical retinoic acid therapy for disorders of the outer eye. *Transactions of the Ophthalmological Society UK*, **104**, 869–874.

Xu, K., Yagi, Y., Toda, I. and Tsubota, K. (1995). Tear function index. A new measure of dry eye. *Archives of Ophthalmology*, **113**, 84–88.

Yetiv, J.Z., Bianchine, J.R. and Owen, J.A. (1980). Etiologic factors of the Stevens-Johnson syndrome. *Southern Medical Journal*, **73**, 599–602.

Zappia, R.J. and Milder, B. (1972). Lacrimal drainage function. 2. The fluoroscein dye disappearance test. *American Journal of Ophthalmology*, **74**, 160–162.

INDEX